The Chemistry of Inorganic Biomaterials

Inorganic Materials Series

Series editors:
Duncan W. Bruce, *University of York, UK*
Dermot O'Hare, *University of Oxford, UK*
Richard I. Walton, *University of Warwick, UK*

Titles in the Series:
1: Pre-combustion Carbon Dioxide Capture Materials
2: Post-combustion Carbon Dioxide Capture Materials
3: Solar Energy Capture Materials
4: Bioinspired Inorganic Materials: Structure and Function
5: Energy Storage and Conversion Materials
6: Two-dimensional Inorganic Nanomaterials for Conductive Polymer Nanocomposites
7: The Chemistry of Inorganic Biomaterials

How to obtain future titles on publication:
A standing order plan is available for this series. A standing order will bring delivery of each new volume immediately on publication.

For further information please contact:
Book Sales Department, Royal Society of Chemistry, Thomas Graham House, Science Park, Milton Road, Cambridge, CB4 0WF, UK
Telephone: +44 (0)1223 420066, Fax: +44 (0)1223 420247,
Email: booksales@rsc.org
Visit our website at www.rsc.org/books

The Chemistry of Inorganic Biomaterials

Edited by

Christopher Spicer
University of York, UK
Email: chris.spicer@york.ac.uk

ROYAL SOCIETY OF CHEMISTRY

Inorganic Materials Series No. 7

Print ISBN: 978-1-78801-753-4
PDF ISBN: 978-1-78801-982-8
EPUB ISBN: 978-1-78801-983-5
Print ISSN: 2472-3819
Electronic ISSN: 2472-3827

A catalogue record for this book is available from the British Library

The Royal Society of Chemistry is a charity, registered in England and Wales, Number 207890, and a company incorporated in England by Royal Charter (Registered No. RC000524), registered office: Burlington House, Piccadilly, London W1J 0BA, UK, Telephone: +44 (0) 20 7437 8656.

For further information see our website at www.rsc.org

Printed in the United Kingdom by CPI Group (UK) Ltd, Croydon, CR0 4YY, UK

Series Preface

Materials chemistry has grown hugely in the past few decades and continues to evolve rapidly. This is especially the case since the subject now finds itself central to many of the major challenges that face global society, such as those concerning energy and the environment; but applications of materials are also found in many everyday situations, with ongoing developments in technology around electronics and the miniaturisation of devices. Although the study of materials is a highly interdisciplinary topic, the role of innovative, fundamental chemistry continues to lie at the heart of the discovery of new materials, the understanding of their atomic-scale structure and the relationship of structure to a useful property.

With the *Inorganic Materials Series*, our aim is to provide a series of themed volumes, each reflecting the diversity of the topic as it continues to develop over the coming years, with many chosen to relate closely to emerging practical applications. The use of the title *Inorganic Materials* is by no means limiting, but instead emphasises the study of the chemistry of all elements in the Periodic Table and how they impact the development of new materials. The chapters in each book will have a pedagogical flavour with a target audience from first-year postgraduate student upwards, but with wide subject coverage ranging from continuous inorganic solids, to molecular and soft matter systems. We will use expert, active researchers as editors and contributors to provide an up-to-date perspective of their field. At the same time our aim is to provide an international perspective, so to reflect the diversity and interdisciplinarity of the now very broad area of the subject.

Duncan W. Bruce, York, UK
Dermot O'Hare, Oxford, UK
Richard I. Walton, Warwick, UK

Inorganic Materials Series No. 7
The Chemistry of Inorganic Biomaterials
Edited by Christopher Spicer

Published by the Royal Society of Chemistry, www.rsc.org

Preface

In recent years there has been increasing interest in the use of biomaterials to treat patients suffering from a wide range of injuries and diseases. Inorganic biomaterials are at the forefront of this movement, being both the most commonly used materials in the clinic, and key components of many of the exciting emerging technologies that are approaching applications in patients. In this book we will look at these materials from a unique perspective, by focussing on their underlying *chemistry*, how this ultimately dictates their properties and effectiveness, and the cutting-edge discoveries that are driving their continued evolution.

Metallic materials are particularly common, serving to fix broken bones, replace damaged joints, and fill dental cavities (Chapter 1). While some of these technologies have now been around for more than a century, as our ability to probe the structure, function, and biological interactions of materials at the molecular- and nanoscales has improved, so has our understanding of the key design principles that must be considered to produce a successful metallic implant.

Alongside improvements in these 'traditional' biomaterial technologies, there has also been an emergence of new and exciting applications of inorganic materials. Cements and glasses that seek to not only fill bone and dental defects, but also to serve as a scaffold for the deposition of new tissue are now prominent in the clinic, and many recent developments in their design promise to increase their relevance in coming years (Chapter 2). Recent efforts to design 'organic–inorganic hybrids' are particularly promising (Chapter 3), harnessing the beneficial properties of both material classes to recreate the structure of bone, a naturally occurring organic–inorganic hybrid. Bioelectronic and neural interface materials that can serve as bioelectronics for the treatment of brain injuries or as neural prosthetics are also at the cutting-edge of modern day medical research, with the

Inorganic Materials Series No. 7
The Chemistry of Inorganic Biomaterials
Edited by Christopher Spicer

Published by the Royal Society of Chemistry, www.rsc.org

potential to impact the lives of people suffering from debilitating conditions worldwide (Chapter 4).

Finally, we will consider a challenge that is common to all of these materials (and indeed all implants that are placed inside the body) – how the chemistry of the material affects its interactions with proteins, biomolecules, and cells, both host and bacterial (Chapter 5). Understanding and controlling these processes is critical to minimise the risk of bacterial infection and implant failure, and therefore ultimately dictates the fate of a biomaterial inside the body.

My thanks go out to the authors of each of the chapters of this book, who have worked tirelessly to produce the fascinating work you will read under challenging circumstances in the midst of a global pandemic. Thank you also to the team at the Royal Society of Chemistry, Robin Driscoll, Connor Sheppard, and Katie Morrey, and to the editors of the series, Duncan Bruce, Dermot O'Hare, and Richard Walton, for all of their help, support, and patience throughout the process. I hope you enjoy reading the book as much as I have enjoyed putting it together.

Chris Spicer

Contents

Inorganic Materials Series No. 7
The Chemistry of Inorganic Biomaterials
Edited by Christopher Spicer

Published by the Royal Society of Chemistry, www.rsc.org

CHAPTER 1

Metallic Implants for Biomedical Applications

O. BAZAKA,[a] K. BAZAKA,*[b] P. KINGSHOTT,[c] R. J. CRAWFORD[a] AND E. P. IVANOVA[a]

[a] School of Science, College of Science, Engineering and Health, RMIT University, Melbourne 3000 VIC, Australia; [b] School of Engineering, The Australian National University, ACT 2601, Australia; [c] Department of Chemistry and Biotechnology, Swinburne University of Technology, Hawthorn 3122 VIC, Australia
*Email: katia.bazaka@anu.edu.au

1.1 Introduction

For over 100 years, metals and their alloys have played a central role in the advancement of medical implant applications across virtually all medical specialties. Early metal implants took some advantage of the highly useful combination of properties pertinent to this class of materials, including their high mechanical strength and excellent toughness, essential for load bearing applications, durability and resistance to wear, good thermal and electrical conductivity, ductility needed for their processing into various shapes, and chemical/biological compatibility. By 1895, scientific reports had been published on the use of metal fixation devices, *e.g.* bone plates, to maintain the correct apposition of fractured bone fragments in the extremities.[1] Still, early designs were fairly simple, suffered from insufficient strength and limited corrosion resistance, and were used to treat only a limited set of conditions. With advances in metallurgy, particularly the development of stainless steel and Co–Cr–Mo alloys, and concomitant progress

Inorganic Materials Series No. 7
The Chemistry of Inorganic Biomaterials
Edited by Christopher Spicer

Published by the Royal Society of Chemistry, www.rsc.org

Table 1.1 Biomedical metals and alloys and their primary use in the medical field.[4]

Type	Primary use	FDA Class
Routinely applied materials		
Stainless steels	Temporary fracture plates, screws, hip nails, *etc.*	II
	Total hip replacements	II
Co-based alloys	Total joint replacement	II
	Dentistry castings	II
Ti-based alloys	Stem and cup of total hip replacements with CoCrMo or ceramic femoral heads	II
	Other permanent devices (nails, pacemakers)	III
Emerging FDA approved materials		
NiTi	Orthodontic dental archwires	I
	Vascular stents	III
	Vena cava filter	II
	Intracranial aneurysm clips	II
	Catheter guide wires	II
	Orthopaedic staples	I
Ta	Wire sutures for plastic surgery and neurosurgery	III
	Radiographic markers for diagnostic applications	II
Materials in clinical trials and research		
NiTi	Contractile artificial muscles for an artificial heart	III
Mg	Biodegradable orthopaedic implants	III

in aseptic surgical procedures, the corrosion and mechanical performance of these implants and patient outcomes continued to improve.[2]

With each passing year, novel alloys, device architectures and functions continued to emerge, a trend that continues today, as reflected by the broad range of functions exhibited by contemporary metallic biomaterials (see Table 1.1), and a great deal of reports that have been published in the scientific literature proposing novel biomedical materials and devices based on metals. However, only a fraction of these devices reach patients. This is because of a stringent regulatory framework that prescribes the kind of benchmarks a device has to achieve in order to be allowed on to the market.

1.2 General Approach to Metallic Implant Design and Manufacturing

For metallic biomaterials, the critical benchmarks include outstanding biocompatibility and mechanical properties, controlled corrosion profile (high chemical inertness and corrosion resistance for permanent implants, and a controlled degradation profile for resorbable platforms), appropriately high resistance to wear (especially for articulating parts), and osteointegration (for implants interfacing with bone tissues). The definitions for these benchmarks are constantly evolving. For example, the descriptors of biocompatibility have evolved to include lack of toxicity, chemical inertness, lack of foreign body or inflammatory response, non-sensitising, non-allergenic and non-tumorigenic nature, and many others. However, with the discovery of various feedback loops within materials–host systems, a more appropriate definition of biocompatibility may be the ability of a material to perform its function with an appropriate host response in a specific application.[3]

The length and complexity of the testing and approval process very much depends on the intended use, with the guidelines defined by the appropriate regulatory body. For example, the US Food and Drug Administration controls three classes of devices used for medical diagnosis and treatment, which are defined by the safety risks their use may pose. Whereas low-risk Class I devices (*e.g.* orthodontic wires) see only general controls, high-risk Class III devices (*e.g.* wire sutures for neurosurgery) have to go through mandatory performance standards, surveillance, and scientific review.[4] For any device, the first step involves lab-based testing to establish materials and device (*e.g.* mechanical or electrical) performance, the propensity to corrode and release debris under physiological conditions, wear performance, and compatibility with different cell lines, to name but a few aspects. Implants that successfully pass this stage may proceed to testing using an animal model, and eventually clinical studies.

Subsequent to their market release, data on the implant performance continue to be collected. This is because the long-term use of metallic implants can bring about the development of local and systemic adverse events, the nature and extent of which, and their effect on host health and implant failure, cannot be predicted at the time of testing and may arise within months to years post implantation. Furthermore, implants approved through a 'fast-track' approval route (*i.e.* on the basis of being substantially equivalent in safety and efficacy to a legally marketed device) may not have been subject to the same level of clinical testing when compared to devices that have gone through pre-market approval, and thus may be missing critically important performance data. For example, over the decades, the US Food and Drug Administration have used the 'fast-track' 510(k) pathway to approve the use of larger-size femoral components and shortened trunnions that connect the head to the stem in metal-on-metal hip replacements with minimal to no clinical testing. Years after their market release, these design modifications have been found to contribute to much greater rates of mechanical failure, increased production of metal ion debris and increased wear.

With respect to metallic implants, the examples of such adverse events include metallosis in response to the release of wear and corrosion particles from total hip arthroplasty systems with metal-on-metal bearings, and the release of soluble metals from coiled wire-based permanent birth control implants made of Ni–Ti and stainless steel. Predicting such adverse events is difficult because in many cases the immune or inflammatory response is only clinically significant in particular groups of predisposed individuals. Nevertheless, the magnitude of these events for the affected individuals can be significant, and include the development of pseudotumours, neurological events, and bone loss and subsequent failure of the prosthesis in the case of the hip replacement system, and systemic sensitivity events in the case of the birth control implant.[5] As in most cases the presence of clinically significant events would necessitate device removal and replacement, the economic and healthcare burden associated with these events is also significant.

The post-market release data collection is also what drives the ongoing device development and optimisation. In spite of the lengthy, multi-stage, and expensive assessment of implant compliance, a growing number of devices are getting cleared and approved for clinical use each year across all application areas. The US Food and Drug Administration has reported a steady growth in approvals of metallic devices over the past decade (see Table 1.2).[6]

1.2.1 Selection of Metals

1.2.1.1 Consideration of Metal Toxicity

Key to metallic implant advancement is the development of new alloys, since a mixture of several different metals and other elements may not only overcome the limitations of single-component materials by taking on the desirable properties of its constituents, but also exhibit unique properties that are distinct from those of the individual components comprising the alloy. Amongst the most common reasons for alloying is to improve the strength, processability or corrosion resistance of a metal. For example, the addition of even a small amount of alloying agents in the metal matrix introduces atoms of sizes different to that of the primary metal; distributed uniformly throughout the crystalline grains (as either substitutional or interstitial elements), where the alloying atoms act as barriers to slip, thus improving strength. Alloying elements can also strengthen the metal through precipitation hardening, where barriers to dislocation are formed by impurity-rich particles of tens or hundreds of atomic diameters in size.

When choosing suitable alloying elements, the individual and cumulative toxicity of elements within an alloy should be low, with a preference given to the essential trace elements Co and Fe (as these elements are already present in heathy tissues of living organisms), and highly corrosion resistant active elements, such as Ti. The majority of the presently used metallic biomaterials are based on Fe-based alloys (*e.g.* stainless steels), Co–Cr alloys, and Ti and its alloys. Where the strength, stiffness and wear resistance of Co–Cr alloys is the greatest amongst the aforementioned alloys; stainless steels offer a favourable combination of cost, cyclic twist strength and ductility; and Ti alloys are considered most biocompatible, with the greatest corrosion resistance and specific strength. This statement is of course a generalisation, as the properties of alloys within each group can differ considerably based on their composition, fabrication method and post-processing treatment, and the environment within which they operate. Biomaterials based on Ta, Nb, Mg and Fe are also becoming increasingly popular due to their bioactivity (in the case of Ta, Mg and Fe), chemical inertness (Nb) or controlled degradation under physiological conditions (Mg and Fe). For some time, Ni was one of the most popular alloying elements due to its

Table 1.2 Growth in the number of devices cleared or approved by the US Food and Drug administration broken down by device type.[6] [a]

Device type	Examples	Application area	Duration	Material	Cleared/approved			Class
					2006	2016	10 years	
Bone fixation	Plates, screws, wires, pins, rods	Orthopaedic, oral, maxillofacial, ophthalmology, ENT	Temporary, Permanent	Ti, SS, Ni–Ti, Au, Ag, Pt alloys	432	608	6573	II
Prosthesis	Sockets, joints, limbs	Orthopaedic, oral, maxillofacial, reproductive, ophthalmology, ENT	Permanent	Ti, Co–Cr alloys, SS, Au, Ag, Pt alloys, Sn–Ag, Ni	389	390	4362	II/(III)
Soft tissue fixation	Clips, sutures, staples	General and specialised surgery	Temporary, Permanent	Ti, SS, Pt, Ta, Ni–Co, Ni–Ti, Au, Pt alloys, Co–Cr alloys	28	34	346	II/(III)
Access and flow control	Catheters, ports, shunts	Vasculature, gastrointestinal, urology, reproductive	Temporary, Permanent	Ti, SS, Ni–Ti, Pt alloys	38	28	364	II/(III)
	Stents, valves, tubal inserts	Vasculature, gastrointestinal, urology, reproductive	Temporary, Permanent	Ti, SS, Ni–Ti, Al, Co–Cr alloys, W, Pt alloys	16	26	246	II/(III)
Sensing and stimulation	Electrical stimulators, receivers, sensors, electromechanical devices	Neurology, cardiology, urology, ENT, endocrinology	Permanent	Ti, SS, Ag, Co–Cr alloys, Pt alloys, Au, Ta	5	12	101	II/(III)

[a]ENT stands for ear, nose, and throat.

ability to enhance the formability, weldability, ductility, and, in some instances, corrosion resistance of stainless steels and Co–Cr alloys. The subsequent discovery of the potential acute and chronic adverse effects of exposure to Ni that leached out from the implants, including delayed hypersensitivity, headaches, vertigo, vision changes, altered iron metabolism and homeostasis of Ca, Mg, Mn and Zn, has led to the reduction in the use of Ni in biomedical alloys. The use of another popular alloying element, Al, has been suspected to cause osteomalacia, hepatic dysfunction, anaemia and neurological effects. While Ni and Al are non-essential metals for the human body, even for essential metals that the body requires for the proper functioning of many of its proteins and enzymes, their quantities are tightly controlled *via* homeostatic regulation, with the presence of excessive amounts of these metals often causing local and systemic toxicity.[7] For example, although vanadium is considered essential for proteins that participate in phosphate and lipid metabolism, and insulin enhancement, the presence of excess vanadium may result in the development of headaches, weakness, tremor and gastrointestinal issues in patients. This holds true for excess Mo, which can cause urinary tract stones, acute renal failure, and myositis, as well as for Co, Cu, Fe, Mn, Zn and Cr essential metals. Undesirable physiological effects have also been reported for other non-essential metals frequently found in biomedical implants, including Ti, Ag, Au, Pt, Pd, Sn, Hf, Zr, W and Ir. Nb, Ta and Zr are often considered among the least toxic alloying elements, yet are expensive. Also expensive, precious metals, including Au and Au alloys with Pt and Pd, Ag and its alloys, and Pt and Pt alloys are used in restorative dentistry, in antimicrobial treatments (Ag) and as electrodes for electrically active devices.

1.2.1.2 Bulk Properties

The bulk properties of metals are defined by their chemical composition, crystalline structure and bond strength. The 3D arrangement of closely packed atoms in metals, where the nuclei of positively charged ions are immersed in a delocalised cloud of freely moving valence electrons, is responsible for the excellent electrical and thermal conductivities of metals, and their high density (in the case of most metals). Greater amounts of delocalised electrons correspond to stronger metallic bonds, *i.e.* stronger electrostatic interactions between the electron cloud and the ions. The strength of the metallic bonds defines the tensile strength of the material, whereas the non-directional nature of metallic bonds is responsible for the malleability and ductility of metals, which allows them to undergo a significant amount of plastic deformation (plastic strain) without failure (cleaving). The strength of the metallic bonds is greater for metals made of atoms of greater nuclear charge, which have a greater number of electrons they can release into the cloud, and of smaller atom size; the greater strength of the metallic bonds is also responsible for the higher melting temperature and hardness of such metals. As such, atoms with half-filled valence shells (such as W and other transition metals with valence-level d electrons) make solids that are stronger than those made of elements with nearly empty or nearly full valence subshells. While the Young's modulus, ultimate tensile strength and toughness define the application of metallic biomaterials,[8] their hardness and melting temperature prescribe the kind of techniques that can be used for their processing into implants and devices.

The atoms in most pure metals tend to naturally assemble into either body-centred cubic (bcc), face-centred cubic (fcc) or hexagonal close-packed (hcp) structures. Bcc corresponds to a crystal organisation in which eight atoms occupy the corners of a cube, with one atom located in the cube centre. In other words, one atom is bonded to eight others, giving a co-ordination number of eight, with two atoms per unit cell. In a fcc structure, eight atoms occupy the cube corners, and six atoms are in the centre of each face (a coordination number of 12, four atoms per unit cell). In a hcp structure, the atoms are organised like stacked hexagons, where six atoms occupy the corners of a hexagon with two atoms located at their centres (6 atoms per cell, a coordination number of 12). Examples of metals that adopt bcc crystal structures include Fe, V, Nb, Cr; fcc structures are adopted by Al, Ni, Ag, Cu and Au; and hexagonal structures by Ti, Zn, Mg, and Cd. The introduction of an alloying element results in the creation of internal stresses within the crystal structure due to the size differences between the atoms; atoms of a larger size apply a compressive force on adjacent atoms, whereas the force the smaller-sized atoms exert is tensile. These internal stresses are often beneficial, enhancing the desirable properties of the material, *e.g.* resistance to deformation, shear and tensile strength, toughness, hardness and ductility. Depending on the choice of the alloying element and its concentration, the bulk density, reactivity, Young's modulus and other physical properties may remain largely unchanged or change greatly from that of the base metal. The crystalline structures of both pure metals and their alloys can also be altered by physical treatment, *e.g.* work hardening. For example, defects can be introduced by subjecting the material to severe plastic deformation, when very large strains (by means of a complex stress state or high shear) are applied to the material to achieve a high density of defects that may act as nucleation sites for precipitates, and change the grain structure to equiaxed ultrafine ($d < 500$ nm) or nanocrystalline ($d < 100$ nm, with some extreme cases of refinement of $d < 30$ nm).[9] These changes will remain permanently within the lattice unless the material is heated and then allowed to re-crystallise. These changes in the structure and texture of the alloy may notably affect its mechanical (strength, ductility) and service characteristics (corrosion resistance). Where grain refinement leads to strengthening of the alloy by inhibiting the dislocation glide, the concomitant transformation of the texture in certain alloys (*e.g.*, Mg1.0%Zn0.3%Ca) may counteract the former strengthening effects by forming an inclined basal texture that promotes basal slip and may also lead to tensile–compression asymmetry.[10]

Certain heat treatment and cooling regimes can also be used to strengthen the alloys and relieve stress that may have accumulated in the material during fabrication. For example, Ti alloys can be subjected to forging at

595–705 °C for 1–2 h, and then air cooled to relieve undesirable residual stresses from non-uniform hot forging, deformation from cold forming and straightening, asymmetric machining, welding and cooling of castings. Single and multistage annealing, and solution treatment and ageing can be used to improve creep resistance and fracture toughness, and attain greater strength, respectively. The latter treatment involves heating the alloy to a specified temperature (defined by the nature of the alloy), followed by quenching at a controlled rate in either oil, air or water, and then reheating to a temperature between 425–650 °C for 2 h.

1.2.1.3 *Withstanding Static* vs. *Cyclic Load*

A combination of the aforementioned techniques may be used to achieve the desired level of fracture toughness, *i.e.* the level of resistance to crack propagation and failure under load, and fatigue strength. The latter is an important determinant of implant performance and lifetime because humans tend to engage in a wide range of repetitive physical activities. For example, hip and knee prostheses, fixation plates and wires will be subjected to cyclic loading during walking and running, dental implants will experience cyclic loading during chewing, and electrodes will be subjected to myocardial activity. Joints will experience cyclic compression and bending with a loading strength of ~50 MPa and ~200 MPa (loading frequency 1 Hz) respectively, with the total number of loading cycles over the lifetime of an implant of 10^7.[11] This type of loading will result in the weakening of the material *via* progressive and localised structural damage and the growth of cracks that start at the point where stress concentration occurs, eventually resulting in implant failure by fatigue. Importantly, failure by fatigue can take place at normal operating loads far lower than the tensile strength of the material.

The microstructure of the material is an important determinant of the maximum intensity of cyclic tension it can withstand prior to failure (see Figure 1.1). Other contributing factors include the nature of the stress (cyclic frequency, load vectors), the corrosivity of the environment, and the surface finish and wear. Fatigue life can be extended through surface modification of the implant material. For example, grit-blasting of dental implants made of commercially pure titanium can produce a surface layer of compressive residual stresses that protects them from failure under cyclical tension-compression loading.

1.2.1.4 *Metal* vs. *Bone*

Among the key drivers behind the development of new alloys is the growing understanding of how different properties of materials affect their behaviour *in vivo*, in particular the interactions between materials and cells. For example, the nature of a cellular response is not only determined by the surface chemistry of the material, but also its surface topography and surface and bulk mechanical properties, including the material's Young's modulus, its specific

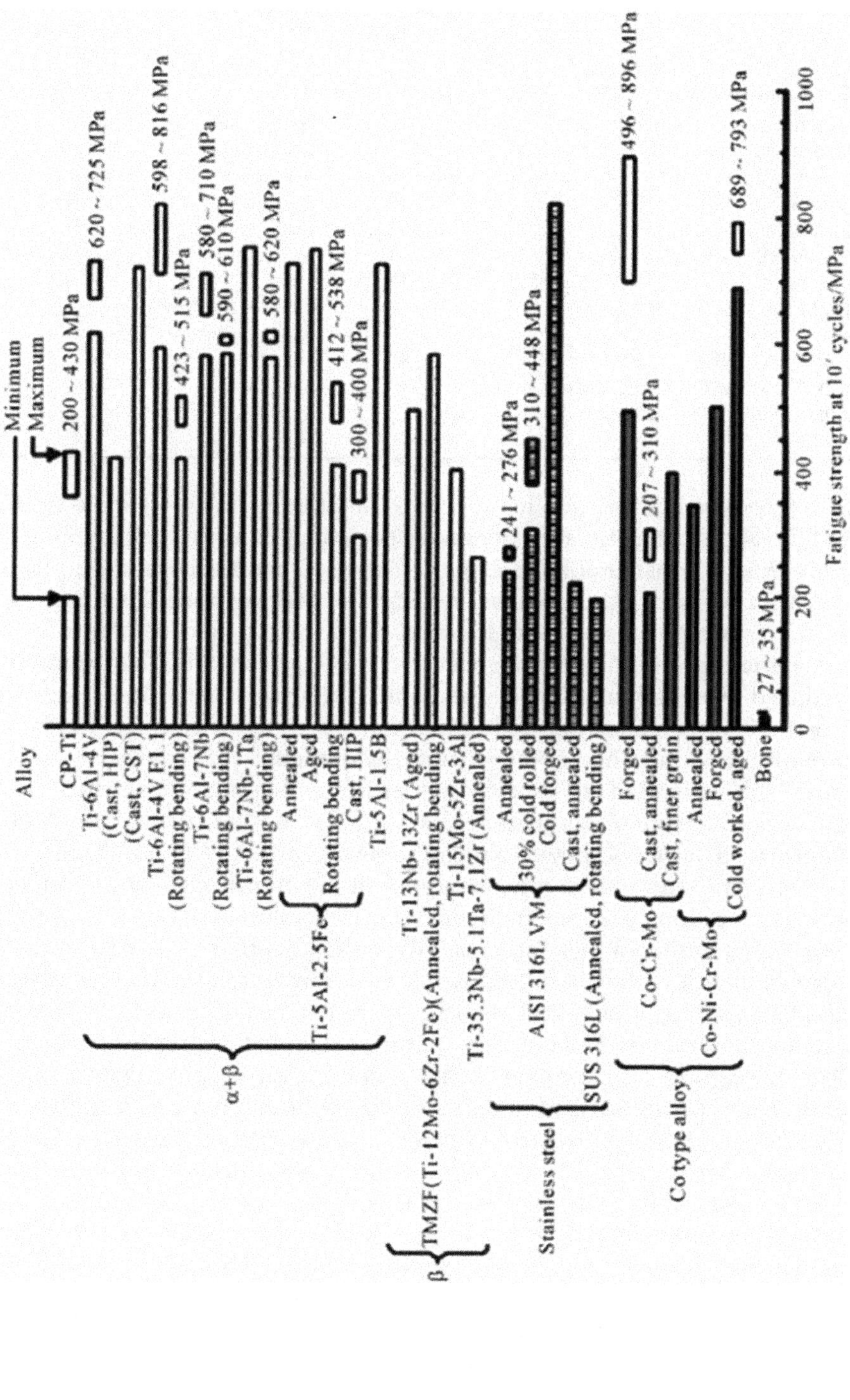

Figure 1.1 Fatigue strength (at 10^7 cycles) for commonly used biomedical metals and alloys, and bone. Data without designation of rotating bending are those obtained from uniaxial fatigue tests in air. Reproduced from ref. 229 with permission from Elsevier, Copyright 2007.

Table 1.3 Mechanical properties of commonly used load-bearing metals and their alloys, and the bone tissue they replace.

Materials	Young's modulus (GPa)	Ultimate tensile strength (MPa)	Fracture toughness (MPa)
Cortical bone	10–30	130–150	2–12
CoCrMo alloys	240	900–1540	~100
316L stainless steel	200	540–1000	~100
Mg alloys	40–45	100–250	15–40
NiTi alloys	30–50	1355	30–60
Ti and Ti alloys	105–125	900	~80
α-type [*e.g.* CP–Ti Grades 1–4]	100–105		
(α + β)-type [*e.g.* Ti6Al(4V,4VELI,7Nb), Ti5Al(2.5Fe,1.5B), Ti(15Sn,5Zr)-4Nb2Ta0.2Pd]	90–115		
β-type [*e.g.* Ti13Nb13Zr, Ti12Mo6Zr2Fe, Ti15Mo, Ti15Mo5Zr3Al, Ti15Mo, 2.8Nb0.2Si0.26O, Ti35Nb7Zr5Ta, Ti29Nb13Ta4.6Zr]	55–80		

strength (expressed as the ratio between its strength and ductility) and others. Most of the alloys currently used in bone tissue reconstruction have a Young's modulus that is considerably greater than that of natural bone tissue (see Table 1.3). This is in part because in order to perform their load-bearing function, they need to have a suitable level of stiffness. The consequences of this difference include non-homogeneity in the implant–bone load transfer and reduced stress stimulation of the bone, resulting in bone loss, limited bone remodelling and sub-optimal implant integration, which affect implant performance and lifetime *in vivo*, and the benefits they deliver to the patient.[12–14] A review of 23 longitudinal studies in implant dentistry that performed a follow-up of 10 or more years to evaluate the survival of ~7700 osseointegrated implants showed a mean survival rate of 94.6% ± 5.97% and a mean marginal bone loss around implants of 1.3 mm ± 0.84 mm, with 14 studies also reporting a cumulative mean success value of 89.7% ± 10.2%.[12] However, in a number of these studies, the probing depth for bone resorption was significantly less than 5 mm, possibly resulting in missed incidences of peri-implant complications. Furthermore, there is a lack of consensus in the field as to the amount of bone resorption considered appropriate for the number of years in the function, health, and success of the implant. Importantly, the measurement of the distribution of stress and strain in dental implants and in bones when the implant is loaded under normal occlusal loading conditions is difficult to conduct *in vivo*, and thus such information often comes from modelling studies[14] or those involving human cadavers.[13]

While Ti and its alloys have relatively low Young's moduli (110 GPa for Ti6Al4V) when compared to 316L stainless steel and Co–Cr alloys (200–210 GPa), they still greatly exceed the Young's moduli of bones (~10–30 GPa). Among the Ti alloys, those that have β-type phase constitution, *i.e.* metals with a bcc structure, are characterised by significantly lower Young's moduli when compared to α-type alloys, *i.e.* those with a close-packed hexagonal crystal structure and (α + β)-type alloys. The phase constitution can be controlled through the selection of alloying elements, with Nb and Ta frequently used to stabilise the β-phase, with some Zr introduced due to its ability to dissolve in the β-phase and increase the strength of the material. As the design of the implant has to take into account economic considerations, some effort has gone into finding suitable affordable alternatives for the use of Nb, Ta, Mo, and Zr, with reports of producing low modulus alloys using Fe (as in Ti–Mn–Fe), Cr (Ti12Cr and Ti–Cr–Sn–Zr), Mn (Ti–Mn), Sn (as in Ti–Cr–Sn and Ti–Mn–Sn), and Al (Ti10CrAl and Ti–Mn–Al).[15–17]

1.2.2 Materials Processing Using 3D Printing

In addition to changing the composition of an alloy, it is possible to change its bulk and surface properties through a careful combination of fabrication and post-processing techniques. Figure 1.2 shows the various stages of metallic implant development where technology-driven innovation can take place.

In parallel to more traditional processing, shaping and forming techniques (the effects of which on the structure and properties of commonly used metals and alloys are discussed in the relevant sections), 3D scanning and printing has recently emerged as a promising means by which to facilitate personalised medicine. Firstly, it allows for the rapid prototyping of implants that fit the unique anatomical features of individual patients, thus improving implant mechanical performance and reducing pain and discomfort for the patient in orthopaedic and dental reconstruction, and improving appearance in craniofacial reconstruction. Because of the use of high-resolution data acquisition tools (*e.g.* computer tomography) in conjunction with computer-aided modelling and design, it is possible to produce more complex implants and correct defects with improved therapeutic outcome. Thus, the time to recovery, incidence of revision surgeries and

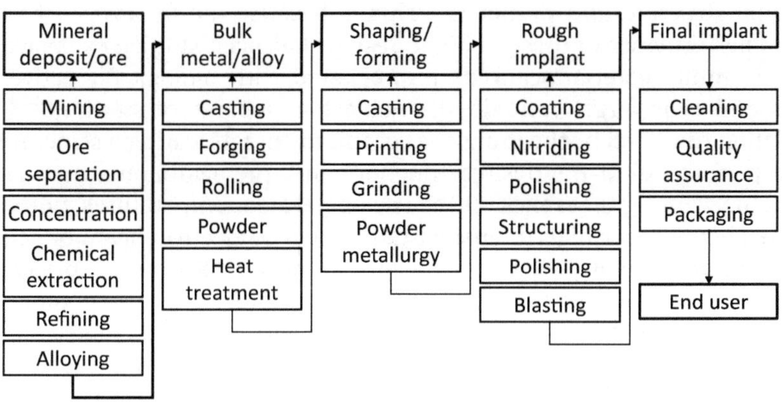

Figure 1.2 General stages of metallic implant development: from mineral ore to an end user-ready product.

the associated radiation burden may also be reduced. More importantly, however, 3D printing allows for the reproduction of the complex structures of the bone the implant is designed to replace, thus facilitating harmonisation of the structural and mechanical properties of the implant to that of the bone, and in doing so, ensuring appropriate load transfer, promoting tissue healing and implant integration. This is because 3D printing allows for the fabrication, and in some instances *in situ* modification, of complex multi-material structures that vary in terms of their chemical composition, density, porosity, and other features, thus reproducing not only the macroscopic features of the tissue it aims to replace, but also the micro- and nanoscale features.[18–21] In metallic implants, this generally means introducing pores while controlling their shape, dimension, orientation and connectivity to promote bone tissue infiltration, blood vessel formation and nutrient exchange, and reduce stress shielding. The latter arises when there is a poor match between the mechanical properties of the implant and the bone, resulting in the typical stress being removed from the bone by the implant. Lack of appropriate levels of stress leads to osteopenia, where the bone density is reduced, increasing the probability of bone fractures and aseptic bone loosening, and compromising the stress transfer between the bone and the implant. In addition to varying the properties of a single material, or of materials of the same type, it may also be possible to use 3D printing to integrate multiple types of materials during implant fabrication. The resulting implant would thus integrate an anatomically-precise metallic scaffold to support mechanical load, polymer and ceramic particles to promote cell attachment and bone tissue formation, active molecules, proteins and growth factors to stimulate cell differentiation and growth, and precursor cells derived from the host from which the tissue would form.

Although the clinical use of 3D printing is certainly growing, owing to the steadily decreasing prices of printers and increasing availability of computer topography and magnetic resonance imaging at hospitals, at present it is limited to the fabrication of patient-specific implants with an otherwise conventional structure and properties. This is because the *in vitro*, *in vivo* and clinical performance of these implant designs has been studied extensively. In contrast, multi-material scaffolds impregnated with biologically active molecules and cells largely remain the subject of laboratory studies,[22,23] with translation hindered by the immense complexity of fabricating a structure that would perform all of the desired functions in a predictable manner. For example, 3D printing that relies on the use of elevated temperatures, potentially reactive chemicals or high-intensity irradiation, *e.g.* electron beam or selective laser melting of metals, will not be an appropriate choice for the fabrication of scaffolds *in situ* seeded with living cells or biologically active molecules. However, the impregnation of such scaffolds with cells and molecules post fabrication is also not trivial. Once cells are within the scaffold, it is necessary to ensure an adequate level of nutrient and gas exchange, however integrating a pathway for vascular network formation into a scaffold is challenging from both technological and biological perspectives. Yet, without an effective means of supply and the removal of molecules and gases, any attempt at tissue development will be futile, whereas dying cells may in fact create significant problems by releasing inflammation-stimulating factors and changing the chemistry of the micro-environment in the pores within the bulk of the implant. The metallic scaffold must also be biocompatible, with a controlled degradation profile and the ability to support cell attachment, migration and the formation of complex tissue architectures. Yet, the increased surface-to-volume ratio and poor circulation of fluids within the pores may create a favourable environment for accelerated corrosion and metal ion release.[24] It should be noted that for certain applications, increased biodegradation may in fact be desirable, as is the case for Fe-based craniofacial implants.[25]

In addition, one should be able to print the scaffold with the high resolution of the desired structure, which the scaffold will then maintain over time. For example, it is not uncommon for the pores to collapse or for the pore networks to be disconnected during scaffold assembly, or, in the case of solid implants, for the pores to be introduced into the structure due to gas entrapment, changing the mechanical properties of the resultant implant.[26] Furthermore, for methods that rely on the use and subsequent removal of the polymer binder, such as the inkjet printing of metal powders, the appearance of defects, shrinkage (and associated delamination and cracking), and difficulty in the removal of unbound powders from the depths of the bulk implant are also of concern.[25,27,28] Examples of defects include spherical voids due to the collapse of vapour cavities, elongated voids due to insufficient fusion, accumulation of partially molten particles, and the balling phenomenon, with the latter two being the result of *e.g.* incomplete melting of powder particles when using insufficiently high power during selective laser melting.[29] This may render the surface less hospitable to cell infiltration and colonisation, as well as affecting the mechanical and chemical properties of the implant. Nevertheless, using electron beam melting and selective laser melting, it is possible to attain a broad range of characteristic crystallographic phases, *e.g.* α (hcp), β (bcc), α' (hcp martensite) and α'' (fc-orthorhombic) and phase microstructures and variable hardness in Ti6Al4V, which would otherwise require complex thermo–mechanical processing if conventional wrought and cast billets were used.[30]

To be a viable option for clinics, the printing process itself should be simple, reliable and affordable. Currently, selective laser sintering, direct metal laser sintering and electron beam melting are used in the printing of metal and alloy powders, including those of titanium, titanium alloys, cobalt chrome, and stainless steel, with fused deposition modelling used for the printing of 3D patterns from eutectic metals (see Figure 1.3).[21] While it is possible to fabricate scaffolds with very good fidelity and resolution of fine features, selective laser sintering is lengthy ($\sim$15 h), requires elaborate and expensive infrastructure, *e.g.* CO_2 lasers, and may expose operators to potentially hazardous dust and nanoparticle condensates (see Table 1.4). Furthermore, as-fabricated implants generally require post-processing to address the surface roughness (*e.g. via* polishing or sandblasting) and accumulated internal stresses (*e.g. via* heat treatment).[20] Unless addressed,

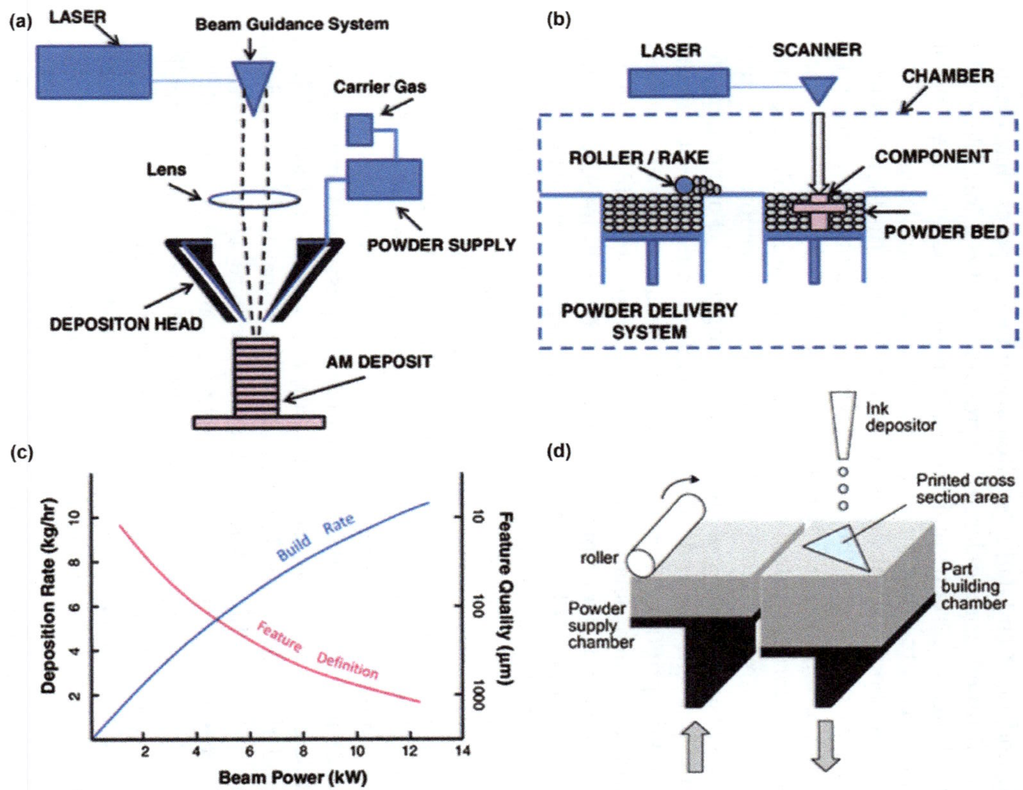

Figure 1.3 (a) Schematic of the general power feed system, such as that used for laser metal deposition (also known as direct laser deposition). Either the work piece or the deposition head may move. (b) General structure of the power bed system, where the beam may be a laser beam (as in selective laser melting) or electron beam (as in electron beam melting). (c) General relationship between the rate of prototype assembly, power of the beam, and feature definition. (d) Inkjet printing using a powder mixed with a polymer binder.
Reproduced from ref. 230 with permission from Springer Nature, Copyright 2014. Reproduced from ref. 28 with permission from Elsevier, Copyright 2010.

Table 1.4 A comparison between selective laser and electron beam melting rapid prototyping. Reproduced from ref. 19, https://doi.org/10.3390/met9070729, under the terms of the CC BY 4.0 license, https://creativecommons.org/licenses/by/4.0/.

Features	Selective laser melting	Electron beam melting
Heat source	Laser beam (up to 1 kW)	Electron beam (60 kW)
Scan speed	Limited by galvanometer inertia	Fast, magnetically driven
Powder size	10–45 μm	45–106 μm
Minimum beam size	50 mm	140 mm
Beam/melt pool dimension	0.5–1.5 μm	2–3 μm
Layer thickness	20–100 μm	50–200 μm
Chamber atmosphere	Argon or nitrogen	Vacuum (+helium)
Environment temperature	Build platform at 100–200 °C	Chamber at 400–1000 °C
Powder pre-heating	Using infrared or resistive heaters	Using electron beam
Surface finish	Excellent to moderate (~20 μm)	Moderate to poor (~35 μm)
Residual stresses	Yes	No

roughness may promote wear in the articulating components of load-bearing systems. Although implants produced using electron beam melting do not generally require thermal post-processing to remove stresses (with beams delivering an energy density of 200 J mm^{-3}), the infrastructure is similarly very expensive, while the resolution of features is generally lower when compared to selective laser sintering.

While the cost of ink-jet printers is lower, the resolution of the features is also lower when compared to the aforementioned approaches due to the spreading of the liquid and undesirable rearrangement of the powder particles due to differences in their wettability.[31] The process is also fairly slow, as the deposition of the next 2D layer of the pattern can only take place once the underlying layer has solidified, with the mechanical properties of the resultant scaffolds being dependent on the chemistry of the binder, and the surface properties and size of the individual metal particles, due to their effect on the distribution of pore sizes and scaffold assembly (see Table 1.5). Electrohydrodynamic inkjet printing produces printed materials with better resolution and speed, as the surface tension effects are overcome by the use of an electric field, and a rapid evaporating solvent is used. Inkjet printing and its modifications may be best for the fabrication of smaller implants, including implantable electrodes and sensors. Although the fabrication of the structure takes place at room temperature, it is followed by high temperature sintering to attain the necessary level of mechanical performance, which will generally lead to implant shrinkage and anisotropy with respect to implant properties, *e.g.* density, and thus load-bearing capacity. With respect to health and safety, issues associated with the handling of nano- and micro-scale powders and the possibility of electrical shorts due to the close proximity of the conductive particles to the circuitry in the printer head warrant consideration.[18]

Table 1.5 Aqueous binders used in the 3D inkjet printing of metallic biomaterials. Reproduced from ref. 18 with permission from Elsevier, Copyright 2017.[a]

Binder	Powder	Mode	Structure
Distilled water + 10% PVA + 0.05% PVP	Titanium : PVA (10 : 1)	Hybrid	Microporous cylindrical scaffolds with designed porosity
Distilled water	Titanium + PVA	Indirect	Microporous cylindrical scaffolds
Distilled water	Titanium + PVA	Indirect	Microporous dental implant
Distilled water	Titanium + PVA	Indirect	Microporous rectangular scaffolds with designed porosity
Aqueous acrylic based binder	316L stainless steel	Direct	Injection moulding tool
Organic-based aqueous binder	Fe–Mn biodegradable alloy	Direct	Rectangular macroporous scaffolds + miniature human femur
Organic-based aqueous binder	Fe–Mn–Ca/Mg alloy	Direct	$10\times10\times4$ mm^3 microporous scaffolds
Distilled water + 20% Maltodextrin + TritonX + NaN_3	Ti6Al4V	Direct	Homogeneous and gradient porous scaffolds + human femoral stem prototype

[a]PVA polyvinyl alcohol, PVP polyvinylpyrrolidone.

1.2.3 Surface Modification

Surface modification is also frequently employed to control the corrosion dynamics and cell–surface interactions between the implant and surrounding issues (see Figure 1.4).[32] Although many metals and alloys will naturally form an oxide layer on their surface, the chemistry and stability of this layer will be determined by the environment to which the implant was exposed at the time of oxide formation. For example, Ca, P, and S may be present in the oxide layer formed on the biomaterial submerged into a biofluid, and be absent when the layer is formed in ambient air. Besides adding new elements, the oxide layer may experience a change in the native oxide composition; for example, a passivation layer grown on Cr–Co–Mo in biofluid was found to be largely composed of Cr oxide, with only a limited amount of Mo oxide.[33] The uniformity, porosity and strength of adhesion to the underlying metal may also be affected. A greater level of control over the final surface properties can be attained *via* implant encapsulation and coating with a protective or bioactive layer, ion implantation to increase wear and corrosion resistance by modifying the top-most layer of the bulk implant, etching and structuring to impart desirable surface topography, and functionalisation to change the surface energy and wettability.

Modification of surfaces by mechanical means, for example using machining, grinding, polishing and blasting is often used to impart specific macro- and micro-scale roughness and texture onto a surface. The fabrication of more intricate patterns and textures, with multi-tier complex topography and

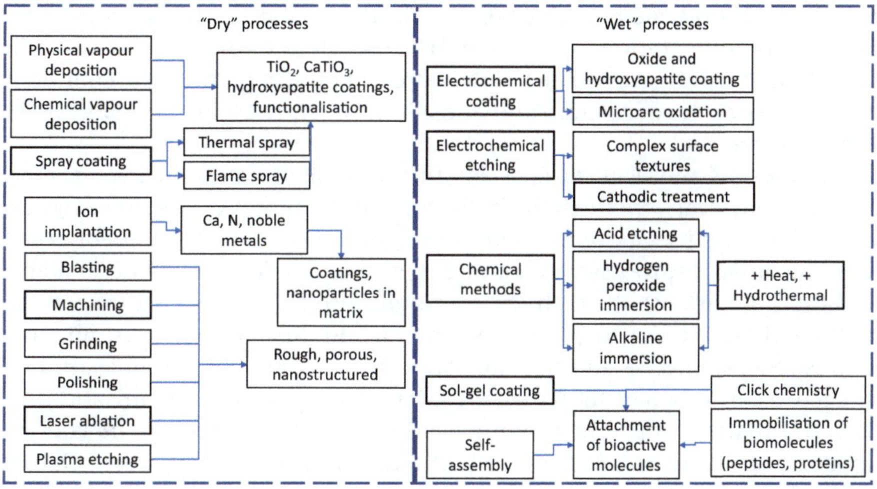

Figure 1.4 Common approaches to surface modification of metallic biomaterials.

well-defined micro- and nano-scale features typically demands the use of more expensive highly-controlled environments and high-energy particles, or the use of self-assembly techniques.[34] The bombardment of surfaces with high energy ions generated in plasmas can be used to manufacture surfaces with high aspect ratio features, including tall thin pillars and deep narrow trenches and pits,[35] with the possibility of fabricating features with aspect ratios of up to 100, deep features with vertical dimensions of >100 μm and radii of <10 μm, or high-aspect ratio features with dimensions below 100 nm. Here, by manipulating the gas composition, and electric and magnetic fields, one can control the intensity, density and direction of the ion flux towards the target surface, thus controlling the shape, density and orientation of the surface features.[36] In addition to etching, plasmas can be used for the growth of patterns of features or smooth continuous films (*e.g.* plasma polymerisation),[35,37] modification of surface chemistry (*e.g.* oxygen plasma treatment),[38] and surface hardening by driving ions into the top layer of the material (*e.g.* plasma nitriding).[39] Both plasma etching and femtosecond laser ablation can also be used to produce surface patterns with high resolution, such as those resembling patterns on natural surfaces capable of repelling bacteria or killing microbial cells on contact.[40–42] These surface patterns may be able to help host cells beat bacteria in their race to the surface, preventing surface biofilm formation and providing host cells with the chemical, physical and mechanical cues that promote tissue formation and implant osseointegration.[43] Exposure of the surfaces to acids, hydrogen peroxide, alkalis, anodic oxidation, sol–gel and chemical vapour deposition will also result in changes to the surface chemistry, and in some instances, surface topography, as in chemical mechanical polishing to obtain ultra-smooth surfaces or the hydrothermal etching of Ti using 1 M KOH to attain a nanostructured surface with strong bioactivity.[43] Electrochemical anodic oxidation has also been used to induce the formation of ordered oxide layers with nano-scale porosities on Ti, Al, Nb, Ta and Zr with the aim of promoting osseointegration;[44–46] this is because such porous oxides show improved mechanical interlocking with hydroxyapatite, a highly osseoconductive crystalline complex of calcium and phosphate that is present abundantly in the osteoid matrix.

Examples of physical surface modification techniques include thermal spraying, physical vapour deposition, ion implantation and sputtering, and pulsed laser and evaporative deposition methods. It is not uncommon to combine multiple processing techniques, and also follow them with biological activation of surfaces through the immobilisation of proteins, attractor molecules and factors that repel microorganisms and/or promote desirable host cell responses. Successful immobilisation is one in which biomolecules can attach in a controlled fashion with respect to density, surface organisation, and 3D conformation, since these biomolecules are the sites that will be later recognised by proteins and cells in the biofluid, and will eventually determine cell attachment preferences. This is generally done by first chemically modifying the surface with *e.g.* carboxylated poly(ethylene glycol), and then using –COOH to graft cell-recognisable biomolecules, including ethane-1,1,2-triphosphonic acid, methylenediphosphonic acid, morphogenetic protein-4, polydopamine, selenocystamine, and the cell–adhesive peptide Arg–Gly–Asp–Cys.[47] On gold and other noble metals, poly(ethylene glycol) and other molecules can be tethered *via* thiol (–SH or –SS–) or amino (–NH_2) groups due to their high affinity for these metals, whereas stainless steel surfaces may need to be first functionalised with a silane-coupling agent and activated by *e.g.* plasma treatment before they can be used for the UV-induced graft polymerisation of poly(ethylene glycol). For metals that form a native oxide, *e.g.* Ti alloys, poly(l-lysine)-*g*-poly(ethylene glycol) co-polymers can be used as they readily adsorb onto the surfaces of the negatively charged TiO_2, $Si_{0.4}Ti_{0.6}O_2$, and Nb_2O_5 *via* the positively charged primary amine groups of the poly(l-lysine), with the hydrophilic, uncharged poly(ethylene glycol) side chains pointing away from the surface.[48] Other strategies for the surface immobilisation of poly(ethylene glycol) and biomolecules include the use of highly-adhesive 3,4-dihydroxyphenylalanine (*e.g. via* conjugation to a single 3,4-dihydroxyphenylalanine residue or to the N-terminus of Ala–Lys–Pro–Ser–Tyr–Hyp–Hyp–Thr–DOPA–Lys);[49] glycose-containing diazirines (*e.g. via* the insertion of singlet carbenes into the H–O bonds of protective oxides, leading to glycosidation);[50] and photoreactive poly(ethylene glycol) (*via* nucleophile-reactive dehydroazepines created by the ring expansion of the photolysed intermediates of aryl azides).[51]

1.3 Key Properties of Major Types of Metallic Implants

1.3.1 Steels

The development of 18/8 (also known as type 304) stainless steel in the 1920s was an early step towards improving the corrosion resistance and mechanical strength of metallic implants, extending the useful lifetime of implants and reducing the incidence of post-operative complications and metallosis (see Figure 1.5). 18 wt% Cr is sufficient for alloy passivation, *i.e.* the rapid formation of a self-healing, chemically inert, protective Cr oxide layer which prevents the diffusion of oxygen to the underlying material, thus effectively providing the alloy with corrosion resistance in reducing acids and against pitting attack in solutions containing Cl^- ions, *e.g.* biofluids. Together with Cr, the 8 wt% Ni allows the alloy to take on an austenitic fcc crystal microstructure, increasing the density, hardness, and abrasive resistance of the steel, thus improving its resistance to oxidation, corrosion and heat. Nevertheless, while it is more resistant to uniform corrosion than carbon steels, 304 steel is susceptible to pitting corrosion when exposed to fluids rich in Cl^- ions and dissolved oxygen (*i.e.* conditions commonly encountered by an implant).

316L steel provides a greater level of corrosion resistance by reducing the fraction of carbon to <0.03 wt%, increasing the proportion of Ni to 10–12 wt% and adding 2–3 wt% Mo. The latter element is frequently introduced into steel to reduce the susceptibility of the alloys to pitting corrosion, whereas the fraction of carbon is intentionally minimised, as it tends to reduce the fraction of Cr available for passivation *via* the formation of Cr_3C_2. 316L steel is also austenitic, with a γ–fcc crystallographic structure. Martensitic and ferritic are the other two possible crystal structures of stainless steel, with the former characterised by a distorted bcc organisation of atoms, and the latter characterised by an α–bcc structure. Martensitic steels can be produced by transforming the fcc structure of the austenite into a body-centred tetragonal crystal structure using quenching. The process of cooling needs to be very rapid to minimise the number of C atoms that can diffuse out of the crystal structure, producing a highly strained, carbon-supersaturated, body-centred tetragonal structure that is very hard due to a significant fraction of dislocations that arise from the shear deformations.[52] The microstructure of martensitic steels makes them more brittle and less corrosion resistant when compared to austenitic steels, and heat treatment can effectively destroy the martensite by relieving the stresses within the crystal structure, making the steel softer. Cold working (rather than heat treatment) is also used to improve strength and toughness in austenitic steel. Although intermittent application of heat may be used during implant manufacturing in order to be able to shape steel into the desired product, excessive heat (*e.g.* during welding) may promote the formation of Cr_3C_2 and reduce the quality of the oxide layer, and as a consequence reduce the corrosion resistance of the alloy. It is often necessary to remove thus-formed suboptimal highly porous coatings using mechanical polishing and/or chemical means.

Efforts have been made to replace potentially toxic Ni with other alloying elements, *e.g.* nitrogen or manganese in Fe–Cr–Mn–Mo–N steels, while preserving the austenitic crystal structure and corrosion resistance of steel. N stabilises austenite and prevents the formation of ferrite, which would

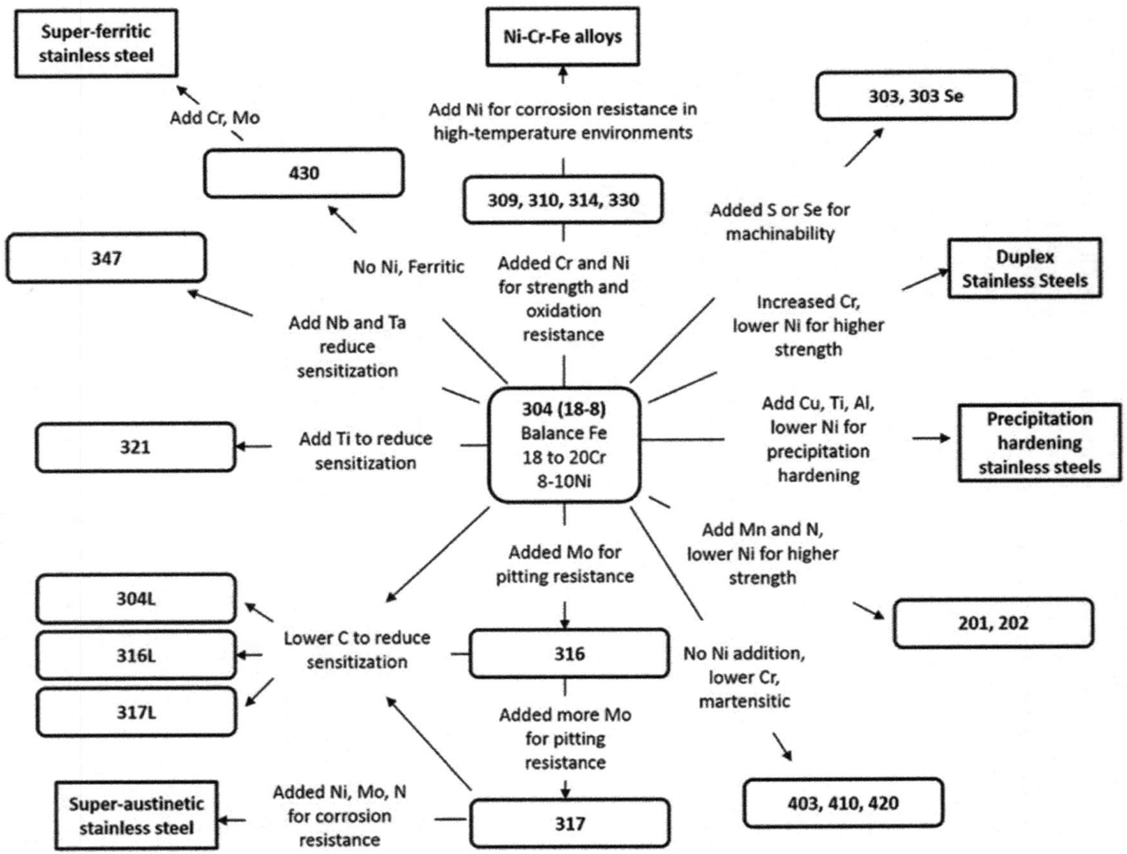

Figure 1.5 Relationships between the composition and properties in stainless steel alloys. Reproduced from ref. 231 with permission from Elsevier, Copyright 2016.

render the material ferromagnetic and reduce its ductility. Adding N to steel also strengthens the alloy, since interstitial N atoms produce large lattice distortions and interact with dislocations by means of electrostatic attractions; dislocation-nitrogen complexes may also drag the dislocations.[53] The changes in the dislocation structure due to the changes in the stacking fault energy, and specifically the planar dislocation arrangement in N-rich Ni-free steels that favours planar dislocation slips over dislocation cross slips, are responsible for the improved fatigue behaviour of these alloys. With respect to corrosion resistance, the formation of NH_4^+ or NH_3 (through the reaction between N and H^+) in pits and crevices may prevent local acidification and thus promote repassivation, whereas the formation of NO_2^- and NO_3^- can inhibit the anion attack due to N enrichment at the interface and the formation of protective Cr_2Mo_3N during passivation.[53] A significant challenge in using higher fractions of N is the resulting increase in the ductile-to-brittle transition temperature, which is proportionate to the amount of N used. This is because the ductile-to-brittle transition temperature is one of the key determinants of the fracture behaviour of the material (along with its impact strength); when the temperature at which steel is used is below the transition temperature, the steel loses its toughness and instead of undergoing plastic deformation it fractures in a brittle mode at the tips of cracks or flaws.[54]

Despite the aforementioned issues with corrosion, due to their significantly lower cost when compared to either Ti or Co–Cr alloys, steels continue to be widely used in medicine, particularly in applications where they are either used for a short time or where the dissolved oxygen-rich conditions prevent rapid corrosion (*e.g.* vascular stents).

1.3.2 Co–Cr Alloys

Co-based alloys owe their superior corrosion performance *in vivo* to the formation of a self-regenerating Cr_2O_3 passivation layer. Compared to steels, Co–Cr alloys are characterised by their better resistance to fatigue, wear and corrosion, and have comparable Young's moduli. They are also sufficiently ductile, for shaping and contouring of the implant, with 8% elongation performance. Mo is often added to reduce grain size and enhance strength by forming a solid solution where the local non-uniformity in the lattice impedes dislocation motion and thus plastic deformation. Depending on their composition (see Figure 1.6), implants made from Co–Cr alloys can be prepared by means of casting (*e.g.* dental Co–Cr–Mo implants), or hot working (*e.g.* wrought Co–Ni–Cr–Mo–Fe and Co–Cr–W–Ni), for example by hot forging (*e.g.* Co–Ni–Cr–Mo load-bearing stems for joint replacement). Wrought alloys display greater fracture toughness and fatigue strength, and are thus capable of withstanding large cyclic loads typical of orthopaedic applications. Yet, their ability to withstand crevice and pitting corrosion and mechanical wear is somewhat lower when compared to cast alloys. Although cold working can in principle be used to improve the strength of forged

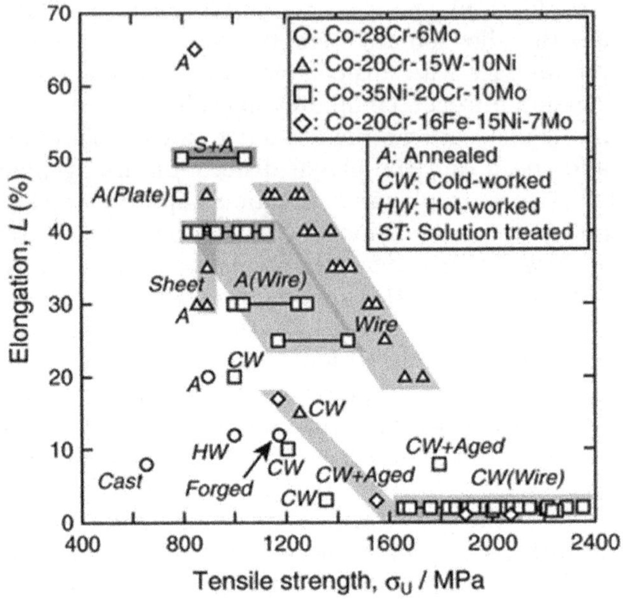

Figure 1.6 Tensile strength and elongation of biomedical implants made of Co–Cr alloys.
Reproduced from ref. 232 with permission from Springer Nature, Copyright 2015.

Co–Ni–Cr–Mo alloys, it is challenging, particularly when implants of a larger size are required. Suboptimal resistance to friction wear limits the use of forged Co–Ni–Cr–Mo as load bearing surfaces in devices.

Cast alloys, however, generally have a larger grain size, and distinct boundary separation, with such defects as holes and cavities present within the body of the alloy. Adding Ni improves the processability of cast alloys, whereas adding a small amount of carbon (~0.25 wt%, *e.g.* Co29Cr6Mo0.23C) makes the alloy easier to cast by reducing its melting temperature and increasing its fluidity. The addition of C also facilitates the precipitation of carbides along the grain boundary and in the interdendritic regions, increasing the hardness (and making the alloy more amenable to work hardening), yield stress, and resistance to wear of the alloy at the expense of its ductility. Solution annealing followed by ageing can increase the ductility by facilitating the dissolution and a more uniform re-precipitation of carbides within the matrix, bringing the austenite/martensite ratio to the desired level.[55]

1.3.3 Ti and Ti Alloys

The attractiveness of Ti as a metal for implants stems from its biocompatibility, and in particular, mechanical biocompatibility (due to a lower Young's modulus), lower density (at 4.5 $g\,cm^{-3}$ compared to >8 $g\,cm^{-3}$ for steels and Co–Cr alloys), and low corrosion under physiological conditions. There are four categories of commercially pure Ti defined by the level of impurities, with grade I having the highest purity of the four. Although 0.7% is the maximum allowable level of impurities for grade IV Ti, the fractions of certain elements is tightly regulated due to their significant effect on the key mechanical properties of Ti. Thus, the maximum allowed content of O, N and Fe should not exceed 0.4, 0.05 and 0.5 wt%, respectively. However, while the fractions of O and Fe should be sufficiently low to attain the desired fracture toughness, creep and stress rupture characteristics, respectively, the fraction of O should also be sufficiently high to achieve the required level of tensile strength. In its pure form, Ti is allotropic, with a close-packed hexagonal crystal structure (α-phase), and transitions to a bcc structure (β-phase) when heated above 885 °C up to 1668 °C. When an alloying element is introduced into Ti, it tends to stabilise either the α- or β-phase. For example, O, Al, N, Sn and C stabilise the α-phase, shifting the temperature at which α-to-β transformation takes place higher. Mn, Cr, Fe, Mo, V and Nb are examples of β-phase stabilising elements, and at sufficiently high quantities may allow the alloy to retain some β-phase when cooled to ambient temperature.

The chemical composition and resulting phase structure will prescribe the types of processing that these alloys can be subjected to in order to tune their properties (see Figure 1.7). Alloys with an α or near-α phase structure, *e.g.* commercially pure Ti, have superior weldability and heat and oxidation resistance, but lower workability and strength. Heat treatment can be effectively used to relieve internal stresses in α-type alloys but will not strengthen the alloy. This is because strengthening often relies on the precipitation of

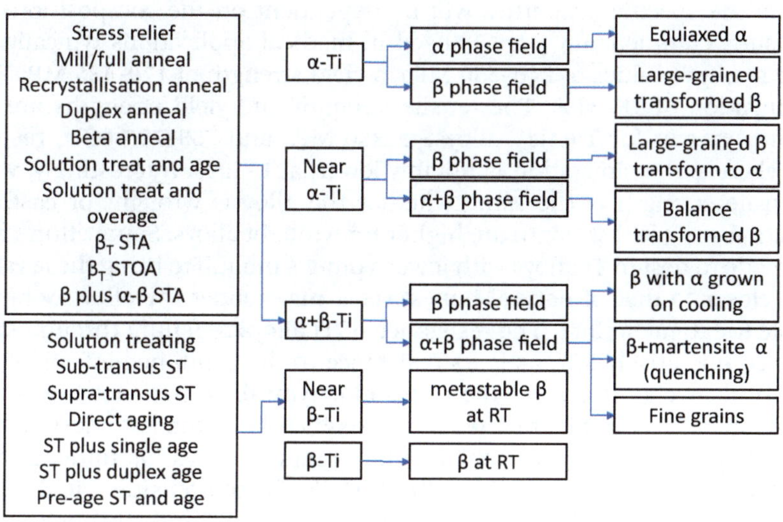

Figure 1.7 Hot working of Ti alloys.

one of the phases within a multi-phase system. For β-phase alloys, exposure to certain temperatures will lead to the loss of this metastable state, and, consequently, strengthening of the alloy. Heat treatment is also used for stress relieving and ageing. Alloys with an α–β two-phase structure, *e.g.* Ti6Al14V, are widely used in medical applications. They contain α-stabilising Al (5.5–6.5 wt%) and β-stabilising V (3.5–4.5 wt%), producing a structure with a β phase dispersed within an α phase. When the alloy is annealed, and then quenched and reheated again, the metastable β phase precipitates as particles, enhancing the strength of the alloy. At greater fractions of V (as in Ti13V11Cr3Al alloys), the β type microstructure is even more clearly pronounced, with annealing leading to a substantial strengthening and at the same time reduction in the ductility of the alloy.

In addition to mechanical properties, adding elements may change the biocompatibility of the alloy by changing the chemical and physical properties of its oxide. For example, adding Zr (as in *e.g.* Ti15Nb4Ta4Zr alloys) not only enhances mechanical strength and corrosion resistance, but also improves bone cell compatibility and bone mineralisation when compared to Ti6Al4V. While the oxide on pure Ti or Ti6Al4V alloy largely consists of amorphous and slightly crystalline TiO_2, with small quantities of Ti_2O_3 and TiO, and an insignificant fraction of Al_2O_3 and hydroxyl in the case of Ti6Al4V, oxides on Ti–Zr alloys have a fraction of zirconium oxides that reflects its fraction in the bulk alloy. Oxidation treatment of Nb-containing alloys, *e.g.* Ti9Nb13Ta4.6Zr, improves their wear resistance due to the formation of Nb_2O_5 particles on the diffusion hardened surface.[56] While in the case of Zr and Nb, the effect is beneficial, as thicker and more stable coatings offer greater protection, and the inclusion of even small quantities of NiO and metallic Ni into oxides on the surface of Ti–Ni alloy may negatively affect its corrosion resistance and biocompatibility.

While the specific properties will be dependent on the composition and processing, commercially pure Ti used in medical applications typically has tensile strength values of 240–550 MPa, a yield strength of 170–485 MPa, and an elongation of 15–24%. The tensile strength and yield strengths are significantly greater for Ti6Al4V alloys, at 860 MPa and 758–795 MPa, respectively, whereas the elongation is notably lower, at 8–10%. Interestingly, while the tensile strength is the same whether the alloy is wrought or cast, the elongation and yield strength are higher for wrought alloys. Some efforts have been made to design Ti alloys with lower Young's moduli to bring these values lower, closer to that of natural bones, yet it may render the alloy more susceptible to fatigue failure. The resistance of Ti alloys to failure through sheer is also comparatively low, as is its resistance to the combined effects of corrosion and mechanical wear, the conditions typically experienced by articulating load-bearing implant systems. Efforts have been made to design more robust alloys, such as those capable of more rapid passivation under fretting and sliding conditions (*e.g.* Ti29Nb13Ta4.6Zr[56]), martensitic Ti6Al4V ELI alloys that are more resistant to adhesive/abrasive wear, and alloys with more refined grain micro- and nano-structures.[9] In parallel, attempts to enhance the surface hardness of alloys through coating with hydrogenated amorphous carbon or nitriding to form a hard TiN/Ti_2N coating,[36,39] however there is a possibility of coating delamination and failure under higher loads.

1.3.4 Noble Metals and Alloys

Noble metals such as Au and its alloys have been used in restorative and conservative dentistry and orthodontics for over 4000 years, because of a favourable combination of corrosion resistance, durability and ease of processing. In its pure form, gold is very soft, with a large elongation (~45%) and a low proof stress of ~30 MPa. For this reason, Au and its alloys can be easily cold worked or cast to fill a cavity, although the softness of Au and some Au alloys (with >85% of Au) makes them unsuitable for load-bearing applications (*e.g.* tooth crowns or cups) as an implant made of such a material would not be able to withstand typical masticatory forces. Alloying Au with small quantities of Cu or Pt of less than 4% can provide additional strength to the alloy, without negatively affecting its corrosion profile. Since in restorative dentistry appearance plays a role, Ag may be added to the alloy to alter its colour. The melting point of Au alloys increases with the addition of greater levels of Pt and decreases when a small amount of Zn is used. Depending on the alloy composition, oxides containing Cu and/or Ag will form spontaneously on the surface of Ag–Pd–Cu–Au, Ag–Cu–Au, and Pd–Ag–In–Sn alloys. Pd-based alloys, such as Pd–Ag and Pd–Ag–Au, tend to have greater elongation compared to Au–Pd and Au–Ag–Pd alloys, whereas their mechanical yield strength is similar.

An alternative to Au tooth filling is an amalgam due to its lower cost, ease of application, strength, and durability. Dental amalgam is produced by mixing liquid Hg with an Ag-based alloy containing Sn (<29 wt%) and solid particles of Cu (<6 wt%), and small quantities of Zn (<2 wt%), Hg (<3 wt%) and sometimes other metals, following the amalgamation reaction $Ag_3Sn + Hg \leftrightarrow Ag_3Sn + Ag_2Hg_3 + Sn_7Hg$. The resultant plastic material is highly deformable, and thus can be used in restorations of larger size or complexity, *e.g.* where the margins of the cavity reside within dentine or cement. After approximately 1 h, the amalgam hardens to 25% of its final hardness level achieved after >24 h. A protective oxide containing SnO_2 encapsulates the fully hardened alloy. Over the years, there has been a lot of debate about the potential toxic effects of Hg, including reports of neurotoxicity (Alzheimer's or Parkinson's diseases), nephrotoxicity and chronic fatigue, yet large scale studies have proven to be inconclusive and this material is still being used today.[57]

1.3.5 Emerging Biomedical Materials

1.3.5.1 Shape Memory Alloys

In recent years, shape memory effects and pseudoelasticity in some alloys (*e.g.* NiTi alloys) have been recognised as valuable properties that can advance a number of surgical procedures. Pseudoelasticity allows the material to undergo large deformation when a load is applied, subsequently recovering is original form once the load is removed. The shape memory effect also allows the material to undergo significant deformation at temperatures below its phase transformation temperature, recovering its original form after the temperature is increased. This is because the application of stress or temperature to an equiatomic (or nearly equiatomic) NiTi alloy leads to a reversible phase transformation between symmetrical B2 austenite and less symmetrical monoclinic B19′ martensite, with the martensite existing in either a stress–induced or stress–free conformation (see Figure 1.8). The diffusionless martensitic phase transformation can be induced by cooling the alloy to below the martensite transformation temperature, T_M, or, at higher temperatures, by applying a load since the application of stress causes an upshift in the value of T_M. Because the lattice size and symmetry of martensite is very different to that of austenite, the phase transformation induced by quenching from high temperature results in the introduction of internal strain energy in the martensitic phase, which is reduced by twinning, with no macroscopic shape deformation taking place. Deforming the twinned martensite to a desired shape will result in the detwinning of the structures, as twins provide a favourable path to accommodate the dislocations when compared to slip, and the material will hold this shape as long as it remains below the temperature for austenite phase transformation, $T_{A\ start}$. Subjecting the material to temperatures above the $T_{A\ finish}$ will recover the B2 structure very rapidly, and subsequent cooling to below $T_{M\ finish}$ returns the material to the twinned or detwinned structure, corresponding to the shape of pre-deformed or deformed martensite, respectively. Both irreversible and reversible shape memory effects have their advantages. For example, a bone plate that contracts at body temperature may be used to exert a compression force onto the bone and promote healing; similarly, NiTi rods can be used to correct the relative position of spine vertebrae as a means to treating scoliosis. However, cycling between distinct shapes may be beneficial for active devices that are designed to respond to changes in their environment. Subjecting the latter type of material to repeated cycles of temperatures or to temperatures above a certain level (determined by the nature of the alloy) will lead to a partial or complete loss of memory.

In pseudoelasticity, stress from the load, rather than the temperature, drives the reversible phase transformation of the symmetrical austenite to twinned martensite and its subsequent detwinning at temperatures well above $T_{A\ finish}$, and the recovery of the original austenite crystal symmetry and macroscopic shape after the load is removed. The stress-induced deformation is highly non-linear since the austenite phase does not lend itself to deformation as well as martensite, due to the lower elastic modulus of the latter phase. Pseudoelastic behaviour is useful for the deployment of stents, since a stent can be severely deformed to facilitate implantation, then, once the conduit is removed, will regain its form. The austenite form is made to be a little greater in diameter than the blood vessel so that there is a force on

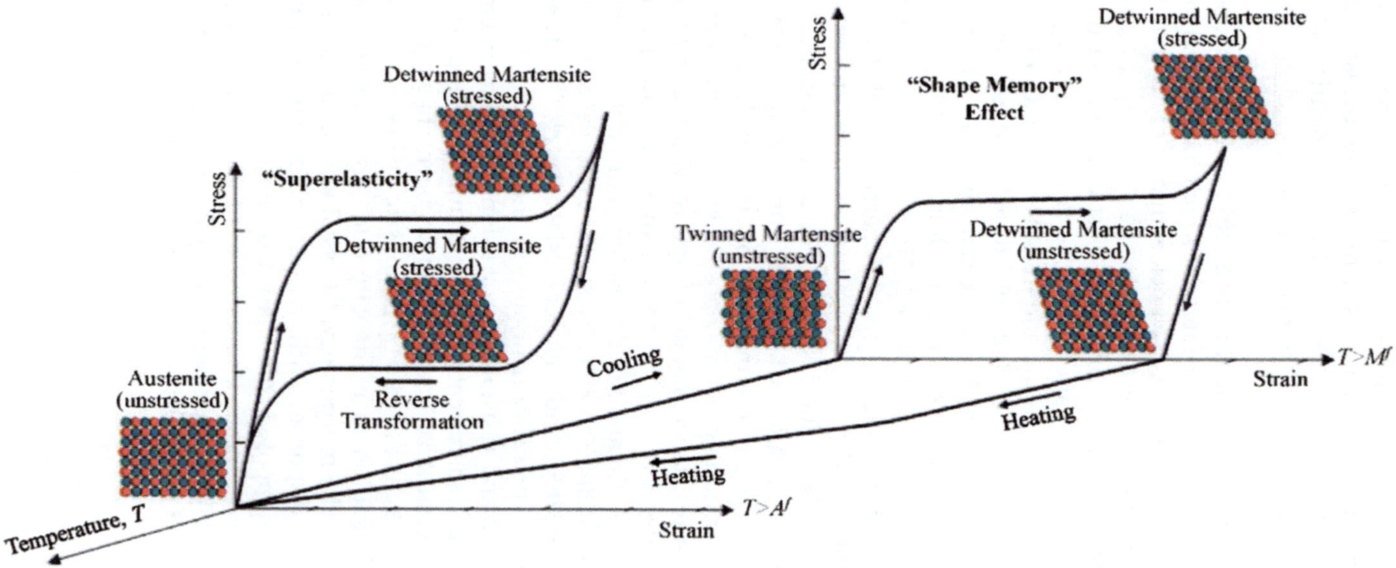

Figure 1.8 Stress–strain–temperature plot that illustrates the superelasticity and shape memory effect in metallic alloys. Reproduced from ref. 58, https://doi.org/10.3390/app5030187, under the terms of the CC BY 4.0 license, https://creativecommons.org/licenses/by/4.0/.

the vessel wall to resist compression due to, for example, muscle contractions. Unlike the case with stainless steel, the stents made of NiTi do not need to be overexpanded to prevent elastic recoil. The forces produced as a result of constrained recovery are also used in orthodontics to treat teeth overcrowding, and in dentistry, where NiTi is used for cleaning and shaping of the root canal.

When used as implants and surgical tools, NiTi and other shape memory alloys are designed so that the temperature at which the material finishes the martensite to austenite transition is well below that of the human body. The alloy composition will have a direct effect on the $T_{A\ start}$, $T_{A\ finish}$, $T_{M\ start}$ and $T_{M\ finish}$ temperatures, the maximum value of strain that can be thermally recovered, and the hysteresis characteristics, with $T_{M\ start}$ and $T_{M\ finish}$ being particularly affected by the nature and relative amounts of the alloying elements. For example, adding Cu (10 wt%) to NiTi reduces transformation and pseudoelastic hysteresis, and limits the sensitivity of T_M to the elemental composition of the alloy; in contrast, the addition of Nb widens thermal hysteresis, and the introduction of Pt, Pd, Hf, Au or Zr shifts the transformation temperatures significantly higher compared to TiNi. The addition of Ag may provide antibacterial activity against common implant-associated pathogens, while also slightly increasing the corrosion resistance, elongation and tensile strength of the alloy, whereas Hf implantation may render TiNi more wear resistant owing to the presence of HfO_2 in the protective oxide layer. Increasing the fraction of Hf may positively contribute towards fatigue resistance. Corrosion resistance may also improve with the addition of Mo.

Apart from NiTi (developed in the 1960s), shape memory behaviour has also been observed in AgCd and AuCd (developed in 1930s), Cu-based alloys (*e.g.* CuAlNi, CuAlBe, CuSn, CuZr), Fe-based FePt and FePd, and others, *e.g.* InTl, NiAl, MnCu, FeMnSi, and magnetic NiMnGa, *etc.*[58,59] However, the manipulation of alloy chemistry is limited by the need for the material to remain biocompatible and resistant to corrosion. The alloy microstructure will also affect both its transformation temperatures and key properties. NiTi alloys with a small grain size (<100 nm) have been shown to have lower transformation temperatures, with a grain size below 50 nm effectively suppressing transformation to martensite due to the effect of nanoscale grains on B19′ martensite morphology.[60] Crystallinity also affects the shape of the stress–strain response, with polycrystalline coarse-grained NiTi showing a plateau, as opposed to a sharp rise that is evident in the nanocrystalline NiTi response. When compared to other metallic implants, the plateau stress of NiTi is on par with their yield strength, but it is more corrosion resistant and biocompatible than stainless steel. In addition to the use of solid NiTi implants as rods, nails, plates, and vertebrae spacers, scaffolds made of porous shape memory alloys are being investigated for their ability to promote bone tissue remodelling and vascularisation while being strong and lightweight.

In addition to cardiovascular and orthopaedic applications, NiTi alloys are also being used in neurosurgery (as coils, stents and guidewires), in the treatment of urethral, oesophageal, rectosigmoidal, and prostatic constrictions, removal of bile duct obstructions, kidney stones and foreign materials, and for fast closing of wounds during surgery. When permanent NiTi implants are subjected to cyclic loads, fatigue may lead to premature failure, even more so than in stainless streel, Ti- and Co–Cr-based alloys. This type of failure has been observed in NiTi stents implanted into peripheral arteries, *e.g.* superficial femoral and femoropopliteal arteries, where they are subject to loads associated with systolic/diastolic pressure cycles and limb movement.[61] The nonlinearity of the pseudoelastic response makes fatigue assessment and alloy optimisation challenging. Care should also be taken when considering NiTi for load-bearing implants with architectures containing multiple fine elements, since the possibility of the structural failure of these devices may be greater. The possibility of the leaching of Ni and associated toxicity needs to be considered for applications that would increase the possibility of implant wear and corrosion.

1.3.5.2 Biodegradable Alloys

In the case of alloys made of Co–Cr, Ti and stainless steel, efforts have gone into rendering them to be as corrosion resistant as possible, Mg- and Fe-based alloys are intended to undergo a controlled break down under physiological conditions. The use of these materials is primarily aimed at applications where tissue healing and regeneration is expected, and can in fact be promoted by gradually removing the mechanical support provided by the implant. Among the obvious examples are plates and screws that hold a bone together after fracture, and stents that provide radial support to a blood vessel during its remodelling. Once the healing process is complete, the presence of the implant can be detrimental for many reasons. Examples range from providing a suitable platform for biofilm formation and interfering with load transfer (*e.g.* bone plate), to platelet attachment and thrombus formation (*e.g.* stent), to releasing potentially allergenic and cytotoxic products of corrosion, or being a source of chronic inflammation, physical discomfort and pain for the patient.

The biocompatibility of Mg, Fe and their alloys stems from their lower density and Young's modulus, which makes them more mechanically compatible with bone, and the ability to form apatites on their surface, which promotes healthy bone formation. However, pure Mg is prone to rapid degradation under *in vivo* conditions, particularly in environments that are rich in Cl^- ions, with the rate of degradation affected by the properties of the implant and that of the environment. Although the products of corrosion (*e.g.* Mg^{2+}) are safely metabolised by the body, the concomitant generation of H_2 and alkalisation of the peri-implant milieu may prevent direct tissue contact, whereas significant loss of material may lead to a rapid reduction in the mechanical strength and implant failure. The principle means to match the degradation rate with the tissue regeneration requirements include varying the alloy composition (see Figure 1.9), controlling the grain structure

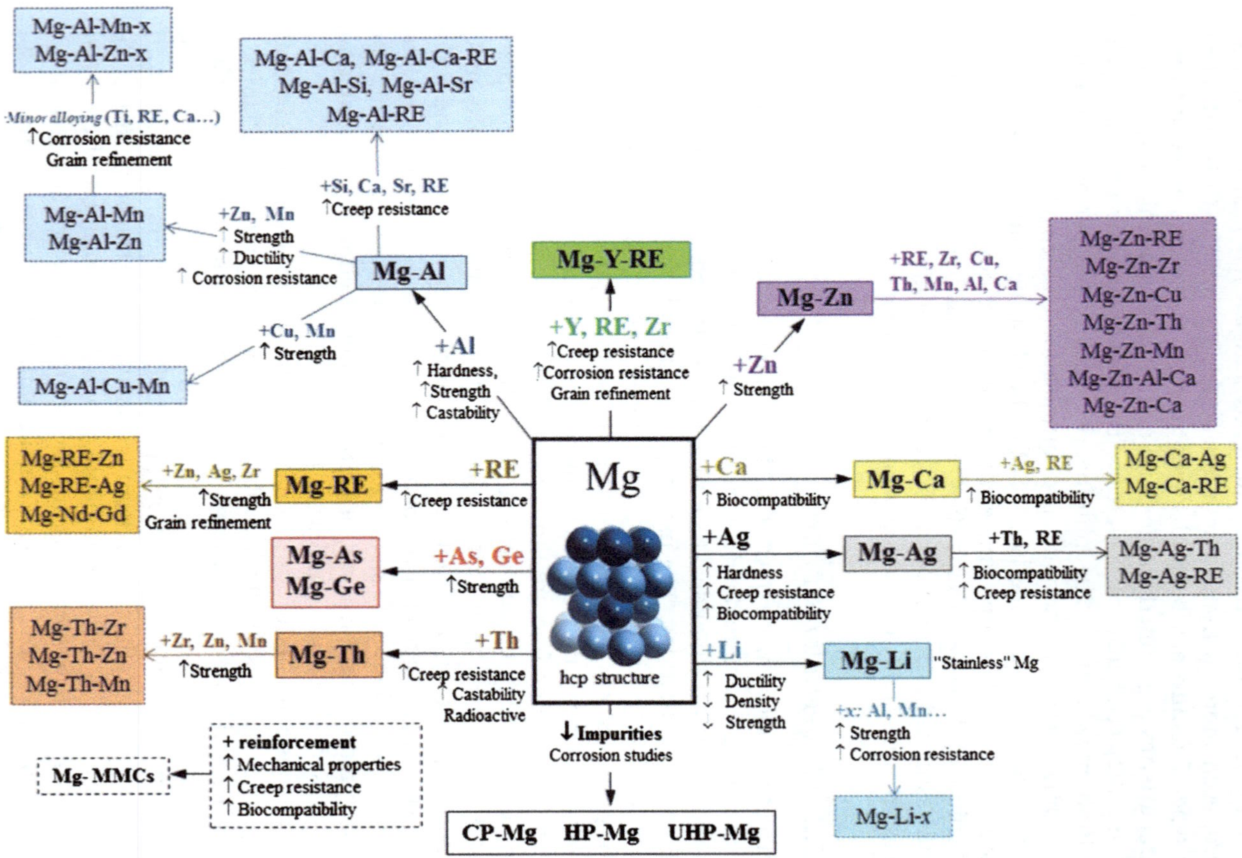

Figure 1.9 Relationships between alloying the elements and properties of Mg-based alloys. Reproduced from ref. 233, https://doi.org/10.1016/j.pmatsci.2017.04.011, under the terms of the CC BY 4.0 license, https://creativecommons.org/licenses/by/4.0/.

or surface modification. For example, adding a small amount of individual rare-earth elements (*e.g.* Gd, Nd or Y), or their combination (*e.g.* Nd, Ce, Dy, La and Pr) can considerably reduce the corrosion rate of MgZnAl alloys. Here, Zn and rare-earth metals are integrated within a protective oxide, increasing their stability when in contact with biofluids by limiting hydration and magnesium egress of the inner layer. In turn, Al may stabilise the hydroxides on the surface of the alloys under chloride-rich conditions, effectively slowing their corrosion. Surface-immobilised amorphous calcium phosphates may further reduce the rate of corrosion.

As a non-essential metal, exposure to Al particles may result in undesirable physiological response, although low doses of Al may be well tolerated. It is worth noting that humans are exposed to potentially large doses of Al and other non-essential elements in their daily lives; for example, Ti and Al are frequent ingredients in sunscreens and antiperspirant deodorants. However, the possible effects of prolonged exposure to rare-earth elements have yet to be fully explored, but La and Ce may be cytotoxic, with Gd and Dy offering the best combination of toxicity and solubility. Some efforts have been made to reduce or fully replace rare-earth elements in Mg alloys with, for example, Ca or Zr.

As an alternative to alloying, a variety of surface treatments have been trialled to either stimulate the spontaneous formation of protective oxides (*e.g.* compact $Mg(OH)_2 + Mg_2Al(OH)_7$ *via* hydrothermal treatment), or to apply a slowly degrading polymer or amorphous SiC coating. To function properly, the coating should be well-attached (including when the implant is subjected to deformation), defect-free and not contain any substances that may be harmful to cells at any point in their degradation.

1.4 Corrosion of Metals *In Vitro* and *In Vivo*

Whether they are used for orthopaedic, dental, or cardiovascular applications, most metallic biomaterials and device components are expected to withstand corrosion, with the exception of biodegradable materials, where controlled corrosion of the metal is part of their function.[62] Corrosion is a coupled oxidation–reduction reaction, whereby the oxidising agent gains an electron and the reducing agent donates an electron. In aqueous environments, *e.g.* biofluids, the process of corrosion involves the establishment of electrochemical corrosion cells and is governed by thermodynamic and kinetic factors that determine the corrosion tendencies and the corrosion rate, respectively.[63]

Traditional inert materials, including base metals and alloys based on cobalt–chromium, titanium, and stainless steel materials, tend to donate electrons in solution, and as such, are all likely to corrode when exposed to fluids in the peri-implant milieu, an inherently corrosive environment rich in oxygen-containing chemical species. As a result of these corrosion reactions, metals undergo dissolution, and metallic ions are produced in the process. The rate and chemistry of these reactions is dependent on the chemistry of the material and the specific characteristics of the environment (see Table 1.6). For example,

Table 1.6 Physical and chemical factors in the body known to affect corrosion.

Environmental condition	Physical parameters	Considerations
Body temperature	36.1–37.2 °C	Increased rate of chemical reactions
pH		
Blood	7.15–7.35	Transient decreases in peri-implant space to ~5.2. Oxygen-containing chemical species drive corrosion. Transient increases in ROS in peri-implant milieu due to inflammation
Intracellular matrix	7	
Cells	6.8	
Dissolved oxygen		
Arterial blood	100 mmHg	Oxygen levels affect formation of protective oxides
Venous blood	40 mmHg	
Intracellular matrix	2–40 mmHg	
Chloride ion		
Serum	113 mM	Monovalent chloride stabilises suspensoid sols preventing the formation of a protective corrosion product layer on metals
Interstitial fluid	117 mM	
Cardiac muscle	20–30 mM	
Mechanical load		
Cancellous bone	0–4 MPa	Trigger metal fracture, stress corrosion cracking
Cortical bone	0–40 MPa	
Arterial wall	0.2–1 MPa	
Myocardium	0–0.02 MPa	
Muscle (peak load)	40 MPa	
Tendon (peak load)	400 MPa	
Load cycling		
Miocardium	5×10^{6}–4×10^{7} per year	Trigger corrosion fatigue
Finger joint	10^{5}–10^{6} per year	
Walking	2×10^{6} per year	
Chemistry of body fluids (Interstitial, synovial, serum)		
Na^{+}	3127–3280 mg l^{-1}	Affect formation of protective oxides and mechanism and rate of corrosion
K^{+}	156 mg l^{-1}	
Ca^{2+}	60–100 mg l^{-1}	
Mg^{2+}	−24 mg l^{-1}	
Cl^{-}	3811–4042 mg l^{-1}	
HCO_3^{-}	1892 mg l^{-1}	
HPO_4^{2-}	96 mg l^{-1}	
SO_4^{2-}	48 mg l^{-1}	
Organic acids	245 mg l^{-1}	
Protein	15 000–4144 mg l^{-1}	

when commercially pure titanium and Ti6Al4V alloy are exposed *in vitro* to the presence of fluoride F^{-} ions at concentrations typical of the oral cavity, the former is subject to pitting corrosion whereas the latter is affected by general corrosion with characteristic surface micro-cracks.[64]

The initial instance of surface corrosion upon first exposure to an *in vitro* or *in vivo* environment is generally followed by the spontaneous formation of a thin barrier film, often an oxide or hydroxide if corrosion takes place in biofluids, which subsequently protects the underlying bulk metal from further corrosion and degradation. Failure to form such a layer, *e.g.* due to unfavourable environmental conditions, or its subsequent distortion, *e.g.* mechanical damage, would render the material vulnerable to corrosion and related failure. The balance between the destruction and reforming of the protective passive layer determines the predisposition of the implanted metal to a specific type of corrosion (see Figure 1.10), *e.g.* pitting, stress corrosion, corrosion fatigue, and crevice, intergranular and fretting corrosion.[63] Given the complexity of chemical and physical processes that can lead to degradation *in vivo*, it is not unusual for biomaterials to show signs of more than one type of corrosion, with factors that affect the outcome varying from those specific to the electrochemistry of the biofluid, to the type of mechanical loading the implant may experience, to the surface finish and geometry of the implant itself.[63]

General corrosion is a process that describes the uniform loss of material across its entire surface-environment interface. In metallic materials used *in vivo*, general corrosion takes the form of passive corrosion, the rate of which is defined as the difference between the rates of passivation *versus* that of passive layer dissolution. Typically, for implants, this rate is low, often not exceeding 1 pm per year, and is unlikely to affect the mechanical properties of the implant. Nevertheless, even at this rate, biochemically non-trivial quantities of potentially toxic Cr, Co and Ni ions may be leached into the peri-implant milieu, affecting tissue health and regeneration. Furthermore, this corrosion rate can increase in the presence of certain chemical species. For example, the corrosion of Ti is greatly accelerated when fluoride ions or hydrogen peroxide (both used in dental care) are added to the solution.[64,65]

1.4.1 Pitting and Crevice Corrosion

A much greater rate of corrosion can occur when the passive layer is removed, *e.g.* due to physical damage or changes in the local pH, resulting in the *localised corrosion* of affected areas. Certain chemical sites in the passivation layer can make it more susceptible to ***pitting corrosion*** due to increased levels of dissolution at these sites, especially when in contact with liquids containing halide ions. For example, impurities in the metal serve as nucleation sites at which pits are formed. Under mild conditions, these formed pits are likely to be repassivated, even though they can propagate in a metastable fashion, remain small (10–100 nm) and do not cause macroscopic damage (see Figure 1.11).[66] Once the conditions reach a certain threshold, termed as the pitting potential, stable pitting corrosion may ensue. It should be noted that the nucleation of pits can take place under a wide range of conditions below the pitting potential. Furthermore, the presence of metastable (repassivated) pits plays a role in stress corrosion cracking, which will be discussed later.

The formation of metal hydroxides and localised drops in pH can be expressed as $M^{n+} + nH_2O \rightarrow M(OH)_n + nH^+$. Local acidification of fluids

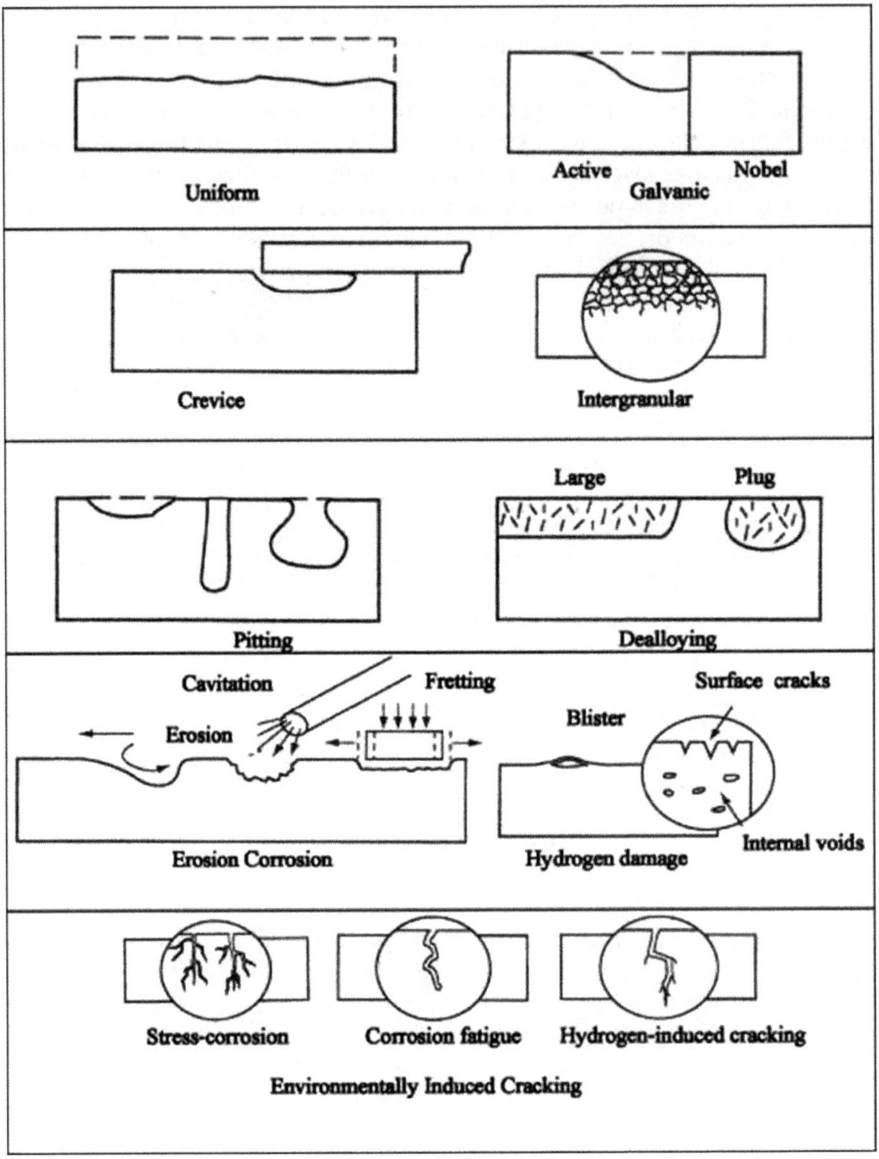

Figure 1.10 Pictorial representation of different types of corrosion. Reproduced from ref. 234 with permission from Elsevier, Copyright 2006.

within the pit is the main driver of corrosion and pit growth, while the geometry of the pit prevents the effective influx of dissolved oxygen from outside of the pit. As a result, cathodic processes that take place on the unaffected passivation layer are physically separated from the anodic processes within the pit. The considerable levels of metal cations generated as a result of

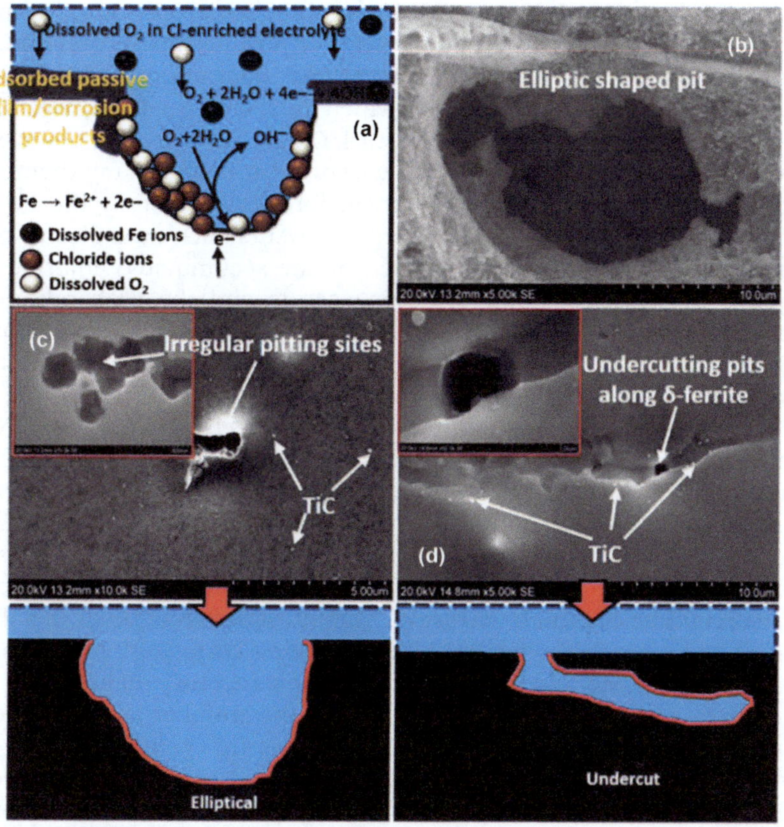

Figure 1.11 Pitting patterns on Ti-stabilised austenitic stainless in NaCl medium. The presence of TiC particles does not promote pitting corrosion of the steel, with pitting taking place along the ferrite lines. (a) Schematic of the pitting patterns in undeformed samples and samples previously subjected to quasi-static loading. (b) Shallow sub-surface pits in undeformed samples. (c) Irregular elliptical-shaped surface pits in previously deformed samples. (d) Undercutting pits at δ-ferrite sites in previously dynamically loaded samples.
Reproduced from ref. 107, https://doi.org/10.1038/s41598-019-48594-3, under the terms of the CC BY 4.0 license, http://creativecommons.org/licenses/by/4.0/.

fast corrosion within the pit cannot effectively diffuse out of the pit, and anions from the surrounding environment, *e.g.* ions of Cl or F, are drawn into the pit instead to maintain electroneutrality. For chloride ions, the relevant reaction can be expressed as $M^{n+}\ Cl_n^{\ -} + nH_2O \rightarrow M(OH)_n + nHCl$.

The resulting inter-pit environment is highly conducive to corrosion processes, and the pits themselves can provide a source of metal ions at potentially toxic concentrations. When considering the impact of pitting corrosion of an implant, it is important to consider the susceptibility of the material to depassivation as well as the frequency of pit formation over time.

For example, where stainless steel is more prone to depassivation compared to Ti and as such has a greater frequency of pit nucleation, over time this frequency reduces to zero as the repeated nucleation and metastable propagation eliminate the sites at which pits tend to form, *e.g.* sulfide inclusions on the surface of stainless steel. On the surface of Ti, however, the pit nucleation remains at a low frequency over time, suggesting that for this metal, nucleation is not driven by inclusions but rather by the inherent properties of the titanium oxide layer, and events such as explosive microscopic rupture of the layer perpetrated by the accumulated chloride at its interface.[66–68] The frequency of pit nucleation and the probability of pit propagation turning from metastable to stable is also dependent on the characteristics of the ambient environment, particularly the concentration of corrosion-promoting ions, reduced levels of dissolved oxygen, the great oxidising potential of the reactive species, and increased temperature and reduced pH. Operating conditions under which implants are exposed to cyclic loads may increase their susceptibility to pitting corrosion, such as is the case with Co–Cr-containing metallic alloys that undergo pitting corrosion under these conditions but not under static load. Materials processing, for example certain types of cold working, may also render the alloy more susceptible to pitting corrosion. The introduction of other metals into an alloy may change its susceptibility to pitting corrosion. For example, a small amount of Mo added to steel renders (as in 316L) makes them more resistant to pitting, substantially reducing the number of pitting-related failures. In contrast, adding elements to Ti (as in *e.g.* Ti6A14V and Ni–Ti shape memory alloys) makes it subject to pitting corrosion under certain medically-relevant conditions, as it creates potential sites of weakness in the passivation layer. Where the pitting is generally superficial in the case of Ti6A14V, the consequences of pitting corrosion in the case of Ni–Ti are more concerning and may result in significant levels of ion release. It should be noted that predicting how the implant will corrode *in vivo* is not trivial, since the outcomes of *in vitro* studies may not always correlate with those later observed *in vivo* in an animal model or in a patient. For example, while type 304 stainless steel undergoes extensive pitting corrosion after days of incubation in 0.9% NaCl (due to localised acidification following the hydrolysis reaction $Fe^{2+} + 2H_2O + 2Cl^- \rightarrow Fe(OH)_2 + 2HCl$), a Sherman plate fabricated from the same alloy and retrieved from a patient after 40 years of implantation showed no visible signs of corrosion.[69] At the same time, other studies have reported significant levels of *in vivo* corrosion in stainless steel, some leading to severe implant failure due to a combination of pitting attack, fatigue striations, and microcracks (see Figure 1.12).[70–72]

Another type of localised corrosion, *i.e. crevice corrosion*, takes place when the inflow and outflow of chemical species is reduced due to the geometry of the space created by two surfaces coming into close proximity with each other, for example at the interface between the head of fixing screws and bone or plate (see Figure 1.13). This is because the formation of the protective oxide relies on the availability of dissolved oxygen, whereas the

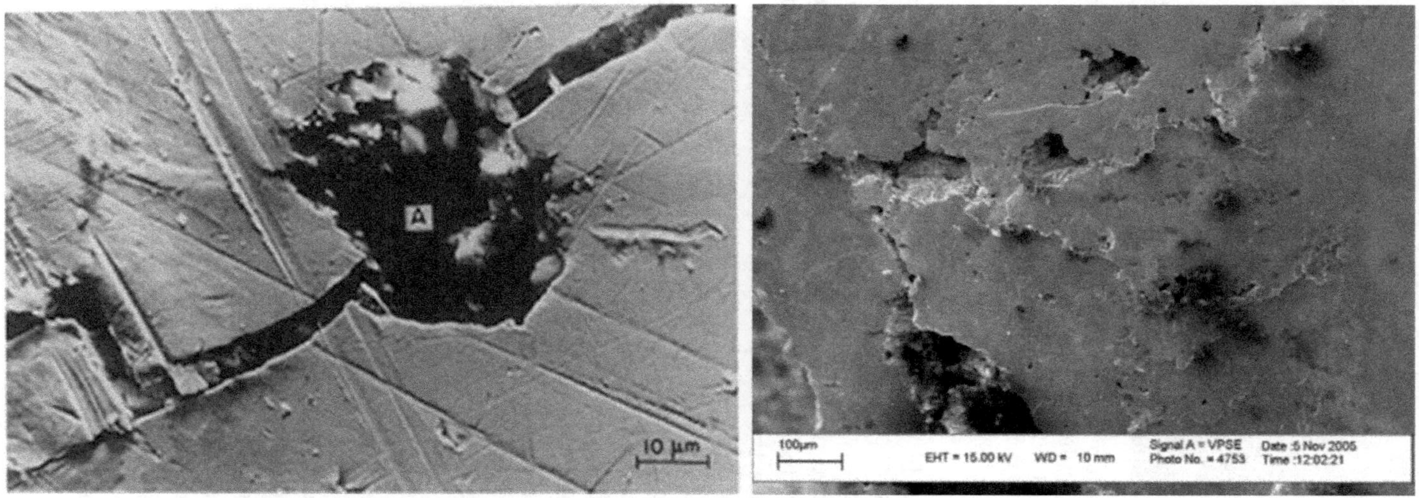

Figure 1.12 SEM visualisation of corrosion pits and associated cracks that develop in orthopaedic stainless steel under applied load. Failed stainless steel plates were removed from patients during revision surgery. Point A shows that the crack was associated with the corrosion pit. Substantial amounts of chloride and sulfur, as well as calcium and phosphorus were the primary corrosion products detected in the pit.
Reprinted from ref. 71 with permission from Springer Nature, Copyright 1992. Reprinted from ref. 70 with permission from Elsevier, Copyright 2006.

Figure 1.13 Microscopic images of the exposed area under the collar showing crevice corrosion in REX 734 steel in a solution of 9 $g\,L^{-1}$ NaCl in deionised water at 37 °C and pH of 5.7 (200× and 7×).
Reproduced from ref. 235 with permission from Elsevier, Copyright 2002.

dissolution processes are promoted by increased concentrations of halide species. Just as in the case of pitting corrosion, the narrow dimensions of the crevice (typically in the order of 3 mm) prevents the convective removal of metallic ions and effective delivery of oxygen, with cathodic and anodic processes spatially restricted to the oxide surface and inner part of the crevice, respectively, since the rate of diffusion-driven mixing is not sufficiently high enough to be effective. In physiological fluids, this typically results in an influx of Cl^- and similar anions into the crevice, and a significant local drop in the pH. The latter will further promote corrosion, and consequently increase the number of metallic ions released into the crevice that will draw even greater quantities of anions into the pit to maintain electroneutrality, reducing the pH even further. Since the chemical mechanism of crevice propagation is very similar to that of pitting corrosion, similar characteristics of the environment will work to promote or hinder it.

In addition to the effects already discussed, certain types of porous implants may also display increased susceptibility to crevice corrosion. For example, even though titanium is generally considered to be immune to crevice corrosion under physiological conditions, oxide films that form on porous titanium have inferior stability compared to those on solid Ti, with the point of initial failure lying within the depths of the pores. In addition to reduced convectional flow of ions and dissolved oxygen, galvanic coupling to the outer surface promotes a fast dissolution of the material within the pores.[73] The geometry and size of the pore, and the degree of porosity and connectivity between the pores also has a notable effect on the corrosion behaviour of cobalt alloys.[74] Furthermore, even materials considered to be relatively immune from pitting corrosion under normal conditions, *e.g.* 316L, 316LVM and austenitic microstructured steel and Co–Cr–Mo, may nonetheless remain prone to crevice corrosion to some degree.[75] When investigating their potential for crevice corrosion, it is important to consider the time over which testing takes place, specifically, the observation period should be sufficiently long for the outcomes of slow processes to come to light, as well as to consider the dynamic nature of the *in vivo* environment – the latter may change over the course of implantation, and vary between patients.

1.4.2 Stress Corrosion Cracking and Corrosion Fatigue

When coupled with mechanical stress, the above described corrosion reactions can lead to crack propagation and eventual failure of the material, a process termed as *stress corrosion cracking* or *corrosion fatigue* depending on whether the loading is static or cyclic in nature. The main issue with both types of failure is that they may proceed slowly (10^{-9} to 10^{-6} ms^{-1}) with little visible evidence up to the point when fracturing occurs. The action of the mechanical and corrosive factors is synergistic in that when applied independently neither of the factors would be significant enough to induce failure. That is, mechanical loading of the pre-corroded metal or alloy at corrosion-generated or manufacturing flaws, at the same level of stress as in stress corrosion cracking, but in the inert environment will not lead to crack propagation. In contrast, the same loading under a corrosion-inducing environment would lead to time-dependent crack propagation. It should be noted that the level of stress concentrated at these surface flaws is below the critical value for stress-intensity factor at which macroscopic mechanical fracture of the material would occur under inert conditions.[76] Both external and residual tensile stresses can cause crack initiation and propagation, although residual stresses that are compressive in nature may provide some degree of protection from this type of failure. The environmental factors that contribute to crack initiation in stress corrosion cracking and corrosion fatigue may differ. Nevertheless, the dynamics of the evolution from localised corrosion to cracking are affected by electrochemical processes that take place at the site of corrosion, the presence of specific species that tend to promote corrosion in a given metallic system, the properties of the material (including its processing, *e.g.* heat treatment), and the nature of the stress–strain system. Superior tensile and fatigue properties in air do not necessarily correspond to reduced susceptibility to stress corrosion cracking or corrosion fatigue. For example, a high-strength low-alloy Mg alloy, MgZn1Ca0.3 (ZX10), showed improved mechanical properties in air when extruded at 325 °C compared to when extruded at 400 °C due to grain-boundary strengthening, yet both materials were found to be susceptible to crack propagation under slow strain tensile and cyclic tension–compression loadings in simulated body fluid.[77]

It should be noted that corrosion processes that take place on a cathode, *e.g.* evolution of hydrogen, absorption, diffusion, and embrittlement, as well as those on an anode, *e.g.* active dissolution and elimination of metal from the crack tip, can participate in crack initiation. Crack propagation relies on interatomic bonds to be broken, which can result from chemical processes (*e.g.* dissolution) as well as ductile and brittle mechanical fracture (of metals and oxides, respectively) induced by such events as the adsorption of chemical species from the peri-implant milieu, reactions at the surface and in the metal ahead of the crack tip, and others.[76] Corrosion sites and flaws, *e.g.* discontinuities, where the tip of the crack is sharp are more likely to undergo crack propagation as stresses can be concentrated at the tip, as in the case of metastable pitting where rapid repassivation ensures that crack tips remain sharp (see Figure 1.14). In contrast, blunt cracks from general corrosion or stable pitting are less likely to result in stress corrosion cracking. For this reason, it is often metals that resist general or active pitting corrosion that succumb to stress corrosion cracking and corrosion fatigue.

In general, cyclic loads may lead to crack propagation at levels below that of static loads. Furthermore, there is no cut-off point, *i.e.* fatigue limit, that needs to be breached for the failure to occur, and the probability of failure due to corrosion fatigue is greater at lower cycle frequencies, which is the case for many biomedical implants, such as those used to restore hip and knee joints. The length of the implantation makes failure due to corrosion fatigue more likely, as it extends the amount of time the stressed material is exposed to the corrosive action of bodily fluids. For some materials, such as steel, reducing the probability of pitting corrosion may have a considerable effect on improving their corrosion fatigue resistance.[78] In addition to alloy chemistry, phase composition and shape also play an important role in

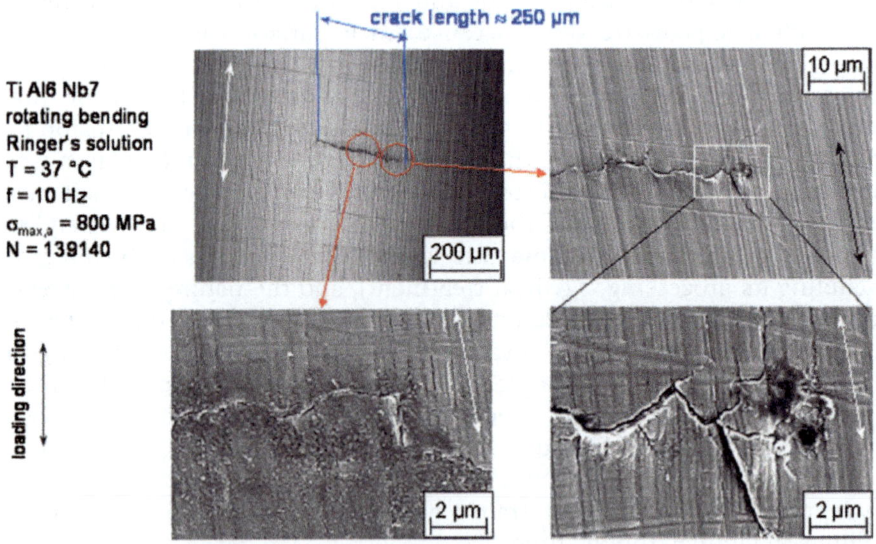

Figure 1.14 SEM images of the surfaces of TiAl6Nb7 subjected to rotating bending in Ringer's solution after test interruption at a defined decrease in the free corrosion potential. In the area of crack origin (bottom left), the surface shows distinct signs of corrosion. It may be assumed that this area was activated due to strain localisation. At the crack-tip, crack branching occurs (top right) and plastic deformation marks are visible (bottom right). Reproduced from ref. 79 with permission from Elsevier, Copyright 2010.

defining corrosion resistance.[79] The extent to which corrosion fatigue and stress corrosion cracking present a real problem for implants remains a subject of debate, with many authors finding no evidence that either of these mechanisms of failure contribute significantly towards loss of performance in real-life implants.[79]

1.4.3 Hydrogen Embrittlement and Fretting Corrosion

Under conditions that favour increased hydrogen evolution at the surface of the intact passivation layer, *e.g.* when the metal in question is galvanically coupled to another metal with greater activity, hydrogen atoms can interact with the metal lattice, leading to the formation of brittle metal hydrides that are prone to cracking under applied or residual stress. Cobalt-based alloys, *e.g.* Zimaloy used in orthopaedic applications, showed a notable loss in ductility when incubated in a saline solution under a cathodic potential that can be attained in a crevice sample after 60 days of incubation.[80] *In vivo* hydrogen embrittlement was suggested to be the cause of severe corrosion attack on the mating interfaces of retrieved hip-implants with Ti6Al4V/Ti6Al4V modular taper interfaces in the stem. The surfaces of the modular connections showed signs of etching, pitting, delamination, and surface cracking, as well as the precipitation of brittle hydrides as a result of crevice and *fretting corrosion*, where the oxide layer was mechanically damaged.[81] Naturally, Ti and other materials that rely on the protective effects of an oxide layer are subject to a considerable reduction in fatigue strength under fretting conditions, with the nature of the environment playing a less important role when the damage from fretting is more severe.[82] Increasing the fretting resistance, for example by implanting nitrogen into the surface of Ti6Al4V, and changes to the alloy grain structure through, for example, heat treatment, may reduce the susceptibility of some materials to fretting corrosion.[78] This is because cracks are observed to propagate along the boundary lines between the acicular α and β phases.

From a chemical standpoint, unlike pitting or crevice corrosion where the removal of the protective oxide is typically brought on and then accompanied by considerable changes in chemistry of the fluid in the immediate proximity of the site, infrequent removal of the oxide by mechanical means only results in a minimum level of corrosion needed to accomplish repassivation. Once the instances of mechanical damage become recurrent due to wear or fretting, the repassivation may fail to occur fast enough, resulting in a considerable loss of material due to corrosion (see Figure 1.15). The mechanical damage may originate from fluid–surface interactions under certain flow conditions, similar to the way flowing rivers erode bedrock, as well as from physical contact between solid surfaces. Where erosion corrosion may in principle take place in biomedical devices used to control the flow of biofluids, fretting corrosion is more likely to be encountered in multi-component load-bearing devices, *e.g.* orthopaedic implants. It is quite possible that in these devices, more than one type of corrosion takes place.

Figure 1.15 Damage due to crevice and fretting corrosion. (Left panel) Localised surface damage on a cemented Ti6Al4V stem region as a result of crevice corrosion. Magnification 70×. (Centre) Fretting scars on the taper neck of a Ti6Al4V cemented stem and a Ti6Al4V head after 17 years *in vivo*, magnification 7×. (Right panel) SEM image of fretting scars on the taper neck of a Ti6Al4V stem and AISI 316L stainless steel head after 14 years *in vivo*. Fretting scars are perpendicular to the machine lines. Reproduced from ref. 83 with permission from Elsevier, Copyright 2007.

For example, in joint implants, a fine layer of fluid typically lubricates the closely fitting surfaces, creating the potential for crevice corrosion events to co-exist with fretting events. The release of metallic particles and ions can be significant under these conditions, affecting the health of the surrounding tissues, with the potential to induce systemic damage to the patient. The loss of material may also compromise the mechanical integrity of the site where the implant is fixed to the bone, leading to implant loosening and compromising the correct load transfer across the interfaces and the bulk of the implant. Thus-stressed materials can be more susceptible to failure due to corrosion fatigue processes. The debris generated in the process of fretting corrosion, *e.g.* oxide particles of various dimensions, tend to accumulate in the peri-implant milieu, thereby increasing stress.[83] If trapped in between the two rubbing surfaces, these particles may enhance damage *via* abrasion.

Although increasing the hardness of the material reduces the damage it is likely to sustain from fretting wear, most Ti and Co–Cr alloys, and stainless steels are subject to fretting due to shearing micro-movements at the implant–cement or implant–bone interface.[84–86] The chemistry and physical characteristics of the environment may also affect the likelihood of fretting corrosion, although the exact effects may differ between different types of materials. For example, when subject to sliding wear, fretting corrosion resistant alloys based on Ti or Co–Cr (namely Ti6Al7Nb and Co28Cr6Mo) show little change when the pH is varied from 2.5 to 6, whereas the fretting corrosion resistance of high-nitrogen Fe22Cr10NiN and AISI 316L stainless steel reduces with decreasing pH.[87] Interestingly, when a 316L SS/Ti6Al4V material with a thin layer of Ringer's solution was submitted to fretting conditions, stainless steel was shown to endure a lower level of fretting damage when compared to Ti6Al4V.[88] The level of damage at each surface is determined by the friction conditions at the contact and the energy that is dissipated at each cycle, as well as the lubricating and corrosive properties of the fluid that separates the two surfaces (see Figure 1.16).[88] Cathodic protection may significantly lessen the extent of damage related to fretting, and

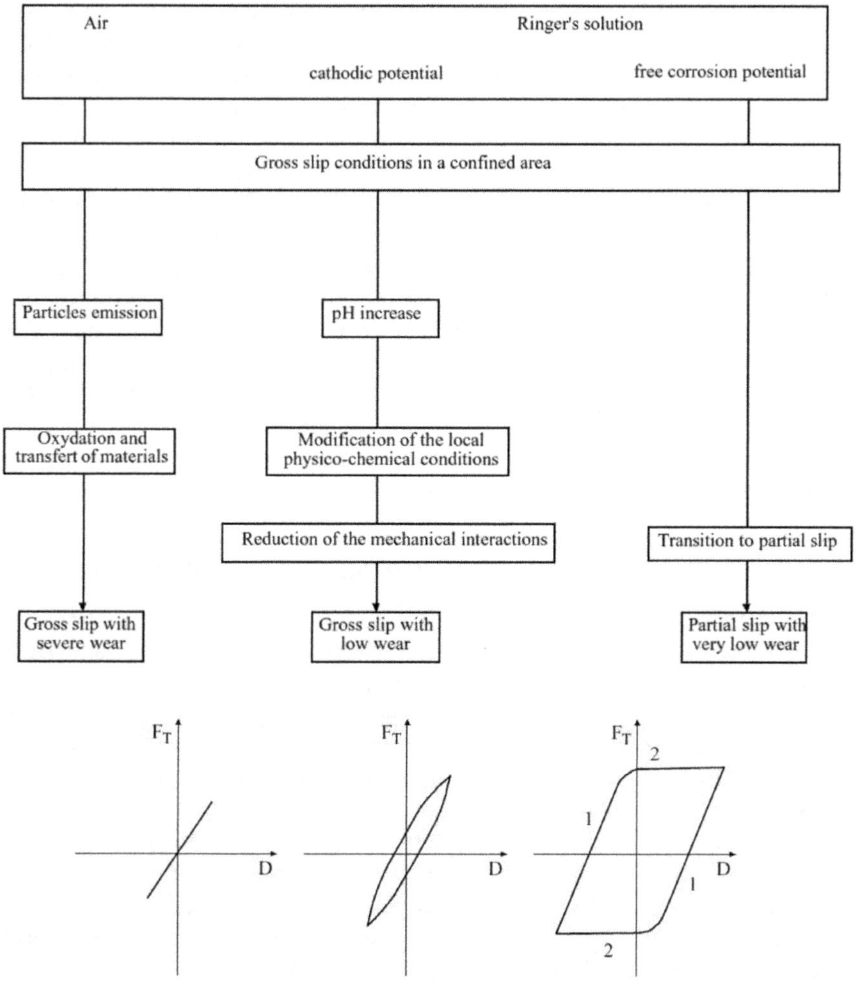

Figure 1.16 (Top panel) Damage mechanism induced by fretting corrosion at the head–neck contact of a total hip joint with a neck in Ti6Al4V alloy and a head in austenitic AISI 316L stainless steel in Ringer's solution. (Bottom panel) Different fretting regimes characterised by specific shapes of tangential force (F_T) *vs.* displacement (D) cycle and representing different wear outcomes. The closed cycle corresponds to the elastic regime without observed wear and no energy dissipation. The elliptic cycle represents partial slip fretting where one part of the surface slips whereas the other part sticks, and which results in failure by crack initiation and propagation. The parallelogram cycle represents a highly dissipative mechanism, with slippage of the entire surface and resulting abrasive wear.
Reproduced from ref. 88 with permission from Elsevier, Copyright 2003.

prevent the transition between fretting regimes. Indeed, the electrochemical potential plays a critical role in defining the fretting corrosion in Ti-based alloys, as was demonstrated using a Ti6Al4V sample in contact with an alumina ball at an amplitude of micro-motions of 50 and 100 μm in a 0.9% NaCl solution.[89] The effect of the applied potential on the wear was particularly pronounced at potentials above −0.2 V Ag/AgCl due to the enhanced oxidation of the third body that is produced as a result of plastic deformation in the metal at higher anodic potentials, requiring less mechanical energy to induce wear damage. As with other types of corrosion, the presence of certain alloying elements may render the material more resistant to chemical dissolution, yet their effect on the mechanical properties of the material may not necessarily be positive, calling for a trade-off. Mixed phase α/β alloys, such as Ti6Al4V and Ti6Al7Nb, may display better characteristics with respect to resistance to corrosion and wear, whereas commercially pure titanium and the near-β Ti13Nb13Zr and β Ti15Mo alloys may show superior corrosion resistance among the group.[90]

1.4.4 Galvanic Corrosion and Intergranular Corrosion

In some applications, *e.g.* in restorative dentistry, it may be necessary to place two dissimilar metallic materials into direct electrical contact. In the presence of a corrosive fluid, *e.g.* saliva rich in acids, salts, proteins and enzymes, *galvanic corrosion* may take place. This type of corrosion is characterised by changes in the rate of the corrosion on these materials, with the corrosion being promoted on a less immune material and suppressed on a more noble material. This is because the difference in the corrosion potential of these materials drives a flow of electric current through the metal/metal junction, as well as through fluids in the peri-implant milieu and in tissues. This can lead to the formation of a galvanic cell that promotes corrosion on the less noble metal. The reactions involved include cathodic reactions that reduce dissolved oxygen in the electrolyte (*i.e.* $O_2 + 4e^- + 2H_2O \rightarrow 4OH^-$ and $O_2 + 4H^+ + 4e^- \rightarrow 2H_2O$) and anodic reactions that dissolve the metal following $M \rightarrow M^{n+} + ne^-$.

Galvanic coupling is held responsible for the intergranular corrosion of dental amalgams, *i.e.* the multiphase alloys commonly used in restorative dentistry. In amalgams, the different phases have distinct electrochemical properties, with corrosion resistance decreasing in the order of Ag_2Hg_3, Ag_3Sn, Ag_3Cu_2, Cu_3Sn, Cu_6Sn_5, and $Sn_{7-8}Hg$.[91] As the most susceptible to corrosion, the latter phase (also known as the γ_2 phase) contributes most towards the observed corrosion in a low-copper amalgam system in an oral environment, whereas the γ and γ_1 phases show greater corrosion resistance. Of most concern is the release of potentially cytotoxic mercury as a result of the reaction $Sn_{7-8} + 0.5O_2 + H_2O + Cl^- \rightarrow Sn_4(OH)_6Cl_2 + Hg$. The free Hg may also interact with the γ phase to generate additional γ_1 and γ_2 phases. In the case of high copper alloys, γ_2 is not present in the final amalgam. However, while the ή phase that is formed instead shows better corrosion resistance than γ_2 and its breakdown does not generate Hg, it is still the most susceptible to corrosion among the phases, interacting with saliva in the manner of $Cu_6Sn_5 + 0.5O_2 + H_2O + Cl^- \rightarrow CuCl_2 \cdot 3Cu(OH)_2 + SnO$.

Outside of amalgams, the passivation of metals greatly reduces the probability and practical consequences of galvanic coupling, since the corrosion potential of the passivation layer on the less noble metal would be that of a more noble metal.[83] Having said that, physical or chemical damage to the protective layer would expose the underlying metal, thereby increasing the probability of galvanic corrosion.[92] For example, when galvanic coupling of spontaneously passivating CoCr, CoCr-c, NiCrTi, gold–palladium and Ti6Al4V dental alloys to titanium grade 2 implants was investigated, the passive domain of the NiCrTi alloy was found to be too narrow in the presence of fluorides to afford the needed level of protection, releasing potentially dangerous levels of Ni ions into the milieu.[93] This study also highlighted the importance of considering multiple environmental factors when evaluating the probability of galvanic corrosion in metallic couples. For example, lowering the pH of artificial saliva from 6.5 to 3 did not result in significant changes in the electrochemical behaviour of TiG2, Ti6Al4V, and CoCr alloys, and their respective couples, with Ti/Ti6Al4V and Ti/CoCr showing the most stable behaviour. However, a combination of low pH (<3.5) with high fluoride content (1000 ppm) triggered the active dissolution of TiG2 and Ti6Al4V, with galvanic corrosion observed in CoCr–c/TiG2 couples.[93] The processing of a material may affect its purity, geometry, grain structure, surface morphology and internal stresses, thus potentially affecting its susceptibility to galvanic corrosion. For example, in the case of cast and machined Ti-coupled oral implants, the ion release is 40 and 25 $ng\,ml^{-1}$ for machined and cast samples, respectively, after 560 h of incubation in saliva.[94] Even though both materials release ions, it is at a much lower concentration when compared to Cr alloy (76.5% Ni, 14.3% Cr, 5.6% Al, 2.3% Mo, 0.8% Si y 0.5% Fe), at 200 $ng\,ml^{-1}$ after only 10 h of immersion. Ti6Al4V/CoCr and Ti6Al4V/CrNi show similar performance when incubated in Ringer's solution.[95] Significant levels of potentially cytotoxic Cu are also released from coupled palladium (73.7% Pd, 14.9% Cu, 11.4% Ga) and gold alloys (50.0% Au, 31.5% Cu, 13.0% Ag, 5.4% Zn) in saliva,[94] whereas a Ti6Al4V/Au couple is likely to resist galvanic corrosion.[95] The galvanic cell phenomenon is often compounded by localised crevice corrosion that is due to the geometry of the assembly.[96]

1.4.5 Modularity as a Promoter of Corrosion

Among the implant designs and features that make them more susceptible to corrosion, *modularity* ranks highly with regard to its potential to induce clinically-relevant negative outcomes. While many of the early implant components were monolithic, contemporary implants for total hip or total knee arthroplasty often feature modular components. This is because modular components such as acetabular liners and femoral necks better support the unique anatomical structures of the patient, optimising the biomechanics and addressing limb length discrepancies, and enabling revision surgeries that only replace damaged components while retaining well-fixed components within the patient. As such, implant modularity can deliver substantial benefits to the patient with respect to implant performance and post-surgery recovery. However, modular junctions create additional crevices where corrosion can take place, not only for materials with dissimilar properties, *e.g.* when Ti contacts Co–Cr alloys, but even in instances where the same material with different surface finishes are used. The junction can vary in shape and size, angles at which they are tapered, surface roughness, tolerances, presence of such features as skirts and so forth. For example, a modular head–neck assembly in femoral total hip replacement prosthesis typically has a tapered conical junction between two metals, or between a metal and ceramic, where the head and stem have a tapered bore and cone, respectively. Stems in these assemblies are frequently made of Ti6Al4V or Co–C–Mo and heads are made of Al_2O_3, ZrO_2, or Co–Cr–Mo.[97] Head–neck taper corrosion can lead to increased levels of Co and Cr in the peri-implant tissues (11.2 and 2.2 ppb, respectively), with patients complaining of adverse local tissue reactions ranging in intensity from tissue discoloration to collection of fluid and lymphocytic infiltrate, soft tissue damage and the appearance of pseudotumours,[98] with some cases requiring surgical revision.[99]

Modularity-related failure is complex and may comprise multiple processes. For example, in assemblies that use titanium alloy stems with Ti alloy or Co–Cr modular necks, both fatigue failure and fretting corrosion processes can be detected (see Figure 1.17).[100,101] The use of neck-preserving stems has been shown to reduce fatigue and fretting corrosion by limiting torsional and bending loads, and thus cumulative stress to which modular taper junctions are subjected.[102] Certain material combinations (*e.g.* using Co–Cr neck adapters in place of Ti when using Ti stems) and assemblies that reduce micro-motion at the stem–neck interface during cyclic loading, as well as strategies that prevent interface contamination, may also reduce the incidence of fretting and corrosion.[103] Improving the tolerances of the interface and application of coatings onto interfacing materials may also positively affect junction stability and corrosion resistance.[104]

1.4.6 Passivation and Formation of Protective Oxides

For the effective protection of the bulk metallic implant, it is imperative that the passivation layer is continuous, non-porous and defect-free, and attaches well to the material so as to prevent diffusion of solutions towards the material bulk. The Pilling–Bedworth ratio (R_{PB}) is frequently used to indicate the probability that the metal will form a sufficiently protective oxide. Defined as the volumetric ratio of the elementary cell of a metal oxide to that of the corresponding metal from which the oxide is formed, a ratio of less than 1 suggests the formation of a thin, discontinuous and/or porous coating with limited protective performance, such as in the case of MgO (R_{PB} of ~0.8).[105] Above an R_{PB} value of 1, the oxide layer is uniform, affording

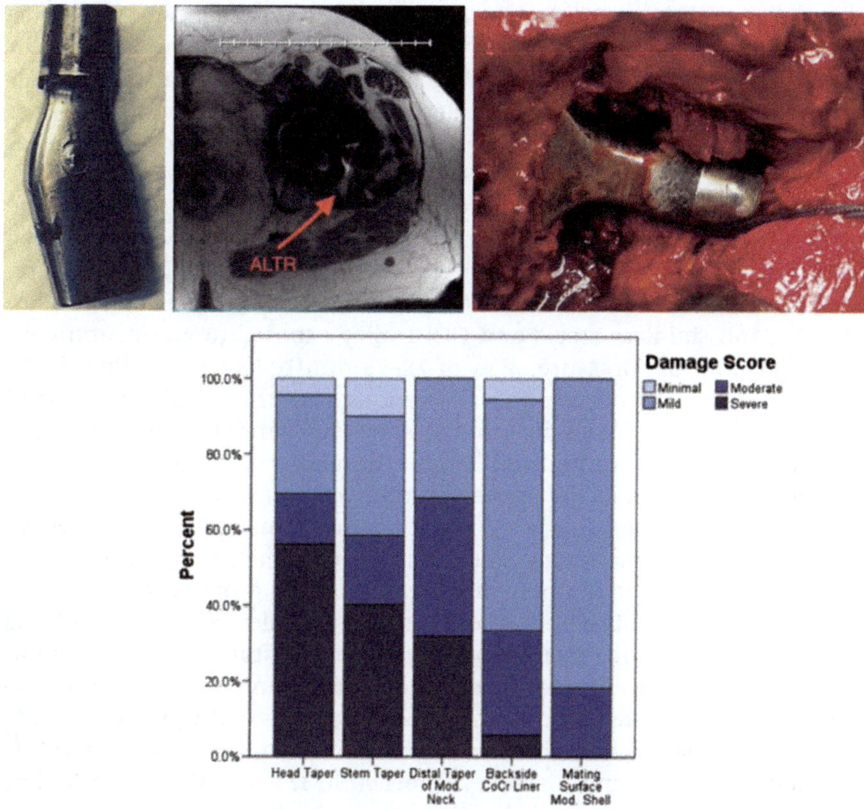

Figure 1.17 Top panel: (left image) Intraoperative image of Zimmer Kinectiv titanium modular neck showing corrosion product at head–neck and neck–stem junctions. (Middle image) Left hip axial proton density weighted magnetic resonance imaging showing adverse local tissue reaction to Zimmer Kinectiv stem. (Right image) Gross corrosion at the typical modular taper visualised as black scaly material. Image taken during revision surgery at a mean of 5.3 years after primary surgery. (Left and middle images) Reproduced from ref. 99 with permission from Elsevier, Copyright 2016 and (right image) Reproduced from ref. 97 with permission from Elsevier, Copyright 2019. Bottom image: distribution of taper damage scores at the head tapers, stem tapers, distal taper of modular necks, backside of CoCr liners, and mating surface of modular shells in 167 surgically retrieved metal-on-metal bearing systems. (Bottom image) Reproduced from ref. 236 with permission from Elsevier, Copyright 2013.

the required protection from subsequent oxidation to metals such as Cr, Ti, and Al. Above an R_{PB} value of 2, the oxide film once again provides limited protection as the mechanical stresses and strains within the coating result in poor adhesion and a tendency of the oxide to flake off the surface.[105] While a useful indicator for certain types of conditions, the behaviour indicated by the Pilling–Bedworth ratio does not always correspond to that under *in vivo* and *in vitro* conditions. Furthermore, for alloys, the sequence of oxide formation is likely to be more complex, and metals that play dominant roles in this process may not necessarily be the ones with the greatest fraction. For example, when the R_{PB} values were calculated for Ni_3Al, NiAl and $NiAl_3$, that all form Al_2O_3 on their surface, they were significantly greater than that for pure Al, at 1.71–1.88, 1.64–1.78, 1.48–1.57, and 1.29, respectively.[105]

The balance between formation and dissolution, and the quality of the passivation layer, depends on the environment. Relatively low levels of dissolved oxygen in bodily fluids are often insufficient to enable efficient re-passivation of the metal surface with an oxide. For example, *in vivo* tests of AISI 316, 316L stainless steels and COP1 alloy exposed to an environment with a typical partial pressure, pO_2, of 28–78 mmHg (corresponding to the values experimentally observed in the venous and arterial blood collected from femoral blood vessels in rabbits) show substantial reduction in the chromium oxide passivation and fatigue durability of these metals, and concomitant increases in the release of ionic iron into the milieu.[106]

Stainless steels rely on their Cr content for the formation of the protective oxide; at the same time, many steels also have a small fraction of carbon introduced with the aim of improving the mechanical properties of the alloy. The presence of both of these elements creates a challenge when the alloy is subjected to thermal treatment, since higher temperatures promote the formation of a chromium-rich $M_{23}C_6$ carbide at the grain boundaries, thus depleting these areas of free Cr, creating local weaknesses in the protective oxide layer and facilitating intergranular stress corrosion (see Figure 1.18).[107] Thermally-treated austenitic stainless steel AISI 321 stabilised with Ti has also been shown to be corroded around TiNs, as these sites participate in galvanic coupling with their surrounding continuous phases, with corrosion enhanced due to threshold chromium concentrations being low enough to form a sufficiently strong protective film.[107]

Chloride ions that are also present in bodily fluids tend to prevent the formation of protective oxides, thereby interfering with this key method of corrosion protection on which many implantable metals and alloys rely. This is because monovalent chloride stabilises suspensoid sols,[108] thus preventing the formation of a protective corrosion product layer on metals. Indeed, the presence of dissolved salts, particularly chloride ions and other halides, in extracellular and intracellular fluids influences implant corrosion *in vivo*. The exact effects and the products that are formed as a result of these interactions are determined by the chemistry of the metal and of the salts, and their concentration. For example, in the case of magnesium-based materials tested *in vitro* and *in vivo*, the presence of $CaCl_2$ leads to the formation of a protective Ca–PO_4 layer on the metal surface, whereas $MgSO_4$ affects the topography of the degradation interface, and $NaHCO_3$ leads to the formation of $MgCO_3$ as a by-product.[109] In the case of Ti6Al4V, the presence of chloride in the solution promotes pitting corrosion, with the degradation process following the initial formation and consequent breakdown of the protective passive layer, subsequent development of unstable and metastable pits and their repassivation

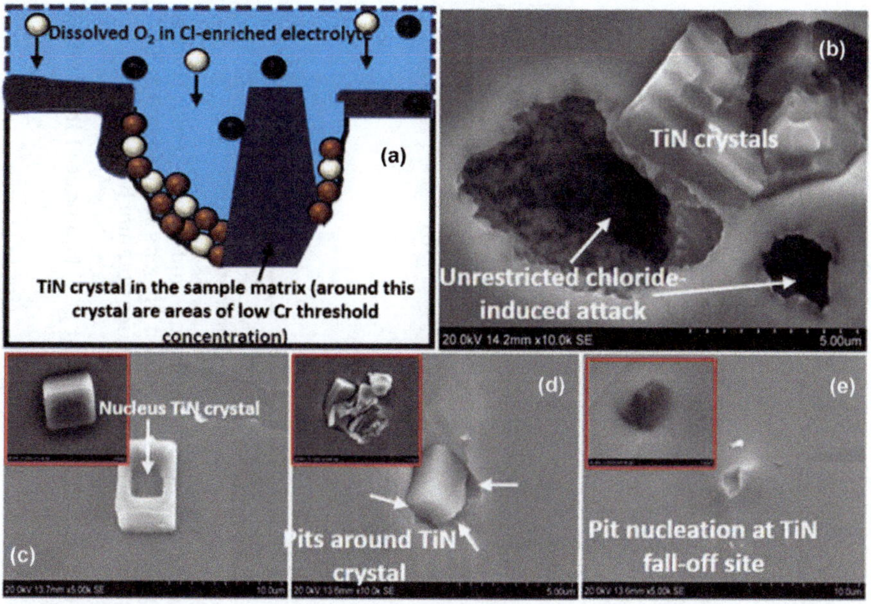

Figure 1.18 Corrosion patterns on Ti-stabilised austenitic stainless in NaCl medium. Galvanic coupling that exists between TiN particles and their surrounding matrix where the concentrations of Cr are reduced leads to the pitting corrosion around TiN particles. (a) Schematic of pit development. (b) TiN are sites where pits deepen due to lower Cr concentration. (c) Depression at the nucleus of smooth crystals. (d) TiN crystals serve as nucleation sites for pits. (e) early stages of pit formation at TiN. Reproduced from ref. 107, https://doi.org/10.1038/s41598-019-48594-3, under the terms of the CC BY 4.0 license, http://creativecommons.org/licenses/by/4.0/.

(see Figure 1.19), and the eventual development of stable pits of greater size and their evolution into other forms of localised corrosion.[110] The high corrosion rate within sites where repassivation has failed to occur is in part sustained by the cathodic electrochemical reactions that take place on the surface of intact oxide films, with the latter typically showing n-type semiconducting rather than insulating behaviour. The availability of complexing ions, particularly Cl^-, in the environment has also been shown to significantly reduce the passive zone of Cr, thereby reducing the *in vivo* performance of stainless steel, since Cr plays a central role in the development of a protective oxide layer on these implants.

1.4.7 Effect of Temperature and pH on Corrosion

The corrosion reaction pathways and their rate are directly linked to the temperature and pH of the environment. In general, an increase in temperature leads to an increase in the rate of chemical reactions. Changes in

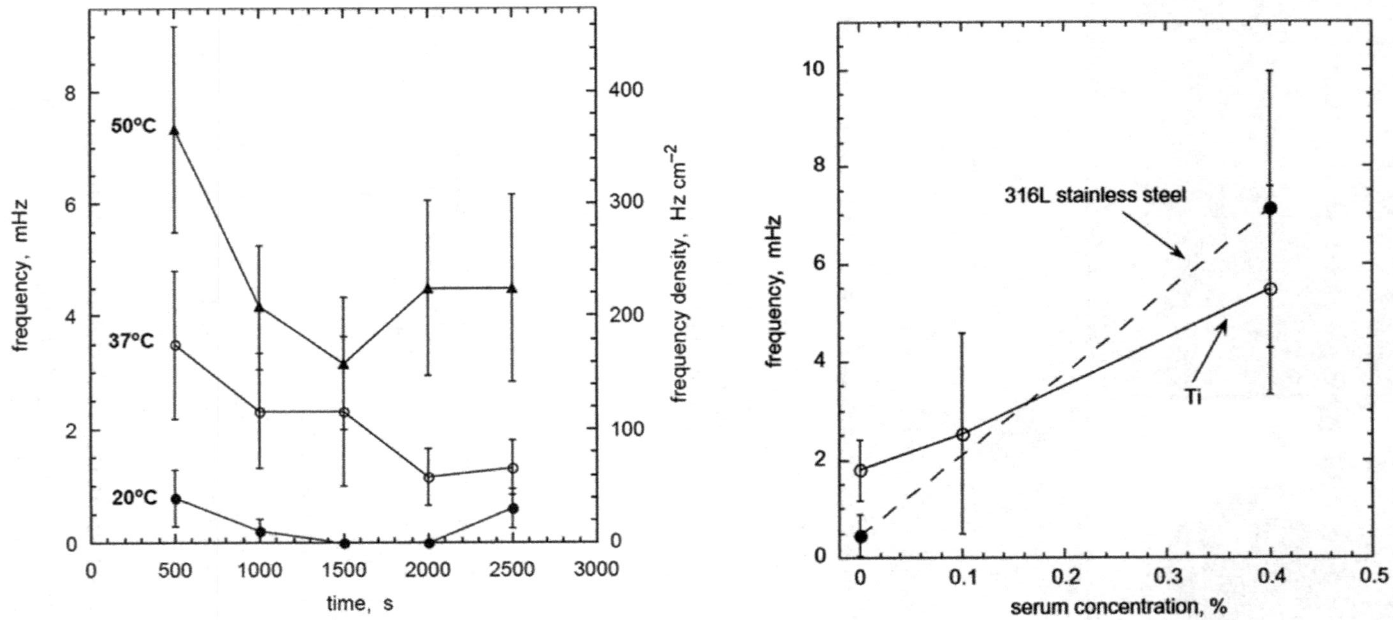

Figure 1.19 The frequency of pit nucleation on titanium in Ringer's solution as a function of temperature and bovine serum concentration. (Left plot) Frequency of pit nucleation on titanium as a function of time of exposure in Ringer's solution at 0.524 V.[21] (Right plot) Mean frequency of pit nucleation events for 306 stainless steel (black points) and for titanium (white points) as a function of the serum concentration in Ringer's solution at 37 °C.[22]
(Left plot) Reproduced from ref. 68 with permission from Elsevier, Copyright 2005 and (right plot) Reproduced from ref. 67 with permission from Elsevier, Copyright 2017.

temperature also affect the equilibrium constant and balance of neutrality, *e.g.* at 25 °C, $[H^+]=[OH^-]$ at a pH value of 7, whereas at 37 °C, neutrality is attained at a pH of 6.81.[111] Even though the human body has several mechanisms to respond to changes in pH, specifically through fast-acting chemical buffers and slower-acting but more potent physiological buffers that involve complex respiratory and urinary response systems,[112] the very act of implantation may perturb pH balance for days and even several weeks. This is because implantation or pathological conditions are often associated with transient changes in the electrochemical state of the environment due to physical trauma (*e.g.* hematomas), inflammation, disruption to tissue blood supply, and thus the level of oxygenation and pH in the peri-implant milieu.[69] For example, the accumulation of hematomas, particularly in patients on anticoagulation or antiplatelet treatment,[113,114] can lead to a significant local acidification of the peri-implant environment (down to a pH of 4.0), resulting in an increased rate of corrosion of the material and severe local damage.[115] Once a deep pit or crevice is formed on the surface, it may be difficult to re-passivate due to the restricted access of oxygen to the bottom of the pit, with a recorded pH of as low as 0.5–2.0 at the root of the pit regardless of the level of oxygenation or pH in the surrounding environment.[116] This is because Ti^{3+} dissolved and trapped within the crevice functions as an anode, producing H^+ *via* hydrolysis and reducing the level of pH within the compartment. In parallel, to preserve electrical neutrality, Cl^- ions migrate into the crevice. Once the pH is sufficiently low, depassivation and active dissolution begins, exacerbated by the presence of chloride ions, and crevice corrosion ensues.

Pourbaix diagrams are frequently used to describe the electrochemical stabilities of materials. These plots show stable phases as a function of electrode potential and hydrogen ion activity of the solution, *i.e.* its pH, with the solution typically being water at 25 °C.[117] The electrode potential, E_H, refers to the voltage potential from the oxidation of molecular hydrogen to solvated protons with respect to the standard hydrogen electrode, as calculated from the standard Gibbs free energies of formation using the Nernst equation. By convention, the standard potential of a reversible hydrogen reaction in a solution, $2H^+(aq)+2e^- \rightarrow H_2(g)$, is set to zero, *i.e.* chemical potentials for protons and hydrogen at unit activity and fugacity equal zero. Gold and other noble metals have higher standard potentials compared to base metals such as magnesium, and thus are far more stable when placed in an acidic solution when compared to base metals, which will corrode unless they are encased in a protective oxide layer. The lines in the plot represent the equilibrium conditions where the activities for the species on each side of that line are equal (see Figure 1.20); and the regions that these lines separate denote the stable species and at times may include such indicators as immunity, corrosion and passivity, each corresponding to the ability of the material to remain stable, corrode or form a protective layer (oxide or another salt) on its surface.

Although Pourbaix diagrams offer some value when one wants to, for example, estimate the possible effects of changing the composition of an alloy

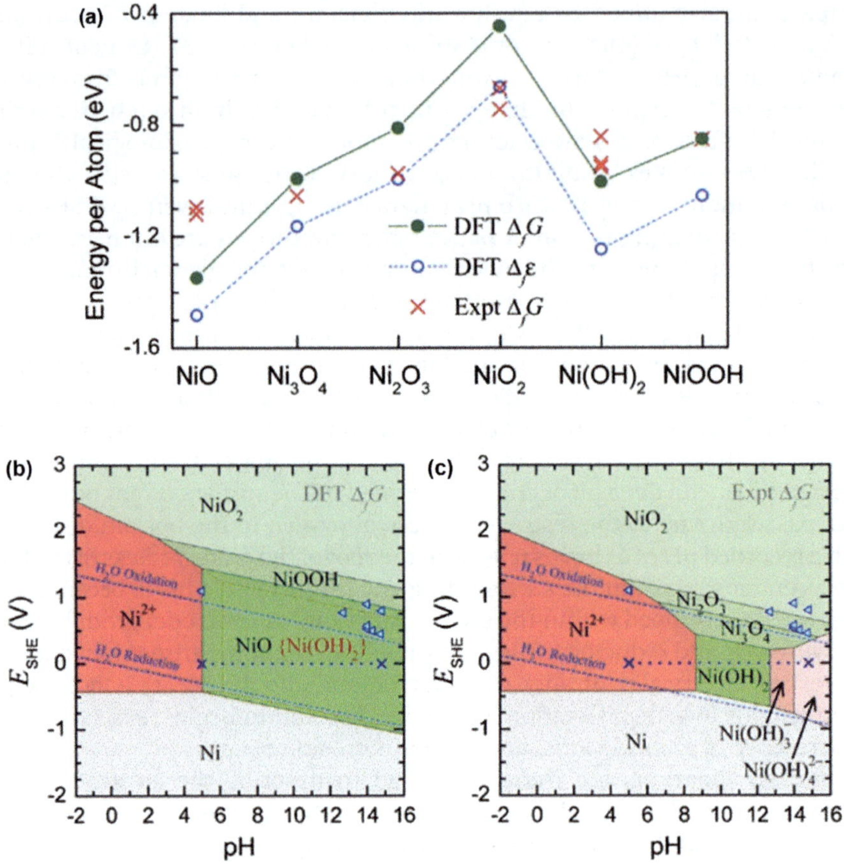

Figure 1.20 (a) The experimental standard Gibbs free energies of formation, $\Delta_f G$, from literature. Ni Pourbaix diagrams produced using density functional theory (DFT, b) are more consistent with direct electrochemical experiments than diagrams produced using experimental (Expt, c) thermodynamic energies ($[I]=10^{-6}$ mol L^{-1}, 298.15 K, 1.0 bar). The blue triangles are the measured oxidation potentials, cross-capped dotted blue lines are stability ranges of $Ni(OH)_2$ and NiO; the blue broken parallel inclined lines show the potentials for the oxidation ($2H_2O \rightarrow O_2 + 4H^+ + 4e^-$) and the reduction ($2H_2O + 2e^- \rightarrow H_2 + 2OH^-$) of water; the red dashed line is the phase domain for the metastable $Ni(OH)_2$. Reproduced from ref. 120 with permission from American Chemical Society, Copyright 2017.

on phase stability, they are generally produced for a standard set of conditions, *e.g.* a concentration of soluble species of 10^{-6} M, a gas pressure of 1 atm, at room temperature, and ignore contributions from kinetic effects. As such, their usefulness in predicting the *in vivo* degradation of metallic implants is limited. For example, while the fraction where gold and other noble metals show immunity corresponds to their real-life corrosion resistance in water, the region of immunity for common metallic biomaterials, *e.g.* Zn and Ti,[118,119] lies below the water stability zone, suggesting that these metals would be readily oxidised when placed in water and therefore should not be used in applications where they may come into contact with water. Yet, in real life, the rapid formation of the protective passivation layer ensures protection of the bulk metal, and only in cases where this layer fails to form do we see rapid corrosion of these metals in water. For Ti, this takes place under extreme reducing and extreme oxidising conditions, as these hinder the formation of the oxide and increase the solubility of hydroxides, respectively. Even though, according to the Pourbaix diagrams, these conditions do not occur within the water stability zone, if we consider the kinetic effects of oxidation/reduction of H_2O_2 at the metal electrode, which in the case of Ti are slow, a lack of dissolved oxygen or presence of H_2O_2 may force the reactions into the negative and positive corrosion domains, respectively.[117]

The importance of considering the kinetic effects is particularly evident when one compares the theoretical and actual electrochemical behaviour of Co and Ni. The apparent similarity of their Pourbaix plots suggests that both should corrode readily over a wide range of conditions. Yet, as Ni tends to acquire a passivation layer more rapidly than Co due to relatively slow kinetics of the dissolution reactions of the former, the conditions under which Ni corrodes are somewhat less extensive.[117] Furthermore, the reported Pourbaix diagrams do not always correlate well with electrochemical observations, possibly due to the inaccurate estimation of experimental free energies of formation ($\Delta_f G$), an issue that may be at least in part addressed by using more advanced approaches to estimating $\Delta_f G$, *e.g.* those based on density functional theory (see Figure 1.20).[120]

1.5 *In Vivo vs. In Vitro* Studies of Implant Degradation

In vitro, biomaterials are commonly tested using artificial biofluids that attempt to reproduce the most critical factors that influence degradation *in vivo*. While the recipes for simpler solutions such as phosphate-buffered saline or Ringer's solution are standard, when it comes to replicating more complex fluids, such as saliva, there is a much wider variety of recipes (see Table 1.7). Yet, none can truly reflect the complex and dynamic nature of saliva that not only differs between patients, but also may differ in one patient as they progress through their life, have diseases and receive medical and physical therapies. This challenge holds true for other biological fluids, too. As a result, the chemistry of simulated body fluids is typically restricted to matching the concentration of inorganic ions encountered in physiological fluids, but rarely do they contain proteins, amino acids, or vitamins that one would expect to find in natural fluids.

As an example, diabetic patients have different glucose levels in their biofluids, and may also undergo antibiotic treatment due to their greater

Table 1.7 Chemical composition of artificial biofluids routinely used for *in vitro* studies.

Component ($g\,L^{-1}$)	PBS	Ringer's	Krebs–Ringer	Hank's	Artificial saliva (1)	Artificial saliva (2)	Glandosane
NaCl	8	8.6	8.66	8	0.4	—	0.84
$CaCl_2$	—	0.33	0.38	0.14	—	—	—
KCl	0.2	0.3	0.402	0.4	0.4	1.5	1.2
KSCN	—	—	—	—	—	0.5	—
$CaCl_2 \cdot 2H_2O$	—	—	—	—	—	—	0.14
$CaCl_2 \cdot H_2O$	—	—	—	—	0.795	—	—
$MgCl_2 \cdot 6H_2O$	—	—	—	0.1	—	—	0.06
$MgSO_4$	—	—	0.15	—	—	—	—
$MgSO_4 \cdot 7H_2O$	—	—	—	0.1	—	—	—
$NaHCO_3$	—	—	—	0.35	—	1.5	—
NaH_2PO_4	1.15	—	0.2415	—	—	0.5	—
$NaH2PO_4 \cdot H_2O$	—	—	—	—	0.69	—	—
$Na_2HPO_4 \cdot 12H_2O$	—	—	—	0.12	—	—	—
K_2HPO_4	—	—	—	—	—	—	0.34
KH_2PO_4	0.2	—	0.042	0.06	—	—	—
$Na_2S \cdot 9H_2O$	—	—	—	—	0.005	—	—
Lactic acid	—	—	—	—	—	0.9	—
Phenol red	—	—	—	0.02	—	—	—
Sorbitol	—	—	—	—	—	—	30
Na carboxymethylcellulose	—	—	—	—	—	—	10
Glucose	—	—	—	1	—	—	—
pH	7.2	7.3–7.4	6.9	7.5	5.5	6.5	6.8

susceptibility to infections. The presence of sugars at different concentrations can affect corrosion. For example, when the stability of AISI 316L stainless steel was evaluated in the presence of a diabetic-serum pool with different concentrations of glucose and an approximately constant concentration of proteins (6.8 $g L^{-1}$) and a temperature of 37 °C, the pitting potential of the material decreased with increasing glucose levels.[121] In contrast, addition of the antibiotics ampicillin or cefradine did not affect the pitting potential. Nevertheless, antibiotics often form part of the therapy, and are also integrated in many of the modern implant coatings and bone cements, *e.g.* SmartSets HV from DePuy International contains gentamicin sulfate at 4.22% w/w. The latter compound has been shown to affect the dynamics of fretting corrosion.[122] Interestingly, 316L stainless steel, commercial purity titanium and CrCo alloy show little change in their corrosion potential when incubated in saline containing one of the antibiotics commonly used for the treatment of orthopaedic infections, *i.e.* tetracycline, tobramycin, clindamycin, cefamandole, bacitracin and chloramphenicol.[123] However, the addition of oxytetracycline leads to anodic polarisation and a subsequent reduction in the corrosion rate of the three types of metallic substrate. This can be attributed to the interactions between the complex ring structures of the oxytetracycline with the metals that could have led to sequestration effects and/or the deposition of an inhibition layer. It should be noted that the corrosion protective *vs.* promoting effect may be dependent on the concentration of the antibiotic in the peri-implant milieu, the presence of other chemical and physical factors, and the chemistry and surface properties of the metal.[123] For example, in another study, the related antibiotics doxycycline and the dihydrate and hydrochloride salts of oxytetracycline were found to promote the corrosion of CoCr alloy, while inhibiting the corrosion of as-received titanium.[124] The corrosion of abraded titanium was inhibited by doxycycline, while for stainless steel inhibition was induced by oxytetracycline dihydrate.

The presence of amino acids that rapidly adsorb onto the surface of the metal can significantly change their polarisation resistance and current densities, with the corrosion outcomes dependent on the composition of the biomaterial. For example, Mg–Ca, AZ31 and AZ91 possess greater corrosion stability in Dulbecco's modified eagle medium (DMEM), a medium containing inorganic minerals and amino acids, than in Hank's solution (which is a mixture of water, inorganic salts and glucose).[125] This can be attributed not only to changes in electrolyte composition (higher $NaHCO_3$ and lower NaCl in DMEM), but also to the presence of alanine, glycine and leucine, which reduce the current density and increase the polarisation resistance of the surface. The protective effect of amino acid adsorption has also been shown for steel and aluminium.[126,127] For reducing pitting corrosion in Al, the order of effectiveness of the inhibitors was found to be arginine > histidine > glutamine > asparagine > alanine > glycine.[128]

The response was less uniform when foetal bovine serum was introduced into the mix. The rate of corrosion increased for Mg–Ca alloy when foetal bovine serum protein was added to DMEM, while the rate was found to decrease for the AZ91 alloy, and follow an initial increase followed by a decrease for the AZ31 alloy.[125] A study on the corrosion resistance of Ti alloys in the presence of whole serum also showed an improved resistance in the Ti6Al4V alloy, and reduced resistance in Ti13Nb13Zr and Ti6Al7Nb when compared to phosphate-buffered saline.[129]

The nature of protein–surface interface reactions is complex and is governed by van der Waals, hydrophobic and electrostatic forces, and hydrogen bonding. In the case of AZ91, the adsorbed protein layer likely changes the electric field on the metal surface, and reduces the effect of changing pH on the corrosion processes on its surface. When compared to the Mg–Ca alloy, it is possible that the proteins adsorb more easily on the Al_2O_3 layer present on the surface of AZ31 and AZ91 alloys and on coarse particles of partially divorced eutectic Mg17Al12 in AZ91, forming a protective layer that inhibits anodic reactions (see Figure 1.21).[125]

In addition to the chemistry of the solution itself, it is important to consider the chemistry of bone cements used in many orthopaedic and dental applications, and the products of degradation that they may release into the peri-implant milieu. For example, an *in vitro* study on the fretting corrosion of cemented Ultima TPS™ femoral stems showed that the chemistry of the poly(methylmethacrylate) cement affected the corrosion regimes at the

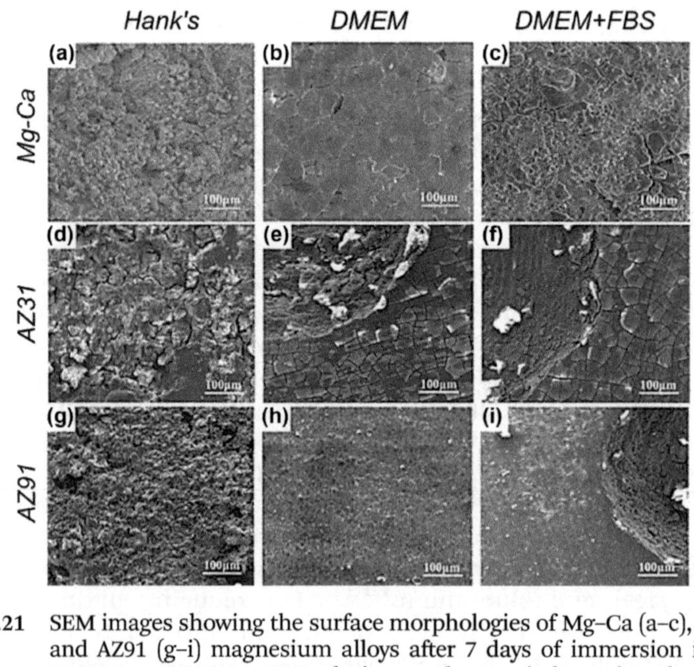

Figure 1.21 SEM images showing the surface morphologies of Mg–Ca (a–c), AZ31 (d–f) and AZ91 (g–i) magnesium alloys after 7 days of immersion in Hank's, DMEM, or DMEM + FBS solutions and potentiodynamic polarisation. Reproduced from ref. 125 with permission from IOP Publishing, Copyright 2009.

stem–cement interface.[122] The use of cements containing the $BaSO_4$ radiopacifier increased the corrosion current under fretting wear, with a greater rate of release of metallic ions when compared to that when ZrO_2-containing cements were used. The corrosion current was further increased under these conditions when a cement loaded with gentamicin sulfate was used. It is possible that $BaSO_4$ and gentamicin sulfate increase the dissolution rate of CoCrMo and 316L stainless steel, as the presence of sulfite ions is known to render the material more susceptible to local corrosion. Although sulfate ions are thought to inhibit the initiation of pitting corrosion (promoted by Cl^-), they are known to accelerate the propagation of localised crevice corrosion after the initiation events have taken place, by reducing the critical pitting temperature, hindering CoCrMo repassivation, and increasing the rate at which H^+ is reduced at the implant surface. The latter may increase the rate at which excess metallic valence electrons are liberated from the surface, and consequently from within the bulk of the metallic matrix. A combination of crevice geometry and elevated temperature at the metal–cement interface, as well as the presence of tiny pores and cracks in the cement, results in a high local concentration of antibiotic-derived sulfates. It possible that barium sulfate particles contribute sulfate ions into the peri-implant milieu through leaching and subsequent dissociation. Their presence may also change the electrochemistry of the environment through other mechanisms, as well as contribute to fretting through abrasion.[122]

1.5.1 Dynamic *vs.* Static Flow Conditions

In addition to the ever changing chemistry of natural fluids, it is important to consider the mechanics of biological flows and how they may affect and be affected by the presence of the implant and the physiological processes in healthy patients and those that may be suffering from relevant medical conditions. Nevertheless, most tests used to evaluate the behaviour of implants *in vitro* do so under static conditions, due to their relative simplicity and reproducibility. For this reason, the results of static studies often do not represent those of the events that take place once the implant is inserted into a patient. Interestingly, *in vitro* studies can both over- and underestimate the *in vivo* performance of a biomaterial. For example, the corrosion rates of gravity-cast AZ91D and LAE442 magnesium alloys were lower when these samples were implanted into guinea pig femura than when tested under standardised immersion and electrochemical tests, with a difference in rate of up to four orders of magnitude.[130] Interestingly, opposite tendencies in the corrosion rates were observed under *in vivo* and *in vitro* conditions.

The importance of selecting relevant testing conditions was highlighted in the study on the degradation behaviour of Mg-based pins under static immersion and interstitial flow (with and without cyclical loading) conditions, using murine (subcutaneous) or canine (tibia) animal models.[131] Subjecting as-drawn pure Mg, as-cast Mg–Zn–Mn, and extruded Mg–Zn–Mn samples to interstitial flow and cyclical loading were both shown to significantly enhance the rate of corrosion, by up to 4–5 and 10 times, respectively, leading to notable reduction in the mechanical strength.[131] The results are similar to those obtained for bioresorbable stents made of AZ31 magnesium alloy and tested under dynamic conditions similar to those experienced by stents in blood vessels, where notable stress- and flow-induced degradation and comparatively high degradation rate are observed.[132] Interstitial flow enables the simultaneous delivery of fresh fluid to, and removal of corrosion products from, the implant site, and promotes the formation of Ca/P complexes on implant surfaces, whereas cyclical loading promotes the detachment of corrosion products, and provides applied stress responsible for crack initiation and propagation. While the effects of flow are fast acting in the case of as-drawn pure Mg and as-cast Mg–Zn–Mn, it takes over two weeks for the corrosion rate of the extruded Mg–Zn–Mn to reflect their contribution, due to a relatively fine-grain boundary of the latter (see Figure 1.22).[131]

As-cast Mg–Zn–Mn pins show considerably different patterns of pitting corrosion when incubated under static immersion and when subcutaneously implanted into mice. Under static conditions, the corrosion begins with the formation of small pits, and then after two weeks evolves into fast localised corrosion due to the multi-phase nature of this material, which leads to microgalvanic coupling. In contrast, under *in vivo* conditions using a mouse model, the corrosion is uniformly localised and similar to that obtained under *in vitro* interstitial flow conditions. When implanted in tibia, the Mg–Zn–Mn pins show more localised degradation as a result of mechanical loading. It should be noted that while the study attempted to use a simulated body fluid containing biomolecules in the *in vitro* tests, DMEM with 10% fetal bovine serum, potential contributions from cells could not be measured, *e.g.* bone and inflammatory cells that could have contributed to the degradation kinetics of the material.

Potentially more accurate degradation data may be obtained if the dynamic immersion tests are conducted not only under constant, but also variable flow rates. This would more accurately reflect the variability in flow rates observed in the human body, *e.g.* those generated in human cancellous bone under the conditions of different physiological activities. Bone homeostasis is affected by physical activity, as the mechanical loading provides pressure differentials that force the fluid to flow at different rates.[133] The magnitude of the loading, and thus the strain in the bone that is induced, varies according to the activity, with stumbling producing approximately three times the loading on hip joints compared to that of walking.[134,135] This results in significantly different pressure differences and marrow flow rates in the bone that can vary from 0.0072 to 1.67 ml min^{-1}. These differences in flow may have considerable implications for the degradation of metallic materials used to replace cancellous bone, particularly for porous scaffolds. For example, the nature of the degradation products that are formed and their accumulation in the pores of pure Mg scaffolds is dependent on the flow rate (see Figure 1.23).[136]

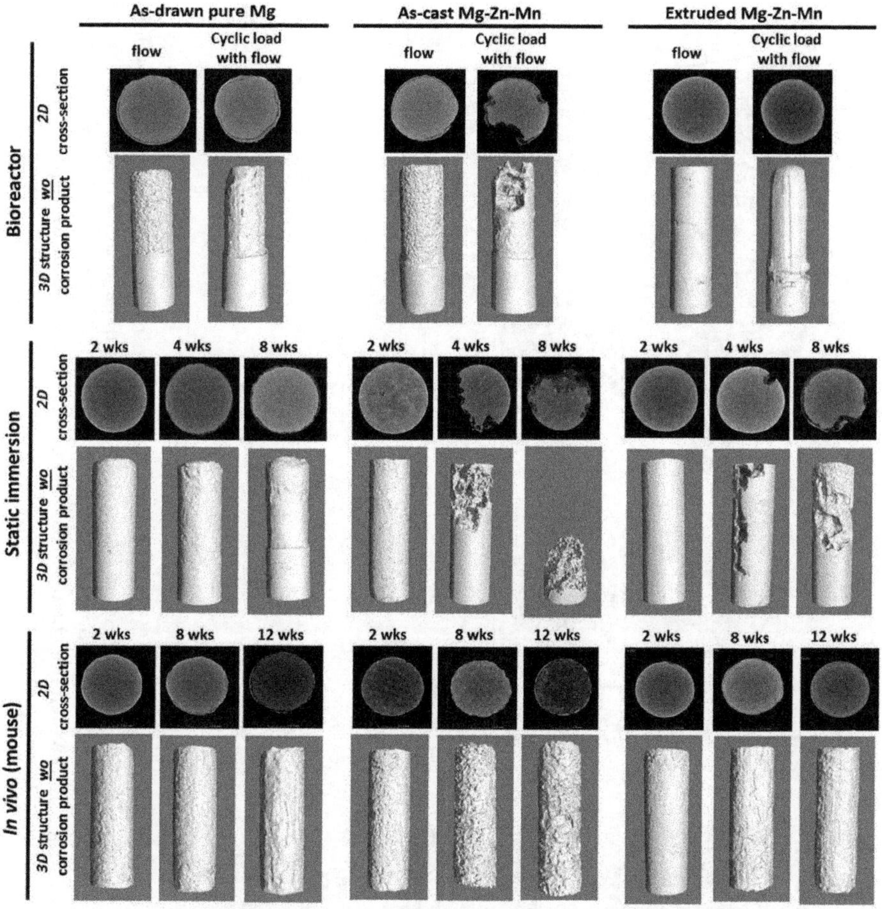

Figure 1.22 Representative cross-sectional 2D micro-computer tomography images (with corrosion products) and 3D surface morphology images (without corrosion products) of the three different Mg-based pins (as-drawn pure Mg, as-cast Mg–Zn–Mn, and extruded Mg–Zn–Mn) in three different models. Top row: bioreactor, simulation under interstitial flow only or cyclic loading with interstitial flow for 2 weeks. Middle row: static immersion after 2, 4, and 8 weeks. Bottom row: *in vivo* (mouse subcutaneous implantation after 2, 8, and 12 weeks).

Where the formation of needle-like clusters containing $MgCl_2$ and $Mg(OH)_2$ took place at different flow rates, the formation of a pore-clogging layer of spherical deposits consisting of oxygen, calcium, and phosphorus (at Ca : P of 1.67) and rod-like structures occurred at a flow rate of 0.025 ml min^{-1}, and of rod clump-, flower- and rod-plate-like structures made of $MgCO_3$ and $CaCO_3$ at flow rates of 0.4 and 0.8 ml min^{-1}.[136] Over the

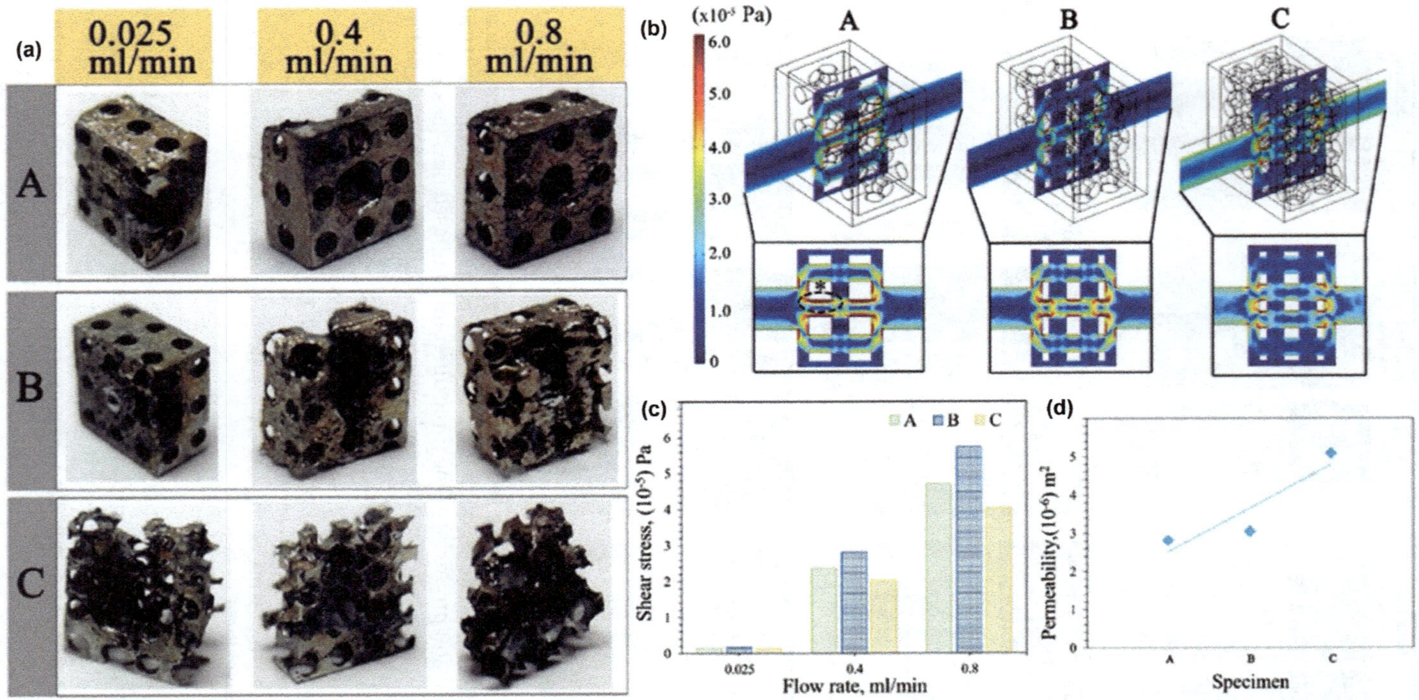

Figure 1.23 (a) Morphology of the pure Mg specimens with porosity of 30% (A), 41% (B), and 55% (C) immersed in simulated body fluid for 72 h under different flow rates. (b) Contour plot of shear stress on the porous Mg specimens under a flow rate of 0.8 ml min^{-1}, (c) average shear stress of the porous Mg specimen under a different flow rate and (d) permeability of porous Mg under different flow rates. * indicates the location in which the shear stress has been taken as similar to the other specimen. Reproduced from ref. 136 with permission from Elsevier, Copyright 2017.

Table 1.8 Comparison of the dynamic and static immersion degradation rates of solid and porous pure Mg. Adapted from ref. 136 with permission from Elsevier, Copyright 2017.

Test condition	Solution	Purity	Porosity (%)	ΔW_m (mg cm^{-2} per day)	P_m (mm per year)
Static immersion	Hank's	HP	0	0.12	0.25
	SBF	CP	0	0.92	1.94
	DMEM	CP	55	0.62–0.73	1.31–1.53
Dynamic immersion	Hank's	HP	0	0.38	0.8
	SBF	CP	30–55	2.3–5.3	4.9–11.2

first 24 h of observation, the weight loss increases with the flow rate. However after 48 h of incubation, samples with a porosity of 30% and 41% show increased corrosion at a flow rate of 0.4 ml min^{-1} compared to that at 0.8 ml min^{-1}, with this trend remaining stable as the time of incubation is increased to 72 h. In contrast, the sample with 55% porosity experiences an increased weight loss at a flow rate of 0.025 ml min^{-1} after 48 h, yet after 72 h, the weight loss increases in line with the flow rate. Table 1.8 shows a comparison between the degradation rates obtained under flow conditions to that under static immersion, highlighting the considerably higher rates reported for the former. This example highlights the importance of studying degradation over a wide range of relevant physiological parameters, with sufficient study length and sampling rate needed to capture the data points necessary to build a comprehensive picture of the likely degradation dynamics under *in vivo* conditions.

1.5.2 Stability of Artificial Bodily Fluids

When trying to untangle the complex nature of implant corrosion and compare the results of different studies, we rely on the assumption that artificial fluids are inherently stable with respect to their pH and ion concentration, both during their storage and *in vitro* testing at physiologically relevant temperatures. Yet, there is a lack of consensus in the literature whether this is indeed the case.[137–140] A recent study on the stability of artificial urine, two artificial salivas and six simulated body fluid samples of various composition, under typical storage (5 °C) and experimental (21 °C and 36.6 °C) conditions, found significant changes in ion concentration and pH, and the formation of precipitate after extended storage (see Figure 1.24).[139] The formation of precipitate in artificial saliva was moderate at 5 °C and substantial at 36 °C, providing further evidence to corroborate the likely change in the ion chemistry of the solution. All fluids had a substantial change in the concentration of calcium ions (decreasing to at least 70% in all tested fluids with the exception of i- and nl-type simulated body fluids), whereas the change was relatively minor in the case of sodium ions (Ca^{2+} and Na^{+} were the only ions tested in the study). Changes in the pH were also statistically significant, with pH increasing over time for all of the fluids. These results suggest that artificial fluids may

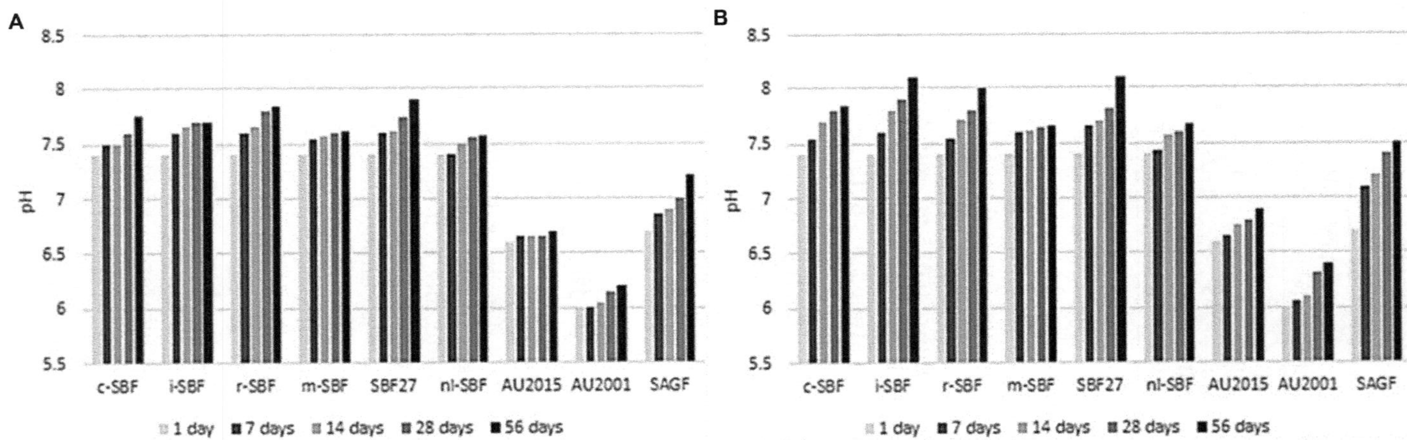

Figure 1.24 The pH value of artificial biofluids at different times of storage at (A) 5 °C and (B) 36.6 °C. c-SBF, i-SBF, r-SBF, m-SBF, SBF27, nl-SBF are simulated blood plasma fluids prepared using different recipes. AU2015 and AU2001 are artificial urine. SAGF is artificial saliva.
Reproduced from ref. 139 with permission from Elsevier, Copyright 2017.

neither be good mimics of *in vivo* fluids (where the concentration of ions and pH may change dynamically), nor can they be considered truly static (where changes in solution chemistry are related solely to the interaction of the solution with the implant). The immersion of the implant into such a solution may further change the nature of the processes that would take place in the absence of the implant, and this effect may be different when the implant is a solid macroscopic material when compared to *e.g.* a highly porous, large surface area structure, a thin film, or a collection of micro- and nanoscale particles.

1.5.3 The Chemical Feedback Loop Between Inflammation and Corrosion

Inflammation can also result from surface corrosion and the consequent release of Ni, Co, Cr, V, and Al species that may induce cellular or systemic toxicity, and cause local allergic reactions or sensitisation as well as changes to the chemistry of the peri-implant milieu. Corrosion may also affect the interface between tissues and implants, leading to implant loosening and consequent changes in how the load is transferred by load-bearing implants. This can not only lead to the mechanical failure of the implant itself, but also affect bone remodelling processes, hindering osseosynthesis and promoting osteoclast activity in the instances of inappropriate bone loading. Tissue acidosis is one of the key hallmarks of inflammatory processes, with substantially increased hydrogen ion concentrations detected in inflamed tissues (pH as low as 5.4), in fracture-related hematomas (pH as low as 4.7), and in malignancies, attributed to a combination of cell injury and death, and infiltration by hyperactive inflammatory cells.[141,142] For example, the presence of HLA-DR+/CD14+/CD68+ macrophages, CD3+ T cells, CD20+ B cells, plasma cells and macrophage granulomas has been detected in periprosthetic connective tissue and femoral heads in response to Co–Cr wear particles released from metal-on-metal hip implants.[143]

Inflammatory cells can release significant quantities of reactive oxygen species, *e.g.* hydrogen peroxide (H_2O_2) and superoxide (O_2^-), and these species can subsequently interact with implant surfaces as well as with metal ions that are released from these surfaces (see Figure 1.25).[144] These interactions are likely to occur through a Fenton reaction, whereby a metal ion (M^{n+}) interacts with H_2O_2 to form a hydroxyl radical ($^{\bullet}OH$), an OH^- ion and a metal ion $M^{(n+1)+}$. These reactions can take place either on the surface of the implant, or in the surrounding fluid. The thus-released hydroxyl radicals can induce further damage in cells through lipid peroxidation and by triggering a number of important biochemical pathways.[145] The extent to which metals partake in hydroxyl radical production is dependent on their chemistry as well as surface topography. For example, *in vitro* studies on Ti, TiO_2, Zr, Fe, Cu, Cr, Ni, Au and Al incubated in the presence of H_2O_2 have shown that metals such as Fe and Cu are associated with greater levels of $^{\bullet}OH$ radical formation compared to Ti, Zr, Au and Al over the pH range of 5.0–8.0.[146] It has been proposed that in the case of Ti, the formation of a hydrated, gel-like TiO–OH matrix on the

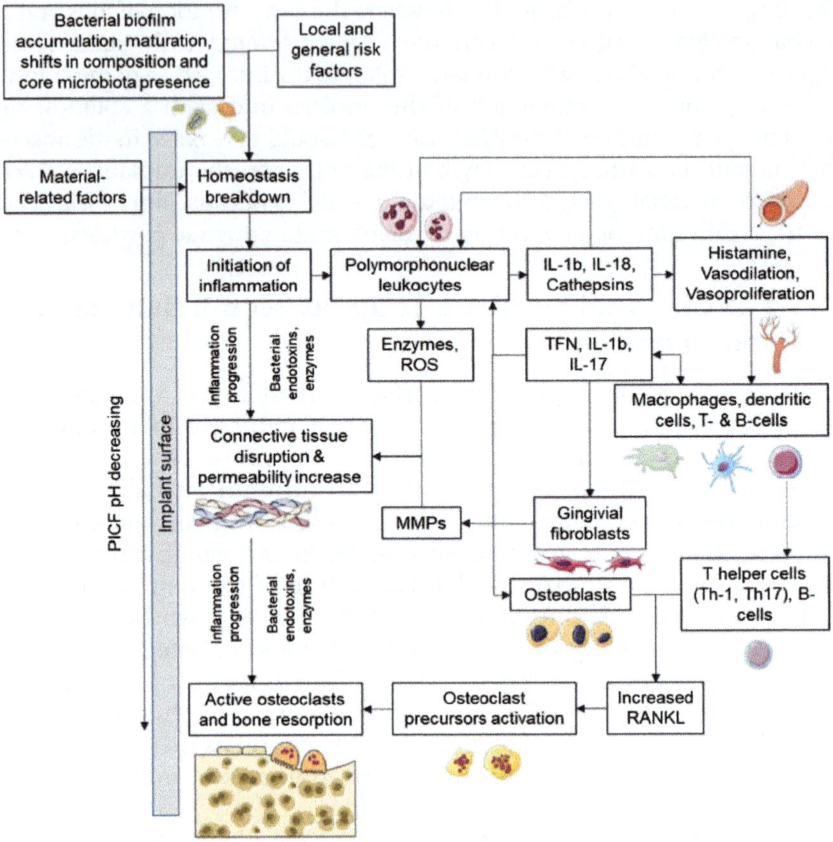

Figure 1.25 Complex interactions between material-, bacterial- and host-related factors affect homeostasis in oral peri-implant milieu. Materials-related factors (roughness, wettability and chemical composition) play a role. Early stages of inflammation due to bacterial products and material-related factors induce microvascular changes and pro-inflammatory cytokine release (IL-1b, IL-8a and cathepsins). Vasodilation and vasoproliferation increase the polymorphonuclear leukocyte infiltrate, cytokine release and recruitment of other immune cell types (macrophages, dendritic cells, T- and B-cells). Rough implants cause greater microvessel density and a higher number of T- and B-cell infiltrate in tissues compared to smooth surfaces. Release of TNF, IL-17 and IL exacerbates the inflammatory response, stimulate neutrophils to produce enzymes, ROS and cause fibroblasts to release metalloproteinase. Bacterial exotoxins and enzymes increase tissue permeability and fibroblast degeneration. The proliferation of epithelium into collagen-depleted areas results in pocket deepening, infection progression and a pH decrease in peri-implant crevicular fluid. TNF, IL-1b and IL-17, and bacterial virulence factors promote the maturation of osteoclast precursors. Decreased pH of the milieu or mechanical injury of the implant surface initiates TiO_2 layer dissolution and corrosion process. The ions/metallic particles of titanium may be phagocytosed by macrophages, causing an additional release of IL-8b when macrophages are coexposed to bacterial lipopolysaccharides.
Adapted from ref. 237 with permission from Dove Medical Press, Copyright 2017.

surface leads to the trapping and oxidation of superoxide radicals, thereby limiting the generation of free hydroxyl radicals through a superoxide dismutase-mediated reaction, $O_2^- + O_2^- + 2H^+ \rightarrow H_2O_2 + O_2$. This phenomenon may, at least in part, account for the favourable biocompatibility of Ti. It should also be noted that the presence of H_2O_2 in solution has also been shown to lead to the formation of a thicker oxide bi-layer, suggesting decreased corrosion resistance under these conditions.[147] The dense inner layer may be responsible for the excellent corrosion resistance of Ti implants, whereas the outer porous layer may promote the precipitation and nucleation of hydroxy-carbonated apatite, and enhance bone cell infiltration and osseointegration by serving as a transition layer between stiff Ti and softer bone tissue.[148–150] Implant surface topography may also play a role in the level of oxygen species produced by cells. It has previously been shown that osteoblasts produce fewer reactive oxygen species, including H_2O_2 and oxygen superoxide, when in contact with machine-prepared smooth titanium surfaces compared to acid-etch-created rough counterparts.[151] Interestingly, roughened Ti surfaces have been associated with lower levels of cytotoxicity and improved biocompatibility, suggesting that the role of reactive oxygen species in wound healing and osseointegration may be more complex.[152]

In vivo and clinical studies have related the levels of short- and long-term pathological changes related to biomaterial implantation to the clinical status of the patient, skill of the surgeon, and the physical and chemical properties of the implant. These include the size, shape, and biochemical activity of its surface, and any degradation products that may be generated at the initial stages of corrosion (*i.e.* prior to oxide layer formation) and subsequently during normal implant operation (*i.e.* associated with normal wear and ageing of the material).

It is important to note that the environment-mediated increase in the degradation rate of the implant material need not be uniform, as the properties of tissues, degree of trauma,[153] inflammation, and thus pH and temperature of the environment, as well as mechanical load may differ over the surface of the implant. This corrosion process and the resultant generation of abnormal electrical currents can effectively transform the metallic implant into an electrode within an electrochemical cell, potentially interfering with the normal functioning of adjacent cells and tissues as the latter receive these extreme electrical signals,[154] as well as affecting the corrosion process itself. Therefore, when judging implant compatibility with tissues, it is important to not only consider the corrosion products, but also the currents of electrons that will be exchanged between tissues and implants, and the reaction products of the ensuing reduction–oxidation processes in which cellular products may be involved.[155]

1.5.4 Corrosion in Electrically Active Devices

Metallic biomaterials are frequently used for the fabrication of devices that deliver electrical stimulation to muscle, cardiac, and neural tissues. For example, with heart failure growing to epidemic levels in industrialised countries, an increasing number of patients are implanted with defibrillators and pacemakers every year.[156] In the United States alone, 5 million patients live with a diagnosed heart failure, with 0.4 million new patients diagnosed annually, costing in excess of \$50 billion in hospital care alone.[157] Once implanted, electrical stimulators pass an electric current through an electrode placed in the proximity of the excitable target tissue, resulting in tissue activation and its improved function. The electrodes are typically made of highly corrosion resistant materials, for example a Co–Cr–Ni superalloy of Co (39–41%), Cr (19–21%), Ni (14–16%), Fe (11.3–20.5%), Mo (6–8%) and Mn (1.5–2.5%), sold under the trade name Elgiloy®. They can also be made from stainless steel, titanium nitride, iridium oxide, or Pt and Pt–Ir alloys, with the latter being the most frequently used electrode materials. The delivery of electrical stimulation can involve Faradaic as well as non-Faradaic processes. In the former, the electrons are transferred between the electrode and biofluid, whereas the latter relies on a rapid change in the concentration of ions at the electrode surface, in the absence of charge transfer between the electrode and biofluid. The Faradaic processes can therefore promote metal oxidation and corrosion, with the subsequent release of soluble metals into the peri-implant environment. By comparison, since the chemistry of tissues and biofluids does not change as a direct consequence of non-Faradaic stimulation, the effect of the stimulation on metal oxidation, corrosion and ion release is significantly lower. Yet, the stimuli that can be delivered using non-Faradaic processes are often not sufficient to attain a clinically significant result, and Faradaic processes are required to deliver the needed level of charge transfer.

Even when corrosion-resistant, chemically inert materials are used, Faradaic corrosion has been reported *in vitro*, *in vivo* and in clinical cases. For example, an X-ray fluorescence microscopy study, using the Australian Synchrotron, on tissues in prolonged contact with multi-channel cochlear implant electrode arrays showed the presence of platinum flakes within macrophages, and particularly in the fibrous tissue surrounding some of the arrays, and the diffusion of finer platinum particles further into the tissues as far as the spiral ligament.[158] The study suggested that the flakes may have been generated as part of the manufacturing process, whereas smaller particles were more likely to originate as a result of electrolysis due to charge imbalance with the bipolar pulses from the implant, as well as corrosion of the electrode material. The mechanism of corrosion that has been proposed, based on the experimental evidence gathered to date, suggests the generation and release of both ions and insoluble particles from the surface, with the extent of degradation dependent on the amplitude and frequency of the stimulation current, and the length of stimulation pulse and stimulation treatment (see Figure 1.26).[6] As with any other type of corrosion, the surface and bulk properties of the material, the geometry of the implant, and the chemistry of the biofluid will also play a role in determining the rate of corrosion and material release. As such, ion and particle release due to

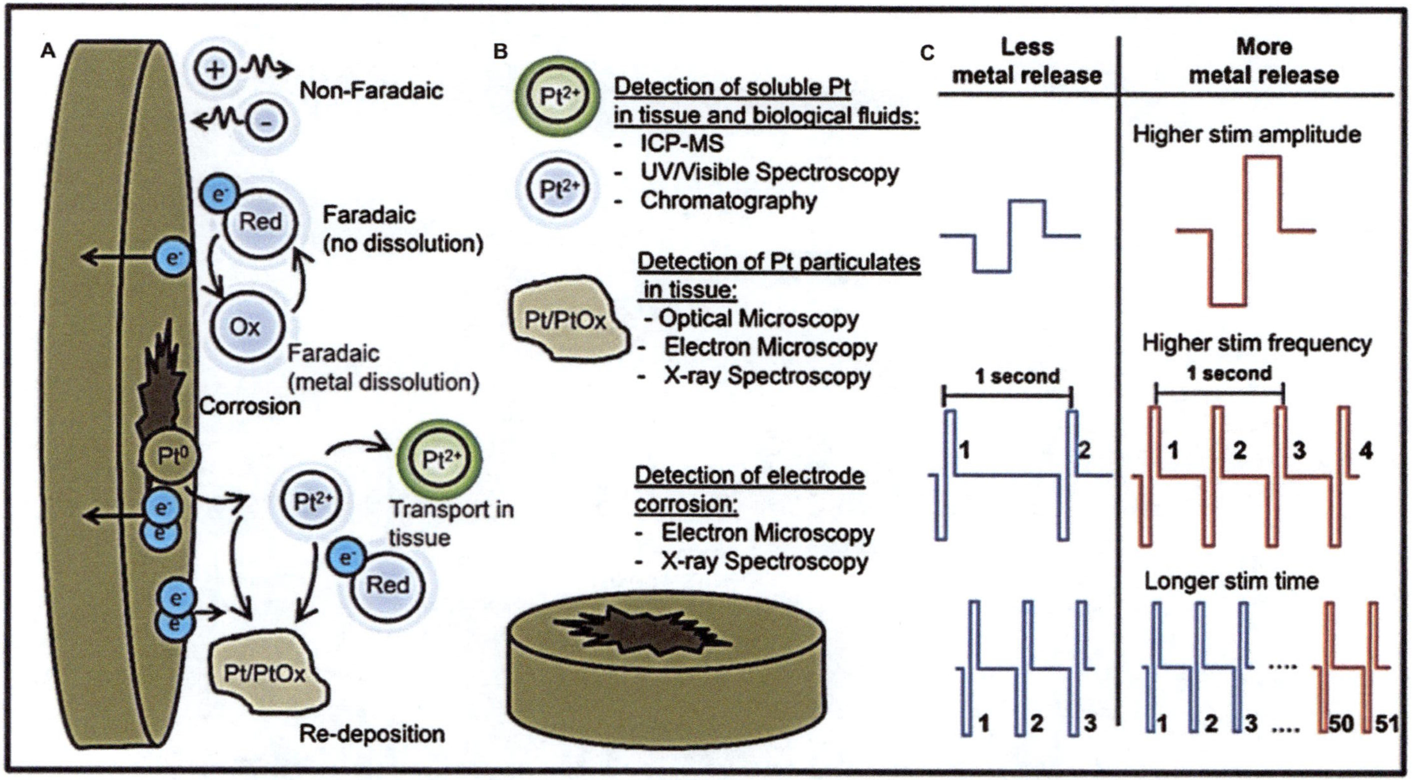

Figure 1.26 Release of platinum from electrically active implants as a function of electrical stimulation parameters. (A) Proposed mechanism of platinum release. (B) Analytical techniques used to detect electrode corrosion and platinum release. (C) Parameters of electrical stimulation that influence platinum release.

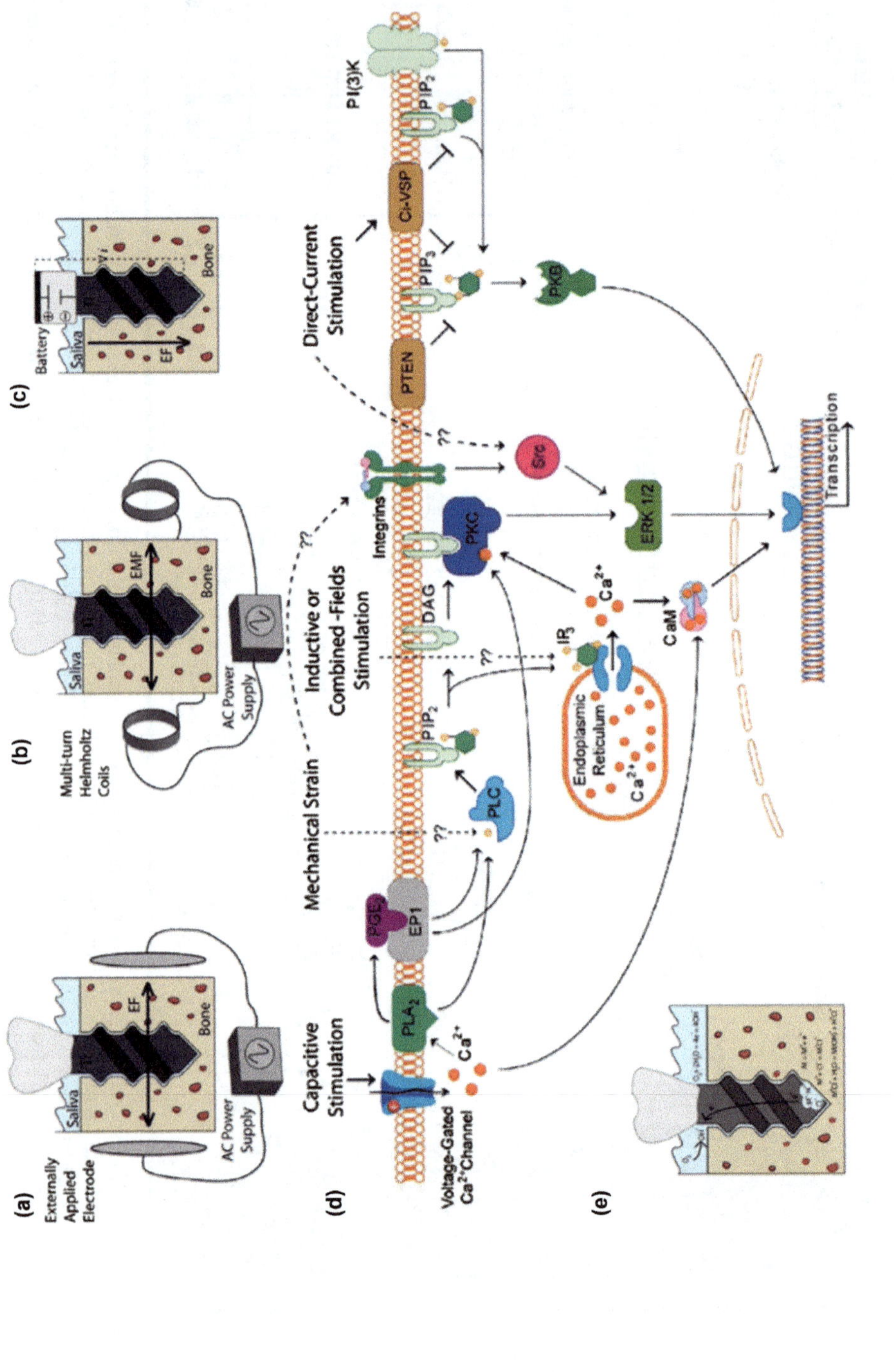
(a)
Externally Applied Electrode
Saliva
EF
Bone
AC Power Supply
(b)
Multi-turn Helmholtz Coils
Saliva
EMF
Bone
AC Power Supply
(c)
Battery
Saliva
EF
Bone
(d)
Voltage-Gated Ca2+ Channel
Capacitive Stimulation
Ca2+
PLA2
PGE
EP1
Mechanical Strain
??
PLC
PIP2
Inductive or Combined-Fields Stimulation
DAG
IP3
Endoplasmic Reticulum
Ca2+
Ca2+
CaM
PKC
Integrins
??
Src
ERK 1/2
PTEN
Direct-Current Stimulation
Ci-VSP
PIP3
PKB
PIP2
PI(3)K
Transcription
(e)

corrosion may become a more significant issue as the size of the electrodes decreases, which effectively increases the charge density applied to the electrode.[159]

1.5.5 Effect of Therapies on the *In Vivo* Corrosion of Implants

In addition to currents that arise due to differences in materials properties or as a result of corrosion processes, external electrical stimulation is often used in clinical practice to reduce pain and inflammation, and promote the healing of injured tissue[160] by activating osseosynthesis processes in the affected bones,[161,162] particularly in the instances of non-union or delayed union of bone fragments, pseudarthrosis and in patients with a history of pathological fractures. For example, direct current, capacitive coupling, and inductive coupling stimulations (see Figure 1.27) have been reported to improve the rate of arthrodesis in patients suffering from non-union complications following spinal fusion surgeries, with the improvement being clinically significant.[163] This is because external electrical stimulation replaces the electro–mechanical effects that arise naturally during the initial stages of bone healing, whereby electrical potentials are generated from a combination of the piezoelectric effect and streaming potential in the deformed bone under applied load.[163] The resulting action currents flowing through the tissues stimulate the enrichment of the facture site with nutrients and bone-forming minerals, and the induction of biochemical processes associated with union activation. In addition to promoting osteoblastic activity, weak electrical stimulation may prevent bacterial colonisation of implant surfaces. For example, in studies conducted *in vitro* and *in vivo*, anodisation of silver or silver-coated Ti electrodes for several days using 8 μA direct current resulted in a significant reduction in the pathogen load of *Staphylococcus aureus*, which lasted for several weeks afterwards,[164] with the inhibitory concentrations of electrically generated silver ions being up to 100 times lower than they are for the clinically approved antibiotic silver sulfadiazine.[165] Subsequent reversal of the poles led to pronounced bone-forming activity. Similar results were observed in a rabbit model, where passing a 200 μA electrical current though a stainless steel

Figure 1.27 Schematics of (a–c) different electrical stimulation systems and (d) the effects of the stimuli they produce on signalling pathways and cellular processes in tissues. (a) Capacitive stimulation set-up, consisting of two externally applied electrodes that generate an electric field. (b) Inductive stimulation set-up, consisting of a pair of multi-turn Helmholtz coils connected in series to generate an electromagnetic field. (c) DC electrical stimulation set-up consisting of a battery that generates an electric field directly through the implant device. The implant becomes the cathode, the anode is exposed to the oral cavity, and the surrounding tissue serves as a path to close the circuit and allow for the flow of current. (e) Initiation and mechanism of the corrosion of a dental implant.
Reproduced from ref. 154 with permission from SAGE, Copyright 2011.

implant was more effective in preventing the colonisation of *S. epidermidis* than standard treatment with doxycycline,[166] and in a goat model, where a low direct current of 100 μA was sufficient enough to hinder the development of clinical infection in the proximity of stainless steel percutaneous pins.[167] The coupling of 15–30 V with anodised and non-anodised nanotubular titanium was also reported to result in a significant reduction in *S. aureus* cell attachment and biofilm formation after 2 days of culture *in vitro*.[168]

The use of external stimulation may have significant implications for implant corrosion, with the outcome dependent on the nature of the stimulation and the properties of the protective layer on the implant. For example, invasive application of a direct current in pulsed mode led to a loss in mass in Cr–Ni–Mo AISI 316L steel implants, even when the latter were encapsulated in a passive diamond coating, with anode screws and cathodes being subject to corrosion damage and surface-limited etching, respectively. Importantly, the corrosion products from the corroded anode were driven into the tissue region, both along the direction as well as in the opposite direction of the current, with measurable levels of Cr, Fe, Mo and Ni detected in the compact bone region.[161] Enhanced corrosion and the infiltration of the products into tissues may compromise the mechanical performance of implants and induce a significant tissue response in surrounding tissues. The latter can result in cell death and severe chronic inflammation, eventually leading to the formation of a thick fibrous tissue capsule around the implant, the role of which is to separate the body from necrotic cells and tissue debris.[169] It should be noted that in the aforementioned study on Cr–Ni–Mo AISI 316L steel implants in tissues subjected to electrical stimulation, the delivery of alternating (sinusoidal) current or direct current in pulse mode using non-invasive and semi-invasive methods was associated with minimal corrosion damage to the implants.[161]

In addition to electrical stimulation, light irradiation, heat treatment and atmospheric pressure plasma treatment may be used to promote wound healing, stimulate tissue regeneration and immune function, prevent pathogenic colonisation and treat malignant growth. Among these, plasma treatment is particularly interesting as it delivers a potent cocktail of UV photons, mild heating, physical agitation, charged and neutral particles, and chemically reactive species to cells and surfaces, changing their behaviour.[170–172] For cells, the treatment may induce a variety of biochemical responses, from activating the differentiation and migration of host cells, to selectively inducing programmed cell death in cancer stem cells and pathogenic microorganisms.[173,174] For biomolecules, plasma treatment may affect their chemistry and biochemical reactivity.[175] For abiotic surfaces, such as those of implants, the treatment may change their chemistry and topography, thus potentially changing the manner in which these surfaces interact with biological agents, *e.g.* proteins and cells, and the environment. The changes in surface properties can be those of material addition (*e.g.* nucleation and growth of films and complex ordered nanostructures on plasma-treated surfaces), material removal (*e.g.* etching of the plasma-treated surface to uniformly remove the top oxide layers, selective etching of a pattern with or without a template, or increasing roughness), or material modification (*e.g.* surface functionalisation to increase hydrophilicity, introduction of defects into nanostructured materials, or ion implantation to increase hardness and wear resistance of the topmost layer of the implant without affecting the mechanical properties of the bulk material).[38,39,176] These addition or removal processes may significantly affect the surface geometry, whereas the modification processes may not.

While many studies have been carried out to investigate the effect of the intentional plasma modification of implants (prior to implantation),[36,37,177] little is known about the consequences of possible unintentional surface modification when plasma is used to treat the peri-implant milieu. In addition to directly modifying the chemistry of the surface, plasma treatment is known to significantly affect the chemistry of treated fluids by introducing large quantities of reactive oxygen and nitrogen species, which then react with molecules within the fluid to produce other chemically active species, including ions of Cl and H_2O_2, and to reduce the pH.[178,179] Mild heating and photons emitted by the plasma may further promote these and other chemical reactions in fluids. Cells, too, respond to plasma-generated hydroxyl radicals ($^{\bullet}OH$), hydrogen peroxide (H_2O_2), ozone (O_3), superoxide (O^{2-}), nitric oxide ($NO^{\bullet}$) and anionic ($OONO^{-}$) and protonated (ONOOH) forms of peroxynitrite by producing micromolar levels of H_2O_2.[180,181] The breakdown of H_2O_2 generated in cells in response to plasma treatment is then catalysed by iron-containing proteins into cytotoxic $^{\bullet}OH$ radicals that can induce cell injury and apoptotic and necrotic cell death, with the outcome dependent on the treatment dose.[180,181] The thus-released H_2O_2 and other cytokines may also trigger lymphocyte activation and immunogenic cell death, although it is difficult to untangle the roles that individual mechanisms may play when plasmas are used to treat animals or humans.[182] The aforementioned changes in the chemistry of the peri-implant milieu may in principle promote material corrosion *via* the previously discussed mechanisms. The treatment may also render cells more susceptible to the uptake and accumulation of particles produced in the process of implant degradation,[183] exacerbating the oxidative stress that cells experience as a result of their interactions with the metallic debris.

1.5.6 Role of Microorganisms in Corrosion and Failure

Most metallic implants are highly susceptible to bacterial attachment and colonisation. As it settles on the surface, a bacterial cell may entrap ions and small molecules between the surface of the bacterial layer and that of the metallic implant, inducing a localised change in the osmotic concentration and pH of the aqueous phase trapped within the confined microenvironment.[184] Subsequently, the presence of infectious agents on the surface of implants can further stimulate pro-inflammatory responses in the patient, as well as alter local pH through pathogens releasing their products of metabolism into the peri-implant milieu. The latter can acidify or alkalise the environment, change its electrochemical potential, enrich it with corrosion-promoting or corrosion-inhibiting chemical species, and directly interact with the protective layer and products of corrosive degradation, with the individual effects depending on the pathogen in question.[185] Before we go on to discuss possible mechanisms and specific examples of microbe-promoted corrosion or corrosion inhibition, it is important to once again stress that *in vitro* findings may not necessarily reflect the events that will take place *in vivo*. For example, if we are to consider the oral cavity, we will find a highly complex and dynamic population of species with great genetic and phenotypic diversity. Furthermore, the chemistry and biochemistry of the oral environment will change with food consumption, oral hygiene practices, pathophysiological processes and so forth (see Figure 1.28).

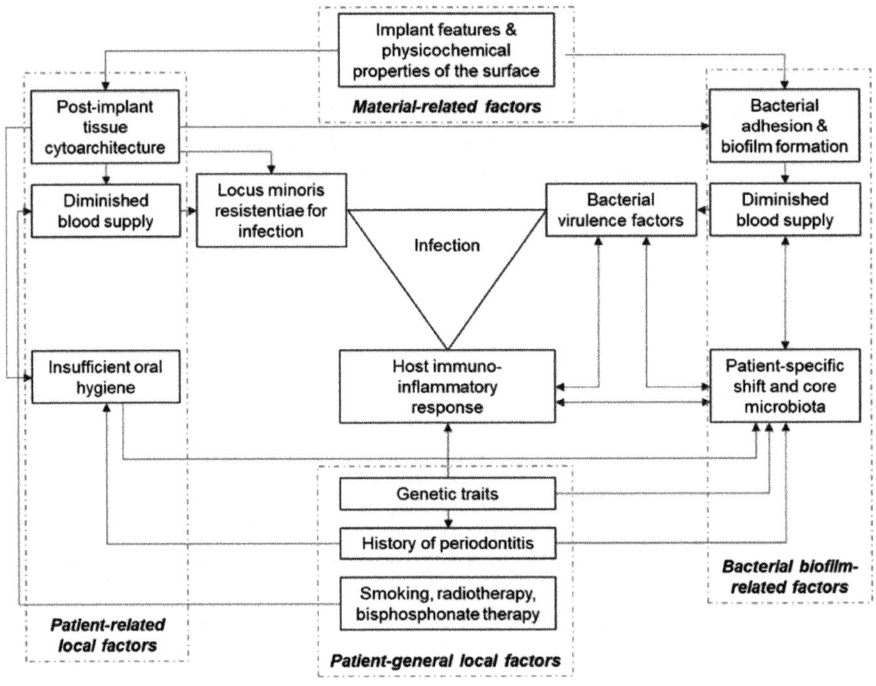

Figure 1.28 Complex interactions between peri-implant infection, implant characteristics and patient-specific local and general factors. The implant characteristics influence peri-implant tissue cytoarchitectonics, which are highly susceptible to bacterial infiltration and constitute a place of less resistance for bacterial infiltration. The implant surface is a favourable site for bacterial adhesion and biofilm maturation. Smoking decreases blood flow in peri-implant capillaries. Insufficient hygiene and periodontitis are risk factors for peri-implant infective disease. Genetic traits define host immunoinflammatory response and subject-specific shifts in the oral biofilm.
Adapted from ref. 237 with permission from Dove Medical Press, Copyright 2017.

In general, most reports in the literature focus on the corrosion-promoting properties of bacteria,[186] especially when it comes to dental applications. As one of the major players in dental plaque formation, several studies have investigated the role of *Streptococcus sanguis* and *S. mutans* in metals and alloys, showing that the former bacterium promotes the corrosion of $Nd_2Fe_{14}B$ and other magnetic materials used as retainers, and the latter increases the rate of corrosion in commercially pure Ti.[187,188] Yet even though the presence of a *S. mutans* biofilm on the surface reduces the corrosion resistance of Ti, it does not prevent the formation of a compact passive film similar to that formed under the same conditions in the absence of bacteria. Furthermore, no pitting corrosion was detected in any of the aforementioned studies, thus challenging previously established associations between biofilms and the development of pits (see Figure 1.29).[189] It is possible that discrepancies in these results arises from differences in experimental protocol, including the length of time the biofilm is allowed to

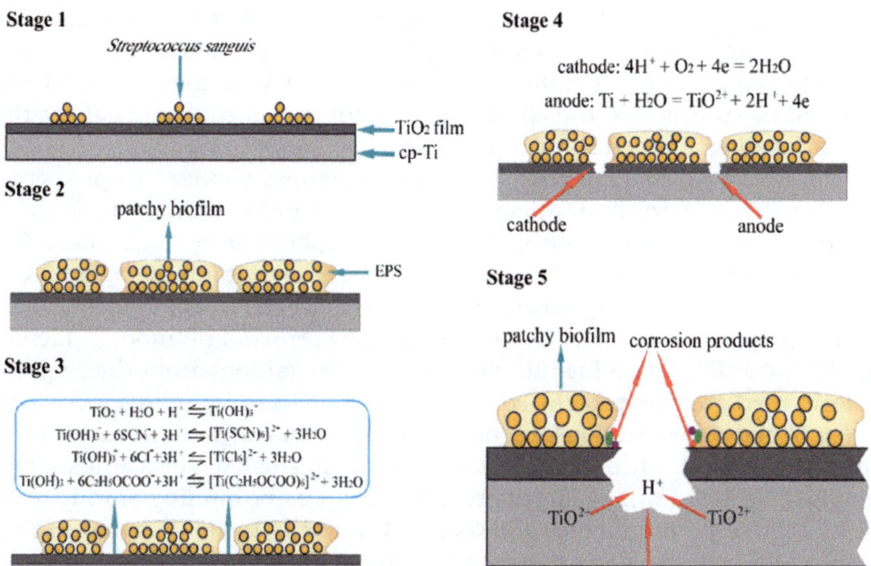

Figure 1.29 Stages of *S. sanguis* biofilm formation on the surface, and its effect on the corrosion of a titanium implant material: (Stage 1) initial attachment of *S. sanguis* cells to the surface of titanium; (Stage 2) formation of a discontinuous biofilm; (Stage 3) dissolution of TiO_2 film in the areas of biofilm discontinuities; (Stage 4) cathodic and anodic reactions at the oxide layer and pit root, respectively; (Stage 5) formation of a pit. Reproduced from ref. 189, https://doi.org/10.3390/ma10030255, under the terms of the CC BY 4.0 license, https://creativecommons.org/licenses/by/4.0/.

grow on the surface, and the presence of specific environmental factors that may the drive the development of biofilms of certain composition, architecture and continuity, or change the properties of the metal.[189–191]

1.5.7 Protein-mediated Mechanisms of Material Degradation *In Vivo*

When implants are first exposed to bodily fluids of a reasonably high ionic strength, their surfaces are likely to be subject to the rapid adsorption of macromolecules, *e.g.* proteins, even in the case where the charge is similar for the macromolecule and the surface.[192] Protein–metal interactions may involve attachment *via* non-specific electrostatic interactions (*i.e.* physisorption) as well as site-specific attachment (*i.e.* irreversible chemisorption). The chemistry, charge and size of the protein, its relative concentration within the fluid, and the characteristics of the implant, in particular its surface chemistry, topography and properties of the oxide layer, will determine the probability of adsorption on any given surface. The physical and chemical properties of the aqueous phase in the peri-implant milieu will also influence the probability of protein adsorption and the eventual desorption of complexes that proteins form with metals, with pH, ionic strength, physical agitation and shear rates shown to play a role.[193] Therefore, such events as inflammation or the release of oxidative species due to cell injury or death may significantly affect the protein binding and detachment processes.

Once adsorbed, proteins may first act in a manner similar to a protective coating to limit corrosion-inducing cathodic reactions and lubricate surfaces to reduce the effects of fretting or friction. Depending on their chemistry and surface conformation, they may also promote or inhibit the attachment of host cells and microorganisms. Over time, however, the protective effects of the protein layer may be overcome by corrosion-promoting factors (see Figure 1.30). These include the release of metal ions from the implant surface, mediated by protein complexation and ligand-binding processes, driven by the concentration-dependent exchange of the protein between the surface and biofluid, and the displacement of smaller surface adsorbed proteins, *e.g.* albumin, by larger proteins with greater binding affinity from the biofluid, *e.g.* immunoglobulin G and fibronectin. It should be noted that metal loss due to the above described Vroman effect is low and thus difficult to capture accurately within a short observation period. Nevertheless, the metal–protein complexation-induced release of Fe, Cr, and Ni due to the Vroman effect was experimentally investigated in both 316L and 303 stainless steels exposed to fluids containing albumin and fibrinogen.[194] The exact manner in which metals are released from the surface may be dependent on the properties of the surface oxide, *e.g.* composition, microstructure, corrosion susceptibility, and the presence of defects, as these may affect the relative strength of the protein–metal bond when compared to metal–oxide or metal–hydroxide bonds. However, these dependencies are not fully understood. A study of protein-mediated metal loss processes in 304, 310,

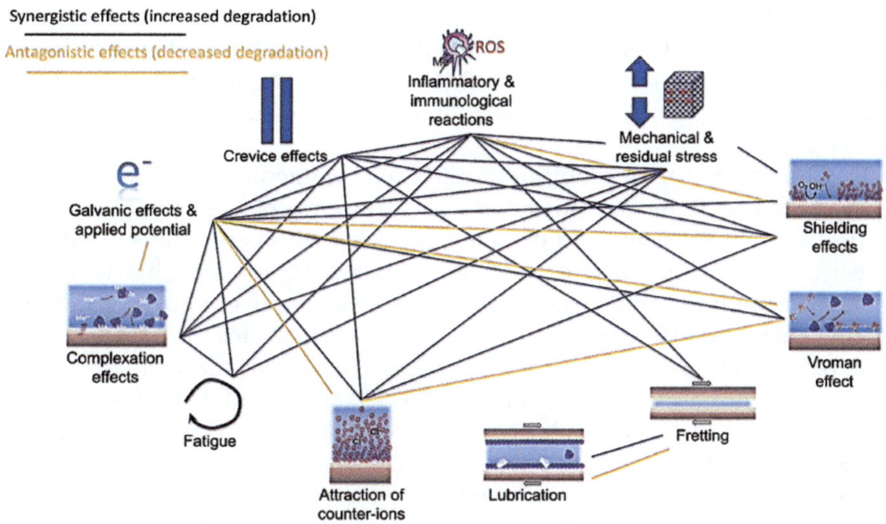

Figure 1.30 Potential synergistic and antagonistic interactions between events that may arise as a result of protein interactions with metallic surfaces, and may affect the degradation kinetics of materials under *in vitro* and *in vivo* conditions. Most events act synergistically to promote material degradation. Antagonistic interactions include adsorbed protein shielding cathodic reactions or reactive oxygen species that decrease the effect of inflammatory conditions; the exchange of proteins by the Vroman effect decreases the local effects of attracted counter-ions; proteins decrease friction by lubrication or fretting-induced corrosion; electrically-coupled different materials or applied potential (*e.g.*, from electronic devices, electrosurgery) change the amount and type of adsorbed proteins.

and 316L austenitic stainless steels, 430 nickel-free ferritic stainless steel, 2205 duplex stainless steel and pure iron found that the complexation ability of the surface oxides differed between these materials.[195] Within 2 h of exposure to albumin-containing phosphate buffer, iron atom–albumin complexes formed on the surfaces of the oxide layer of 304 steel, resulting in a notable enrichment in Cr in the surface oxide. This then acted as a passivation-like coating that hindered the rate of release of metal and increased the polarisation resistance of this alloy when compared to other alloys in the study. The changes in the polarisation resistance of alloys were attributed to the preferential complexation to Fe over Cr atoms. The complexation and release of Mn was found to be significantly more dependent on the pH of the biofluid compared to Fe, Ni, or Cr, whereas its dependence on complexation mechanisms was the lowest amongst these atoms. While poor protective performance of the iron oxide layer meant that pure Fe had a significantly greater rate of Fe loss when compared to the tested alloys, the addition of albumin to the buffer resulted in a significant decrease in the quantities of Fe lost from its surface, since the binding of the protein acted to partially block sites where the reduction of oxygen may otherwise take place.

In addition to the availability of particular elements on the surface, the microstructure of the outermost surface, surface-to-volume ratio (in the case of powders), crystallinity and presence of a large number of defects will also affect the rate of metal release.[196] It is interesting to note that while the greater crystallinity of oxides is linked to greater rates of metal oxidation during corrosion, the rate of protein-mediated metal loss is greater in the case of amorphous oxides. This is possibly because of the increased probability of metal complexation and complex detachment due to weaker metal–oxide bonds in the latter oxide structures. The presence of reactive oxidative species, particularly H_2O_2, can significantly increase the susceptibility of the metallic implant to protein-mediated metal loss; this phenomenon can have significant clinical ramifications, since increased levels of H_2O_2 are often detected at sites of infection, inflammation or injury. An *in vitro* study of metal loss from a corrosion-resistant TiAlV alloy when exposed to simulated physiological fluids with varying concentrations of H_2O_2 and albumin showed significant changes in thickness and composition of the surface oxide, with the layer becoming depleted of V and thicker and enriched in Al as a result of H_2O_2 exposure alone, and depleted of Al when exposed to a combination of H_2O_2 and albumin.[197] Interestingly, accelerated corrosion studies over the same period of 24 h did not show any significant changes, with the albumin-catalysed dissolution of the corrosion product layer and consequent formation of a thinner oxide layer detected when the study was run for 120 h.[198] An even longer *in vivo* investigation of the same alloy implanted in rat tibia, which would expose the surfaces to similar concentrations of H_2O_2 and albumin, reported the eventual stabilisation of the level of metals in the peri-implant tissue after 1.5–6 months. These results highlight the significance of considering both corrosion and protein-mediated degradation mechanisms and selecting appropriate timelines when testing the stability of implant materials.

In addition to H_2O_2, conditions that damage the protective oxide or prevent its repair, *e.g.* fretting, are likely to facilitate direct protein–metal interactions, which may result in an increased rate of material removal due to the synergistic effects of dissolution and corrosion. Fretting corrosion rates of stainless steel plates and screws in fluids containing proteins of different charge, and adjusted to a pH between 3 and 5, were found to be similar when the alloys were exposed to an albumin-containing solution with a pH below the isoelectric point of the albumin.[199] Interestingly, Ni was preferentially released in protein-rich fluids when compared to the metal release in saline. In contrast, the rate of corrosion was decreased when the alloys were exposed to negatively charged proteins in solutions with pH above their isoelectric points, confirming that protein charge plays an important role in defining its effect on corrosion. The charge of the adsorbed proteins may also affect the nature of the charged species these proteins can attract to the surface, with net-positively-charged lysozymes attracting negatively charged Cl^- ions, and net-negatively-charged albumin attracting H^+ ions. The surface-localised changes in ion concentration may further destabilise the protective oxide layer, increasing the local rate of corrosion.

Besides the aforementioned chemical effects of protein adsorption, the layer may entrap wear debris and metal ions within its matrix, positively contributing towards material deterioration *via* friction and wear.[200] The wear debris may also interfere with protein–protein interactions, leading to protein aggregation around the wear particle, preventing its degradation and potentially leading to increased inflammation,[201] thus creating favourable conditions for cell damage and increased material degradation *via* other mechanisms of corrosion.

1.6 Physiological Implications of Corrosion and Wear

Whether it is produced through wear or corrosion, metallic debris may significantly undermine the function of the implant system and pose a significant threat to the health of tissues in the peri-implant milieu, and to host systems in general (see Table 1.9). The probability of metallic debris generation is particularly high in systems containing articulating components, *e.g.* orthopaedic implants. Yet, the use of these devices is on the rise, with the annual number of total hip arthroplasty surgeries exceeding 1.4 million worldwide, with 0.6 million in Europe alone, and is expected to continue to grow.[19] Total hip replacement surgeries are also on the increase, with 332 000 procedures conducted annually in the United States alone.[5] In part, this is driven by a rapidly ageing population that wants to remain active for longer, as well as by the needs of younger patients suffering from painful and often debilitating joint conditions, including, but not limited to, osteoarthritis, rheumatoid and traumatic arthritis, avascular necrosis, acetabulum defects, bone fractures, noncancerous and malignant abnormal bone growths.[202] Greater levels of physical activity, however, are often associated with a greater incidence of short-term failures of these implants, necessitating replacement surgery. Adverse reactions to metal debris, also known as metallosis, pseudotumour or aseptic lymphocytic vasculitis-associated lesions, are one of the leading causes of replacement surgery. A recent retrospective study using data from the National Joint Registry for England, Wales, Northern Ireland and the Isle of Man showed that of 873 188 primary hip replacements implanted over the observation period, 0.38% required a revision surgery due to adverse reactions to metal debris.[5] While metal-on-metal hip implants were associated with the greatest risk of revision due to metallosis (3.7%), other implant designs were not immune,

Table 1.9 Local and systemic toxicity of small-sized debris particles after total hip replacement using implants containing metal bearings. Adapted from ref. 238 with permission from John Wiley & Sons, Copyright 2012 John Wiley & Sons, Ltd.

System	Particle	Action
Haematopoietic	Al	Erythropoiesis impairment
	Ni	Decreased erythrocyte thermostability, deformability, and the rate of O_2 release
	Co, Ni	Reduction of human endothelial cell number within 24 h
	Co	Impairment of growth and differentiation of bone marrow derived $CD34^+$ human haematopoietic progenitor cells
Immune	Co, Cr, Ni, Al, *etc.*	↓CD8 (+) cells levels (T-cytotoxic/suppressor) ↓leucocytes, myeloid cells, lymphocytes, CD16 cells ↔ CD3, CD4, CD8, and CD20 (+) cells Alterations in spleen architecture (capsule and medulla), depletion of T4 & B cells Inhibition of the rapid release of antimicrobial ROS by neutrophils
Hepatobiliary	Cr	Hepatic malfunction, potentially severe hepatic lesions, hepatocellular necrosis and possibly disseminated intravascular coagulation
	TiO_2	Fatty degeneration of rat hepatocytes
Renal	Cr, Cu, Al	Tubular necrosis and interstitial cell damage
Respiratory	Pt salts, Ni, Cr, Co	Acute chemical pneumonitis, pulmonary oedema, acute tracheobronchitis Bronchial asthma and allergic sensitisation
	TiO_2	Follicular lymphoid hyperplasia with inflammatory cells aggregated around bronchia
Nervous	Al, Co, Ni	Retinal degeneration
	Al	Amyotrophic lateral sclerosis, Alzheimer's disease
Cardiovascular	Co	Cardiomyopathy, impaired left ventricular function
	TiO_2	Accumulation of dispersed and aggregated brown particulates in interstitial fascicle, cytoplasm and nucleus of vascular cells of ventricular endocardium
Musculoskeletal	Al	Osteomalacia, pathological fractures, impaired bone remodelling, impaired response to vitamin D, proximal myopathy
Skin	Co, Ni, Cr	Dermal reactions, positive skin-patch testing to Co, Ni and Cr (>50% in patients with loose total joint replacement)

Table 1.9 (*Continued*)

System	Particle	Action
Endocrine and reproductive	Co	Goitre and myxoedema
	Co, Cr, Ni, Al, *etc.*	Altered production or circulation of reproductive hormones, impaired sperm and ova quality, decreased fertility
	Cr	Testicular toxicity, disrupted spermatogenesis (monkey)
	Cr, Ni, Co, V, Al	Translocation of particles through the maternofoetal circulation and lactation to offspring
	Cr, Ni, Co, V, Al	Affected conception, foetal implantation and later teratogenicity
Carcinogenesis	Ionic Cr, Co, Ni, V, Al, Ti	Mutagenic actions on cells *in vitro* mediated by direct action, free-radical mediated DNA breaks, inhibition of DNA repair
	Co, Cr, Ni, Al, *etc.*	Mutagenic damage in bone marrow and peripheral blood lymphocytes
		Aberrations in chromosomes in patients with joint implants
	Co, Cr, Ni	Cr(VI) and Ni(II) considered carcinogenic, metallic Ni and soluble Co as possibly carcinogenic, and metallic Cr, Cr(III) compounds and implanted orthopaedic alloys as unclassifiable
		Increased incidence of different cancers in patients with metal-on-metal articulating implants

with metal-on-polyethylene, ceramic-on-ceramic, ceramic-on-polyethylene, ceramic-on-metal, and metal-on-ceramic implants having non-trivial levels of risks of 0.024%, 0.055%, 0.023%, 0.69%, and 1.97%, respectively. Importantly, while the mechanisms underpinning metal particle generation in metal-on-metal implants are somewhat understood, the case is less clear with the other types of hip joint prosthesis. In metal-on-metal implants, multiple mechanisms of wear particle generation are considered to be at play, including corrosion at modular implant junctions, specifically at the junction between the femoral head and neck and between femoral neck and stem, and articulating mixing of alloys that make up metal femoral heads and titanium femoral necks.[97,203,204] The body responds to these particles with a T lymphocyte-mediated type IV hypersensitivity reaction and associated tissue damage, which is often but not always found to be dose-dependent. The rate and pattern of wear are design-dependent, with greater wear rates reported for metal-on-metal implants with significant edge loading (*e.g.* in shallow acetabular cups or those implanted at abnormal inclination angles). Metallic particle generation due to increased trunnion wear in ceramic-on-ceramic and metal-on-polyethylene bearings has also been reported, with certain design features, such as the greater size of the head in metal-on-polyethylene bearings, associated with greater risk of wear and corrosion in modular junctions due to greater transmitted torques.[205,206] In addition to material- and design-specific factors that affect debris generation and the pathological events that ensue, the installation of the implant and skill of the surgeon, the goodness of the fit to the anatomy of the patient, and the patient's level of physical activity will all play a role. For example, the likelihood of adverse implant reactions and implant failure due to metallosis is typically greater for patients with a pre-existing metal sensitivity, although some patients will develop sensitivity as a result of implantation, making data collection and analysis challenging.[207] Furthermore, patients with certain genetic conditions, chronic diseases, and diet and lifestyles may be more susceptible to premature implant wear and failure.

When a sufficient number of wear particles enter the peri-implant milieu, they trigger a cell-mediated inflammatory response in the tissues in contact with the implant, leading to bone loss (*i.e.* periprosthetic osteolysis), and subsequent loss of implant integration (*i.e.* aseptic loosening). The latter will affect the transfer of mechanical stresses from the implant to the bone, as well as between the implant components, eventually leading to implant failure. Even in the absence of obvious mechanical failure of the implant, the performance of loosened implants will be sub-optimal, with local inflammation, pain and reduced function. Understanding the mechanisms of debris-initiated inflammatory processes and bone lesion formation is not trivial due to the complexity and interconnectivity of biochemical and biophysical processes that these particles can trigger, both locally and at the systemic level. We have already discussed some of the challenges in understanding the kinetics of material failure due to chemical and physical events, and the role biomolecules and cells may play in these processes.

It should be noted that in addition to metal particles, chemical degradation and wear of load-bearing implants, especially those with articulating surfaces, is generally accompanied by a release of fragments of bone cements used to fix the implant to the bone, and significant quantities of polymer particles, including poly(methyl methacrylate) and polyethylene debris with ultrahigh molecular weight, as these polymers are commonly used in polymer-on-metal and polymer-on-ceramic bearings. These particles may also induce adverse cell responses, inflammation, bone loss and changes in the chemistry of the peri-implant milieu, thereby affecting the corrosion rate of metals. The loss of cement may also facilitate the ingress of biofluids from the milieu into the bone–metal interface, where the physical confinement creates favourable conditions for the corrosion of metallic components. These particles may also get entrapped between interfaces that are articulating or subject to micromotion, promoting surface wear. Generally, the generation of these particles has been found to be reduced when using polymer-on-ceramic bearings, highly crosslinked polymers in polymer-on-metal bearings, and cement-free metal-on-metal and ceramic-on-ceramic systems.

When compared to polymer particles generated by joint implants, the metallic debris are more numerous, and generally have a smaller dimension and more uniform size distribution. While the volume of the wear particles produced by metal-on-ultra high molecular weight polyethylene polymer implants is estimated to be 40–100 greater than that by metal-on-metal implants, the number of released polymer particles is notably lower, at an average of 5×10^{11} particles per year, compared to an average of 6.7×10^{12}–2.5×10^{14} metal particles per year released by metal-on-metal implants.[208] Their size allows metallic debris to migrate more easily away from the peri-implant milieu when compared to large molecular weight polymer debris that accumulates within peri-implant tissues. In doing so, metallic particles can reach distant organs, and activate a systemic response. There have also been reports on metallic particles becoming attached to such proteins as albumin, alpha-1-antitrypsin, and apolipoprotein, acting as a hapten and eliciting an adaptive immune response by activating specific cell surface receptors and initiating the inflammatory cascade.[209] With a size in nanometres (often, few tens of nanometres for individual particles, and few hundred nanometres to few hundred micrometres for agglomerates) and a large surface-to-volume ratio, metallic particles are also more biochemically reactive compared to larger polymer particles (*e.g.* ultra-high and high molecular weight polyethylene) or less reactive ceramic particles (*e.g.* alumina), being increasingly prone to rapid corrosion and releasing considerable quantities of ions into physiological fluids. As a result, the release of the inflammation-related factors TNF-α, IL-6 and IL-1β may be greater when cells are exposed to metallic particles when compared to ceramic particles, as demonstrated by exposing primary macrophages and THP-1 monocytes to Ti and TiO_2 rutile particles,[210] even when the rutile particles are significantly smaller (0.45 ± 0.26 μm for TiO_2) than Ti particles (3.32 ± 2.39 μm). Having said that, end-stage inflammatory responses and associated bone resorption have also been reported in response to metal oxide particles, *e.g.* alumina. Furthermore, some studies suggest that polymer particles with a small size (0.3 to <5–10 μm) and irregular shape may elicit greater proinflammatory responses than either metallic or metal oxide particles, yet this concept remains a subject of debate.[209,211] Nevertheless, it is generally accepted that smaller particles (<1 μm) tend to simulate a greater inflammatory response, with particle concentration, shape, surface area and surface energy controlling the nature and extent of the histological response.

It is worth noting that although debris generation is generally discussed with reference to load-bearing articulating implants, metals and alloys used in, for example, dental implants and cranial anchorage devices, will also deteriorate over time. While their degradation is more commonly associated with the release of metallic ions, loss of integrity in the protective passivation coating due to micro-motion associated wear and/or corrosion may also result in the release of particles. For example, commercially pure cranial anchorage devices used in bone-anchored hearing aids, and thus not subjected to macroscopic wear processes, have been shown to release Ti metallic and oxide particles and ions into the surrounding soft tissue, leading to peri-implant tissue inflammation.[212] Where orthopaedic implants are generally fabricated from titanium grade V alloys, it is not unusual for titanium grades II–IV to be used for the fabrication of dental and anchorage devices. The differences in interstitial elements and processing renders these grades different with respect to the mechanical strength, chemical reactivity, and biocompatibility of the bulk materials and products of their degradation. When exposed to suitable conditions, *e.g.* acidic pH levels or increased concentrations of H_2O_2 in the fluid, these materials can undergo significant corrosive degradation at the interface between metallic components of the devices, even under minimal load-bearing and wear conditions.

1.6.1 Interactions Between Macrophages and Metallic Wear Debris

In vivo, a response to an adverse stimulus, such as the presence of a metallic particle, involves resident and recruited cells of different types, including macrophages and giant cells, bone cells (osteoblasts and osteoclasts), cells of the extracellular matrix (fibroblasts) and lymphocytes. This response is mediated by a large number of signalling molecules that cells may express as a result of direct contact with metallic particles or after being activated as part of the inflammation cascade, including chemokines, growth factors, cytokines, eicosanoids, degradative enzymes, and reactive oxygen species; their release may promote or suppress inflammation and trigger tissue loss and regeneration.[208] When the contact event between the cells in the peri-implant milieu and debris is transient, the recruitment and stimulation of inflammatory cells and acute inflammation-related bone loss is followed by increased osteoblastic activity, tissue regeneration and healthy bone formation. In contrast, when cells in the peri-implant are persistently challenged by particles and ions, the inflammation becomes chronic, and the tissue regeneration stage is often compromised due to increased activity of osteoclasts and a concomitant decrease in the activity of osteoblasts.[211] Unless addressed, this can lead to a considerable loss of bone tissue in the immediate proximity of the implant, and the formation of fibrous pseudo membranes at the interface between the bone and the implant, *i.e.* fibrosis. The membrane effectively prevents the in-growth of bone tissues into the implant (in the case of porous implants) or the formation of a strong bond between the bone and the implant. This lack of proper integration leads to increased micro-motion of the implant against the tissues, causing tissue injury and further stimulating inflammation.[203]

Among the aforementioned cells involved in debris-induced inflammation, inflammatory events mediated by macrophages are among the best studied, in part due to the ability of these cells to migrate towards areas where the concentration of metallic debris is high. Once at the site, macrophages will attempt to remove the particles through endocytotic or pinocytotic uptake (in the case of smaller particles, <150 nm) or phagocytosis for particles with a size of 150 nm–10 μm.[213] It should be noted that some of the particles will also be removed from the peri-implant milieu by other cell types, *e.g.* osteoblasts, fibroblasts, and endothelial cells. Due to their ability to stimulate the inflammatory response *via*, for example, the release of signalling molecules and cell-to-cell stimulation of monocytes, the activity of macrophages generally widens the zone of soft-tissue damage and inflammation.

The signalling chemicals released by mononuclear macrophages upon particle ingestion can also trigger mononuclear phagocytes to fuse and subsequently differentiate into cells capable of degrading larger objects. The biochemical stimuli include factors TNF-α, IL-1β, IL-6, PGE2 and GM–CSF. For example, debris from Co–Cr–Mo alloys, including soluble Co, Cr, Mo, and Ni ions and particulates, have been shown to activate the inflammasome-mediated pathway in macrophages, causing caspase-1-induced cleavage of intracellular pro-IL-1β into its mature form, which in turn causes the secretion of highly pro-inflammatory IL-1β and induction of a broader proinflammatory response.[214]

In the case of debris-releasing bone implants, the fusion of cells from the monocyte-macrophage lineage is likely to have two outcomes: chronic inflammation and the presence of foreign particles will trigger the programme for macrophage fusion and differentiation into multinucleated giant cells, whereas events at the bone will trigger the programme for macrophage fusion and differentiation into osteoclasts. The multinucleate giant cells are capable of ingesting foreign fragments with a size greater than 10 μm, whereas osteoclasts resorb the bone tissue. Interestingly, the mechanism by which the degradation of a material is achieved is very similar between these two cell types. Both rely on the cells attaching strongly to the substrate (an oversized debris particle, the surface of the bone or that of the implant), and then releasing lysosomal enzymes and protons into the thus-formed sealed space to degrade the material.[215]

The internalisation of metallic debris (both particles and ions) can induce DNA damage, with the extent of damage dependent on the cell type and properties and concentration of the particles. For example, at low concentrations of <0.01 mM, the *in vitro* exposure of osteoblasts, fibroblasts, and lymphocytes to fluids containing ions of Al, Co, Cr, Fe, Mo, Ni, and V has little effect on viability. At greater concentrations of up to 1 mM, exposure to ions of Co, Ni, and V negatively affects cell proliferation, viability and morphology, whereas it is necessary to increase the concentration of ions >5 mM for fluids containing Al, Cr, Fe, and Mo to obtain an effect of similar magnitude. At lower concentrations of below 1 mM, exposure to ions of Al, Cr, Fe, and Mo induces changes in the shape of the cell, and loss of filopodia and lamellipodia, suggesting that the migration ability of the cells may also be affected across the three cell types.[216] An important conclusion from this study was that soluble Co and V were able to induce notable toxicity in all three cell types at concentrations below that detected in the peri-implant milieu in animal and clinical models that use Co- and Ti-based alloys. Metal ions have also been shown to be more effective in inducing allergies to metallic implants when compared to metallic particles, as they are effective at both stimulating macrophage secretion of pro-inflammatory cytokines IL-1β, IL-6, and TNFα, similar to particles, as well as upregulating CD80, CD86, and ICAM-1 T cell costimulatory molecules in human monocytes and macrophages.[217] Ions of Co have also been shown to induce cellular injury through nitric oxide-induced protein nitration to form nitrotyrosine, and apoptosis in macrophages through activation of the caspase-3 pathway.[218]

Chemical reactions between metallic particles and biomolecules present in biofluids can produce significant quantities of reactive oxygen and nitrogen species. These molecules play important roles as signalling molecules, capable of activating macrophages without the cell needing to ingest the metal particle. Due to their considerable chemical reactivity, they can also cause oxidative damage to cellular components. For example, Ti debris can generate radicals capable of inducing the peroxidation of linoleic acid in the plasma membrane to malondialdehyde, and to induce the NFkB signalling pathway by triggering NSmase activation *via* the hydrolysation of sphingomyelin in the membrane.[219] It is important to note that all cells are susceptible to radical-mediated damage from the exposure to metallic debris, and oxidative stress can arise even from the exposure to nanoscale particles of noble metals. For example, the inhalation of gold particles by rodents causes metabolic and structural changes in the lung tissue, with animals suffering from interstitial pneumonia, fibrosis, chronic inflammatory cell infiltrates of small lymphocytes, congested and dilated blood vessels, scattered dense extravasation of red blood cells, and foci of hemosiderin granules.[220] The extent of the damage suggests that particles not only induces oxidative stress, but also interferes with the antioxidant machinery of the exposed cells through interactions with relevant protein structures.

Unlike pathogenic microorganisms, metallic particles cannot be fully digested by macrophages, and return to the extracellular environment once the macrophage dies where they can once again interact with other cells. Interestingly, experimental evidence suggests that, at least *in vitro*, phagocytosis by macrophages renders Ti particles unable to stimulate TNF-alpha secretion by macrophages without changing particle size or shape, most likely through changing the chemical composition of their surfaces.[221] As previously mentioned, the surface properties of nanoparticles, *e.g.* their chemistry, free surface energy, topography, are important determinants of their biochemical reactivity and toxicity. Such a mechanism may offer some protection in the case of a one-off release of particles, however, the continued release of particles due to corrosion or wear may limit its efficacy.

The presence of certain types of molecules attached to the surfaces of metallic nanoparticles may increase their cytotoxicity. These molecules can originate from activated or lysed cells, where cells can be host cells, *e.g.* alarmins or danger signals from injured or dying macrophages or osteoblasts, or they can be microbial cells, *e.g.* extracellular polysaccharides and virulence factors, including liposaccharides, peptidoglycan, lipoteichoic acid, and teichoic acid. In principle, particles can also harbour attached bacterial cells or even their biofilm communities. Microbial cells and particles that attach to wear particles may originate from infected implants, with microbial patterns remaining even on implants that have been sterilised, or be present in the peri-implant milieu.[222] The human body is well equipped to deal with a bacterial threat, and once recognised by immune cells, particle-bound bacterial fragments and microbial-associated molecular patterns may trigger strong pro-inflammatory responses. It should be noted that such a response may be mounted by the host body in the absence of clinical infection,[222] although it is possible that conventional diagnostic methods simply fail to detect microbial populations. Importantly, the response to low quantities of bacterial fragments bound to particle surfaces may far exceed that to the particles themselves. This has been shown experimentally *in vitro* and *in vivo* by using Ti particles with and without bacterial fragments on their surface. The inflammatory response that particles free of endotoxins are able to elicit in bone marrow cells from mice and human monocytes is significantly lower than that induced by particles with endotoxins. The exposure to the former particles is also associated with a much lower incidence of osteolysis (by 50–70%).[223] When endotoxins are re-introduced into the fluid containing the endotoxin-free nanoparticles, the endotoxins rapidly attach to the particle surface, with the resulting opsonised particles regaining their ability to evoke cytokine release and stimulate osteoclast differentiation. This is because endotoxins have strong affinity for metallic biomaterials, and this results in it also being difficult to remove endotoxins using conventional methods of cleaning. The significance of endotoxins in the ability of wear particles to induce inflammation was also confirmed by using an antibiotic capable of inactivating bacterial lipopolysaccharides, where the use of polymyxin B suppressed the pro-inflammatory activity of Ti particles coated with bacterial lipopolysaccharides.[224] The presence of bacterial toxins may also explain the pro-inflammatory response induced by wear particles of inert materials, as these endotoxins ligate TLR2 and TLR4 toll-like receptors, the pattern-recognition receptors on the cells of the innate immune system responsible for the detection of the lipopolysaccharides of Gram-negative bacteria and the lipoteichoic acid and peptidoglycan of Gram-positive bacteria, respectively, stimulating macrophage infiltration of the peri-implant space. TLR2 and TLR4 also respond to endogenous ligands derived from host tissues or cells, including fragments of cells, induced gene products, extracellular matrix components, *e.g.* fibronectin, heparan sulfate, biglycan, fibrinogen and hyaluronan breakdown fragments.[225]

Many of the materials-based strategies to prevent inflammation focus on the prevention of wear debris generation, *e.g.* through the development of wear-resistant coatings or selecting wear-reducing combinations of materials for interfaces that are subject to articulation, and on mitigating bacterial attachment and removal of implant-attached bacterial motifs. Other approaches include immunomodulation strategies, *e.g.* through interrupting macrophage activation mediated by toll-like receptors (TLRs) and factor-kappa B transcription factor, inhibiting macrophage migration,

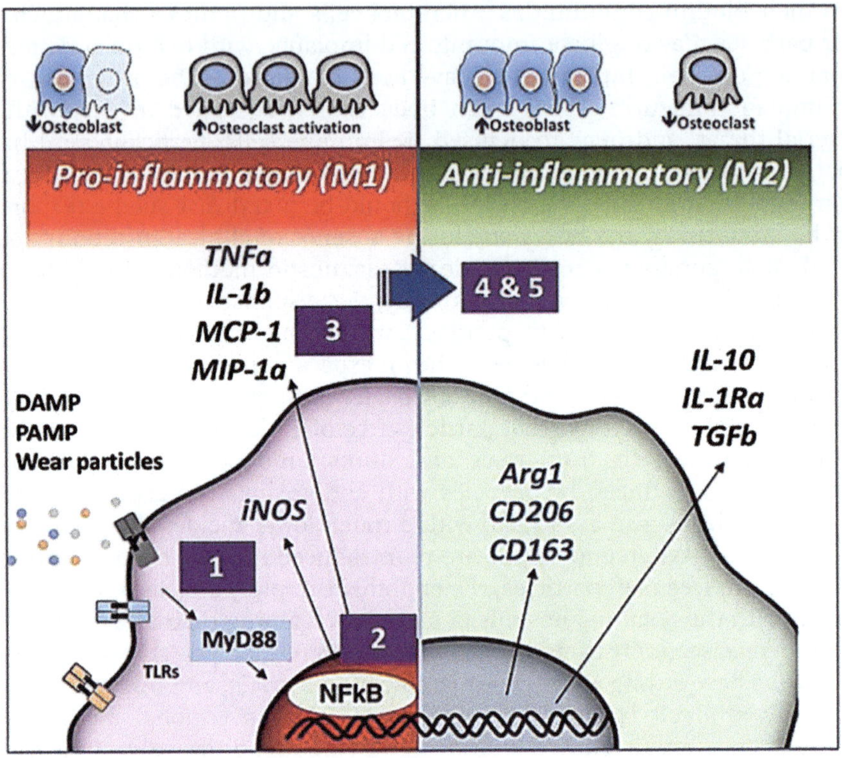

Figure 1.31 Strategies for immunomodulation to mitigate periprosthetic osteolysis induced by wear particles. Wear debris, adherent pathogen-associated molecular patterns (PAMPs) and damage-associated molecular patterns or DAMPS released by injured and dying cells are recognised by toll-like receptors (TLRs) on macrophages, activating downstream pathways, including the key transcription factor nuclear factor-kappa B (NFκB) and induce the expression of inducible nitric oxide synthetase (iNOS) and cytokines/chemokines including tumor necrosis factor alpha (TNFα), interleukin 1 beta (IL-1b), macrophage chemotactic protein 1 (MCP-1), macrophage inhibitory protein 1 alpha (MIP-1α), and others. These events lead to periprosthetic osteolysis trough reduced osteoblast and increased osteoclast activity. Mitigation strategies include inhibiting (1) the TLR pathway; (2) NFκB activation; or (3) macrophage migration, and polarisation of pro-inflammatory macrophages (M1) by (4) interleukin 4 (IL-4) treatment or (5) genetically modified or preconditioned mesenchymal stem cells (MSCs) into an anti-inflammatory, pro-tissue repair macrophage (M2) phenotype. M2 macrophages produce interleukin 10 (IL-10), IL-1 receptor antagonist (IL-1ra), and transforming growth factor beta (TGFβ). Reproduced from ref. 211, https://doi.org/10.3389/fbioe.2019.00230, under the terms of the CC BY 4.0 license, https://creativecommons.org/licenses/by/4.0/.

or by polarising pro-inflammatory macrophages *via* interleukin 4 treatment to shift them towards an anti-inflammatory, pro-tissue repair macrophage phenotype (see Figure 1.31).[211]

1.6.2 Effect of Metallic Wear Debris on Tissue Regeneration

In the previous section, the ability of wear debris to activate macrophages and trigger their fusion and subsequent differentiation into bone-resorbing osteoclasts was mentioned. However, in addition to stimulating osteoclast activity, wear particles and soluble metals have been reported to depress the differentiation and bone-reforming activity of osteoblasts, hindering implant integration at the initial stages of implantation, and subsequently compromising the continuing repair of the periprosthetic bed needed for the implant to perform its function properly over extended periods of time.[209] This is because metallic particles interfere with mesenchymal stem-cell differentiation into functional osteoblasts, reducing their ability to express genes associated with collagen production and proliferate; depending on the particle concentration, they may also initiate programmed cell death in osteoblasts. For example, exposure of Saos-2 osteoblasts to particles from the degradation of Ti and its alloys affects their attachment apparatus by altering their cytoskeletal structure, specifically by interfering with the production and arrangement of ventral stress fibres, β-tubulin and acetylated α-tubulin fibres, and inducing changes in the phosphorylation states of both FAK and Pyk2 tyrosine kinases at focal contacts.[226] The stage of osteoblast maturation and the properties of the particle, *e.g.* their size, significantly affects the type of events and the extent of damage in response to the debris, with more mature cells showing lower susceptibility to particle-induced apoptosis, and larger particles stimulating the activity of enzymes responsible for the degradation of extracellular matrix proteins during bone formation, *e.g.* matrix metalloproteinases (MMP)-2 and -9.[227]

Mitigating the damaging effects of wear particles on bone formation requires better understanding of the events and processes triggered by particle exposure. Yet, the design of *in vitro* models that would reflect the complexity of the peri-implant milieu, and capture the rich variety of feedback mechanisms and timing of signalling events that take place between cells, is not a trivial matter. Furthermore, peri-implant inflammation and osteolysis may arise due to reasons other than the effects of the wear debris, and these events may be difficult to decouple. For example, both bone resorption and formation are affected by the mechanical pressure that bone cells experience. Osteocytes, the star-shaped cells typically found in mature bone, continuously experience mechanical cues from the surrounding bone (lacunae) and neighbouring osteocytes, osteoblasts and osteoclasts which they contact by means of long cytoplasmic extensions, *i.e.* processes that reside in pericellular interstitial fluid-filled canals termed canaliculi. Introduction of the implant changes the nature of the peri-implant environment, affecting the flow of fluid within the lacuno-canalicular porosity, and the pressure that this network of cells experience. For example, the micromotion of the implant may be sufficient to drive the joint fluid into the implant–bone interface, and thus increase the pressure in the peri-implant space. In response to these flow and pressure changes, osteocytes release a host of chemical signals, including nitric oxide, which alter osteocyte apoptosis, and osteoblast and osteoclast activity. The effect of mechanical instability on osteoclast differentiation and bone loss have been shown to depend on the type and magnitude of mechanical stimulation, with supraphysiological loadings of MLO-Y4-osteocytes enhancing osteoclast differentiation by 1.9-fold when compared to physiological loading, with the stimulatory effect on the membrane-bound receptor activator of NF-kappa B ligand (RANKL) and nitric oxide production also being greater in the case of supraphysiological loading when compared to stress shielding.[228] These results suggest that wear particle-induced inflammation and bone resorption may proceed in parallel or be secondary to mechanical instability-induced events.

1.7 Concluding Remarks

Despite tremendous progress in the development and application of metallic biomaterials, and the significant benefits that their use brings to patients, there are still gaps that prevent us from taking full advantage of this class of materials. Our knowledge of how these materials truly behave *in vivo* is still incomplete, which limits intelligent design and optimisation, as well as our ability to predict and respond to the short- and long-term consequences of their clinical use. We still lack *in vitro* models and characterisation tools that would allow us to capture the complexity of events that take place in the body after implantation, and the intricate links that may exist between different mechanisms of corrosion, wear, and host response. Data from implant clinical use are a result of multiple factors, including those arising from the choice of material, device design, the method of implantation, skill of the surgeon, treatment regime, and a great diversity of patient-related factors. Furthermore, there is great variability with respect to how *in vitro*, *in vivo* and clinical studies are designed, performed and interpreted, making comparison challenging. Ideally, a systematic approach based on a unified experimental framework and set definitions would certainly help, however there is a lack of consensus as to what such a framework should look like.

References

1. W. A. Lane, *Br. Med. J.*, 1895, **1**, 861–863.
2. H. K. Uhthoff, P. Poitras and D. S. Backman, *J. Orthop. Sci.*, 2006, **11**, 118–126.
3. D. F. Williams, *Definitions in Biomaterials: Proceedings of a Consensus Conference of the European Society for Biomaterials, Chester, England, March 3-5, 1986*, Elsevier Science Limited, 1987.
4. C. F. Van Eck, A. F. Chen, B. A. Klatt, J. D'Antonio and F. Fu, *J. Long Term Effects Med. Implants*, 2009, **19**(3), 185–193.
5. G. S. Matharu, H. G. Pandit, D. W. Murray and A. Judge, *BMC Musculoskeletal Disord.*, 2016, **17**, 495.

6. Biological responses to metal implants, *U.S. Food and Drug Administration Centre for Devices and Radiological Health*, 2019, retrieved from https://www.fda.gov/media/131150/download.
7. D. C. Carvalho, L. M. Coelho, M. S. M. Acevedo and N. M. Coelho, in *Handbook of Mineral Elements in Food*, ed. M. de la Guardia and S. Garrigues, 2015, pp. 109–122.
8. M. Long and H. Rack, *Biomaterials*, 1998, **19**, 1621–1639.
9. V. K. Truong, R. Lapovok, Y. S. Estrin, S. Rundell, J. Y. Wang, C. J. Fluke, R. J. Crawford and E. P. Ivanova, *Biomaterials*, 2010, **31**, 3674–3683.
10. N. Martynenko, E. Lukyanova, V. Serebryany, D. Prosvirnin, V. Terentiev, G. Raab, S. Dobatkin and Y. Estrin, *Mater. Lett.*, 2019, **238**, 218–221.
11. M. Sumita, T. Hanawa, I. Ohnishi and T. Yoneyama, in *Comprehensive Structural Integrity*, ed. I. Milne, R. O. Ritchie and B. Karihaloo, Pergamon, 2003, pp. 131–167.
12. V. Moraschini, L. d. C. Poubel, V. Ferreira and E. dos Sp Barboza, *Int. J. Oral Maxillofac. Surg.*, 2015, **44**, 377–388.
13. J. Du, J. H. Lee, A. T. Jang, A. Gu, M. Hossaini-Zadeh, R. Prevost, D. A. Curtis and S. P. Ho, *J. Biomech.*, 2015, **48**, 3486–3494.
14. D. Robinson, L. Aguilar, A. Gatti, J. Abduo, P. V. S. Lee and D. Ackland, *J. Adv. Prosthodont.*, 2019, **11**, 169–178.
15. M. Nakai, M. Niinomi, X. Zhao and X. Zhao, *Mater. Lett.*, 2011, **65**, 688–690.
16. P. F. Santos, M. Niinomi, K. Cho, M. Nakai, H. Liu, N. Ohtsu, M. Hirano, M. Ikeda and T. Narushima, *Acta Biomater.*, 2015, **26**, 366–376.
17. P. F. Santos, M. Niinomi, H. Liu, K. Cho, M. Nakai, Y. Itoh, T. Narushima and M. Ikeda, *J. Mech. Behav. Biomed. Mater.*, 2016, **59**, 497–507.
18. S. Barui, S. Mandal and B. Basu, *Curr. Opin. Biomed. Eng.*, 2017, **2**, 116–123.
19. L. Dall'Ava, H. Hothi, A. Di Laura, J. Henckel and A. Hart, *Metals*, 2019, **9**(7), 729.
20. A. Dawood, B. M. Marti, V. Sauret-Jackson and A. Darwood, *Br. Dent. J.*, 2015, **219**, 521–529.
21. Y.-G. Park, H. S. An, J.-Y. Kim and J.-U. Park, *Sci. Adv.*, 2019, **5**, eaaw2844.
22. N. E. Putra, M. J. Mirzaali, I. Apachitei, J. Zhou and A. A. Zadpoor, *Acta Biomater.*, 2020, **109**, 1–20.
23. D. Han, C. Yang, N. X. Fang and H. Lee, *Addit. Manuf.*, 2019, **27**, 606–615.
24. Y. S. Hedberg, B. Qian, Z. Shen, S. Virtanen and I. O. Wallinder, *Dent. Mater.*, 2014, **30**, 525–534.
25. D.-T. Chou, D. Wells, D. Hong, B. Lee, H. Kuhn and P. N. Kumta, *Acta Biomater.*, 2013, **9**, 8593–8603.
26. W. E. King, H. D. Barth, V. M. Castillo, G. F. Gallegos, J. W. Gibbs, D. E. Hahn, C. Kamath and A. M. Rubenchik, *J. Mater. Process. Technol.*, 2014, **214**, 2915–2925.
27. E. Sachs, E. Wylonis, S. Allen, M. Cima and H. Guo, *Polym. Eng. Sci.*, 2000, **40**, 1232–1247.

28. F. E. Wiria, J. Y. M. Shyan, P. N. Lim, F. G. C. Wen, J. F. Yeo and T. Cao, *Mater. Des.*, 2010, **31**, S101–S105.
29. T. DebRoy, H. Wei, J. Zuback, T. Mukherjee, J. Elmer, J. Milewski, A. M. Beese, A. Wilson-Heid, A. De and W. Zhang, *Prog. Mater. Sci.*, 2018, **92**, 112–224.
30. L. E. Murr, S. A. Quinones, S. M. Gaytan, M. I. Lopez, A. Rodela, E. Y. Martinez, D. H. Hernandez, E. Martinez, F. Medina and R. B. Wicker, *J. Mech. Behav. Biomed. Mater.*, 2009, **2**, 20–32.
31. S. F. S. Shirazi, S. Gharehkhani, M. Mehrali, H. Yarmand, H. S. C. Metselaar, N. Adib, Kadri and N. A. A. Osman, *Sci. Technol. Adv. Mater.*, 2015, **16**, 033502.
32. T. Hanawa, *Sci. Technol. Adv. Mater.*, 2012, **13**, 064102.
33. Y. Bao, T. Kudo, S. Cao, A. I. Munoz and S. Mischler, *J. Bio- Tribo-Corr.*, 2020, **6**, 58.
34. I. Levchenko, K. Bazaka, M. Keidar, S. Xu and J. Fang, *Adv. Mater.*, 2018, **30**, 1702226.
35. K. Bazaka, M. V. Jacob, R. J. Crawford and E. P. Ivanova, *Acta Biomater.*, 2011, **7**, 2015–2028.
36. O. Baranov, K. Bazaka, H. Kersten, M. Keidar, U. Cvelbar, S. Xu and I. Levchenko, *Appl. Phys. Rev.*, 2017, **4**, 041302.
37. K. Bazaka, M. V. Jacob, V. K. Truong, F. Wang, W. A. A. Pushpamali, J. Y. Wang, A. V. Ellis, C. C. Berndt, R. J. Crawford and E. P. Ivanova, *Biomacromolecules*, 2010, **11**, 2016–2026.
38. K. Bazaka, O. Baranov, U. Cvelbar, B. Podgornik, Y. Wang, S. Huang, L. Xu, J. W. M. Lim, I. Levchenko and S. Xu, *Nanoscale*, 2018, **10**, 17494–17511.
39. O. Baranov, I. Levchenko, J. M. Bell, J. W. M. Lim, S. Huang, L. Xu, B. Wang, D. U. B. Aussems, S. Xu and K. Bazaka, *Mater. Horiz.*, 2018, **5**, 765–798.
40. E. Fadeeva, V. K. Truong, M. Stiesch, B. N. Chichkov, R. J. Crawford, J. Wang and E. P. Ivanova, *Langmuir*, 2011, **27**, 3012–3019.
41. E. P. Ivanova, J. Hasan, H. K. Webb, G. Gervinskas, S. Juodkazis, V. K. Truong, A. H. F. Wu, R. N. Lamb, V. A. Baulin, G. S. Watson, J. A. Watson, D. E. Mainwaring and R. J. Crawford, *Nat. Commun.*, 2013, **4**, 2838.
42. E. P. Ivanova, J. Hasan, H. K. Webb, V. K. Truong, G. S. Watson, J. A. Watson, V. A. Baulin, S. Pogodin, J. Y. Wang, M. J. Tobin, C. Löbbe and R. J. Crawford, *Small*, 2012, **8**, 2489–2494.
43. V. J. Wandiyanto, K. V. Truong, M. Al Kobaisi, S. Juodkazis, H. Thissen, O. Bazaka, K. Bazaka, J. R. Crawford and P. E. Ivanova, *Materials*, 2019, **12**(10), 1575.
44. H. Tsuchiya and P. Schmuki, *Electrochem. Commun.*, 2005, **7**, 49–52.
45. V. S. Saji, H. C. Choe and W. A. Brantley, *Acta Biomater.*, 2009, **5**, 2303–2310.
46. I. Sieber, H. Hildebrand, A. Friedrich and P. Schmuki, *Electrochem. Commun.*, 2005, **7**, 97–100.

47. H. Chouirfa, M. D. M. Evans, P. Bean, A. Saleh-Mghir, A. C. Crémieux, D. G. Castner, C. Falentin-Daudré and V. Migonney, *ACS Appl. Mater. Interfaces*, 2018, **10**, 1480–1491.
48. G. L. Kenausis, J. Vörös, D. L. Elbert, N. Huang, R. Hofer, L. Ruiz-Taylor, M. Textor, J. A. Hubbell and N. D. Spencer, *J. Phys. Chem. B*, 2000, **104**, 3298–3309.
49. J. L. Dalsin, B.-H. Hu, B. P. Lee and P. B. Messersmith, *J. Am. Chem. Soc.*, 2003, **125**, 4253–4258.
50. M. Weber, A. Vasella, M. Textor and N. D. Spencer, *Helv. Chim. Acta*, 1998, **81**, 1359–1372.
51. Y. Ito, H. Hasuda, M. Sakuragi and S. Tsuzuki, *Acta Biomater.*, 2007, **3**, 1024–1032.
52. G. Krauss, *Mater. Sci. Eng. A*, 1999, **273–275**, 40–57.
53. K. Yang and Y. Ren, *Sci. Technol. Adv. Mater.*, 2010, **11**, 014105.
54. J. W. Morris, *Science*, 2008, **320**, 1022.
55. P. V. Muterlle, M. Zendron, M. Perina, R. Bardini and A. Molinari, *J. Mater. Sci.*, 2010, **45**, 1091–1099.
56. S. J. Li, R. Yang, S. Li, Y. L. Hao, Y. Y. Cui, M. Niinomi and Z. X. Guo, *Wear*, 2004, **257**, 869–876.
57. R. A. Bernhoft, *J. Environ. Public Health*, 2012, **2012**, 460508.
58. J. Seo, Y. C. Kim and J. W. Hu, *Appl. Sci.*, 2015, **5**, 187–208.
59. İ. Özkul, M. A. Kurgun, E. Kalay, C. A. Canbay and K. Aldaş, *Eur. Phys. J. Plus*, 2019, **134**, 585.
60. T. Waitz, T. Antretter, F. D. Fischer and H. P. Karnthaler, *Mater. Sci. Technol.*, 2008, **24**, 934–940.
61. D. Allegretti, F. Berti, F. Migliavacca, G. Pennati and L. Petrini, *Shape Mem. Superelasticity*, 2018, **4**, 186–196.
62. P. K. Bowen, E. R. Shearier, S. Zhao, R. J. Guillory II, F. Zhao, J. Goldman and J. W. Drelich, *Adv. Healthcare Mater.*, 2016, **5**, 1121–1140.
63. J. Kruger, Fundamental Aspects of the Corrosion of Metallic Implants, in *Corrosion and degradation of implant materials*, ed. B. C. Syrett and A. Acharya, ASTM International, West Conshohocken, PA, 1979, pp. 107–127.
64. J. C. M. Souza, S. L. Barbosa, E. A. Ariza, M. Henriques, W. Teughels, P. Ponthiaux, J.-P. Celis and L. A. Rocha, *Mater. Sci. Eng. C*, 2015, **47**, 384–393.
65. C. Fonseca and M. Barbosa, *Corros. Sci.*, 2001, **43**, 547–559.
66. G. Burstein and S. Vines, *J. Electrochem. Soc.*, 2001, **148**, B504–B516.
67. G. T. Burstein and C. Liu, *Corros. Sci.*, 2007, **49**, 4296–4306.
68. G. T. Burstein, C. Liu and R. M. Souto, *Biomaterials*, 2005, **26**, 245–256.
69. D. J. Blackwood and B. P. Pereira, *J. Mater. Sci.: Mater. Med.*, 2004, **15**, 755–758.
70. H. Amel-Farzad, M. T. Peivandi and S. M. R. Yusof-Sani, *Eng. Failure Anal.*, 2007, **14**, 1205–1217.
71. M. Sivakumar and S. Rajeswari, *J. Mater. Sci. Lett.*, 1992, **11**, 1039–1042.
72. M. Sivakumar and S. Rajeswari, *Steel Res.*, 1995, **66**, 35–38.

73. D. J. Blackwood and S. K. M. Chooi, *Corros. Sci.*, 2002, **44**, 395–405.
74. B. S. Becker and J. D. Bolton, *Powder Metall.*, 1995, **38**, 305–313.
75. J. Pang and D. J. Blackwood, *Corros. Sci.*, 2016, **105**, 17–24.
76. R. H. Jones and R. Ricker, *Stress-Corrosion Cracking, Material Performance and Evaluation*, ASM International, 1992, p. 1.
77. S. Jafari, R. S. Raman, C. H. Davies, J. Hofstetter, P. J. Uggowitzer and J. F. Löffler, *J. Mech. Behav. Biomed. Mater.*, 2017, **65**, 634–643.
78. J. Yu, Z. J. Zhao and L. X. Li, *Corros. Sci.*, 1993, **35**, 587–597.
79. C. Fleck and D. Eifler, *Int. J. Fatigue*, 2010, **32**, 929–935.
80. B. Edwards, M. Louthan and R. Sisson, Corrosion and Degradation of Implant Materials: Second Symposium, 1985.
81. D. C. Rodrigues, R. M. Urban, J. J. Jacobs and J. L. Gilbert, *J. Biomed. Mater. Res., Part B*, 2009, **88B**, 206–219.
82. R. Waterhouse and M. Dutta, *Wear*, 1973, **25**, 171–175.
83. S. Virtanen, I. Milošev, E. Gomez-Barrena, R. Trebše, J. Salo and Y. T. Konttinen, *Acta Biomater.*, 2008, **4**, 468–476.
84. S. Brown, C. Flemming, J. Kawalec, H. Placko, C. Vassaux, K. Merritt, J. Payer and M. Kraay, *J. Appl. Biomater.*, 1995, **6**, 19–26.
85. J. L. Gilbert, C. A. Buckley and J. J. Jacobs, *J. Biomed. Mater. Res.*, 1993, **27**, 1533–1544.
86. M. Khan, R. Williams and D. Williams, *Biomaterials*, 1999, **20**, 765–772.
87. M. Windler, J. E. MacDougall and R. Schenk, in *Stainless Steels for Medical and Surgical Applications*, ASTM International, 2003.
88. L. Duisabeau, P. Combrade and B. Forest, *Wear*, 2004, **256**, 805–816.
89. S. Barril, S. Mischler and D. Landolt, *Wear*, 2005, **259**, 282–291.
90. M. Khan, R. Williams and D. Williams, *Biomaterials*, 1996, **17**, 2117–2126.
91. M. Fathi, M. Fathi and V. Mortazavi, *J. Res. Med. Sci.*, 2004, **9**, 42–51.
92. T. Chaturvedi, *Indian J. Dent. Res.*, 2009, **20**, 91–98.
93. A. Mellado-Valero, A. I. Muñoz, V. G. Pina and M. F. Sola-Ruiz, *Materials*, 2018, 11, 171.
94. M. Cortada, L. Giner, S. Costa, F. J. Gil, D. Rodriguez and J. A. Planell, *J. Mater. Sci.: Mater. Med.*, 2000, **11**, 287–293.
95. H. Arslan, H. Çelikkan, N. Örnek, O. Ozan, A. E. Ersoy and M. L. Aksu, *J. Appl. Electrochem.*, 2008, **38**, 853–859.
96. L. Reclaru and J. M. Meyer, *J. Dent.*, 1994, **22**, 159–168.
97. D. R. Plummer, R. A. Berger, W. G. Paprosky, S. M. Sporer, J. J. Jacobs and C. J. Della Valle, *J. Arthroplasty*, 2016, **31**, 264–268.
98. I. P. S. Gill, J. Webb, K. Sloan and R. J. Beaver, *J. Bone Jt. Surg., Br. Vol.*, 2012, **94-B**, 895–900.
99. R. Shah, C. Talmo and S. Nandi, *Arthroplasty Today*, 2019, **5**, 145–147.
100. T. McTighe and D. Brazil, Memorandum - Modular Necks, *Reconstr. Rev.*, 2012, **2**(1), DOI: 10.15438/rr.v2i2.22.
101. T. McTighe, D. Brazil, L. Keppler, J. Keggi and E. McPherson, Metallic Modular Taper Junctions in Total Hip Arthroplasty, *Reconstr. Rev.*, 2015, **5**(2), DOI: 10.15438/rr.5.2.108.

102. T. McTighe and D. Brazil, Modular Necks and Corrosion - Review of Five Cases, *Reconstr. Rev.*, 2016, **6**(3), DOI: 10.15438/rr.6.3.163.
103. S. Y. Jauch, G. Huber, E. Hoenig, M. Baxmann, T. M. Grupp and M. M. Morlock, *J. Biomech.*, 2011, **44**, 1747–1751.
104. S. A. Yerby, J. K. Taylor, J. Park and J. F. Shackelford, *J. Arthroplasty*, 1996, **11**, 157–162.
105. C. Xu and W. Gao, *Mater. Res. Innovations*, 2000, **3**, 231–235.
106. M. Morita, T. Sasada, I. Nomura, Y. Q. Wei and Y. Tsukamoto, *Ann. Biomed. Eng.*, 1992, **20**, 505–516.
107. A. A. Tiamiyu, U. Eduok, J. A. Szpunar and A. G. Odeshi, *Sci. Rep.*, 2019, **9**, 12116.
108. S. W. Dean, in *Encyclopedia of Materials: Science and Technology*, ed. K. H. J. Buschow, R. W. Cahn, M. C. Flemings, B. Ilschner, E. J. Kramer, S. Mahajan and P. Veyssière, Elsevier, Oxford, 2001, pp. 5930–5938.
109. N. A. Agha, F. Feyerabend, B. Mihailova, S. Heidrich, U. Bismayer and R. Willumeit-Römer, *Mater. Sci. Eng.: C*, 2016, **58**, 817–825.
110. R. W.-W. Hsu, C.-C. Yang, C.-A. Huang and Y.-S. Chen, *Mater. Sci. Eng.: A*, 2004, **380**, 100–109.
111. N. Eliaz and E. Gileadi, *Physical Electrochemistry: Fundamentals, Techniques, and Applications*, John Wiley & Sons, 2019.
112. N. Goraya and D. E. Wesson, in *Metabolic Acidosis: A Guide to Clinical Assessment and Management*, ed. D. E. Wesson, Springer New York, New York, NY, 2016, pp. 1–6.
113. N. Ilov, N. Ilov, A. Nechepurenko, A. Abdulkadyrov, D. Paskeev, E. Damrina, E. Kulikova, M. Terent'eva, D. Stompel and D. Tarasov, *J. Atrial Fibrillation*, 2016, **9**, 1391.
114. O. Beton, E. Saricam, H. Kaya, H. Yucel, O. Dogdu, O. O. Turgut, O. Berkan, I. Tandogan and M. B. Yilmaz, *BMC Cardiovasc. Disord.*, 2016, **16**, 73.
115. N. Eliaz, *Materials*, 2019, 12, 407.
116. M. Prikryl, S. C. Srivastava, G. R. Viviani, M. B. Ives and G. R. Purdy, *Biomaterials*, 1989, **10**, 109–117.
117. D. J. Blackwood, K. H. W. Seah and S. H. Teoh, in *Engineering Materials for Biomedical Applications*, World Scientific, 2004, vol. 1, pp. 3-1–3-56.
118. B. Beverskog and I. Puigdomenech, *Corros. Sci.*, 1997, **39**, 107–114.
119. W. G. Cook and R. P. Olive, *Corros. Sci.*, 2012, **58**, 291–298.
120. L. F. Huang, M. J. Hutchison, R. J. Santucci, J. R. Scully and J. M. Rondinelli, *J. Phys. Chem. C*, 2017, **121**, 9782–9789.
121. T. Moura e Silva, J. M. Monteiro, M. G. S. Ferreiraa and J. M. Vieira, *Clin. Mater.*, 1993, **12**, 103–106.
122. M. Bryant, R. Farrar, K. Brummitt, R. Freeman and A. Neville, *Wear*, 2013, **301**, 290–299.
123. J. A. von Fraunhofer, N. Berberich and D. Seligson, *Biomaterials*, 1989, **10**, 136–138.
124. J. A. von Fraunhofer and S. H. Stidham, *J. Biomed. Eng.*, 1991, **13**, 424–428.
125. X. N. Gu, Y. F. Zheng and L. J. Chen, *Biomed. Mater.*, 2009, **4**, 065011.

126. H. Ashassi-Sorkhabi, M. Majidi and K. Seyyedi, *Appl. Surf. Sci.*, 2004, **225**, 176–185.
127. H. Ashassi-Sorkhabi, Z. Ghasemi and D. Seifzadeh, *Appl. Surf. Sci.*, 2005, **249**, 408–418.
128. A. El-Shafei, M. Moussa and A. El-Far, *J. Appl. Electrochem.*, 1997, **27**, 1075–1078.
129. M. A. Khan, R. L. Williams and D. F. Williams, *Biomaterials*, 1999, **20**, 631–637.
130. F. Witte, J. Fischer, J. Nellesen, H.-A. Crostack, V. Kaese, A. Pisch, F. Beckmann and H. Windhagen, *Biomaterials*, 2006, **27**, 1013–1018.
131. Y. Koo, H.-B. Lee, Z. Dong, R. Kotoka, J. Sankar, N. Huang and Y. Yun, *Sci. Rep.*, 2017, **7**, 14710.
132. Y. Koo, T. Tiasha, V. N. Shanov and Y. Yun, *Sci. Rep.*, 2017, **7**, 1173.
133. L. M. McNamara and P. J. Prendergast, *J. Biomech.*, 2007, **40**, 1381–1391.
134. G. Bergmann, F. Graichen and A. Rohlmann, *J. Biomech.*, 1993, **26**, 969–990.
135. E. Birmingham, G. Niebur, L. McNamara and P. McHugh, *Ann. Biomed. Eng.*, 2016, **44**, 1191–1203.
136. A. P. Md Saad, R. A. Abdul Rahim, M. N. Harun, H. Basri, J. Abdullah, M. R. Abdul, Kadir and A. Syahrom, *Mater. Des.*, 2017, **122**, 268–279.
137. A. Oyane, H. M. Kim, T. Furuya, T. Kokubo, T. Miyazaki and T. Nakamura, *J. Biomed. Mater. Res., Part A*, 2003, **65**, 188–195.
138. F. Grases and A. Llobera, *Micron*, 1998, **29**, 105–111.
139. M. Pietrzyńska and A. Voelkel, *Microchem. J.*, 2017, **134**, 197–201.
140. M. Pietrzyńska, R. Tomczak, K. Jezierska, A. Voelkel and J. Jampílek, *Mater. Sci. Eng.: C*, 2016, **68**, 70–77.
141. A. Mainnemare, B. Megarbane, A. Soueidan, A. Daniel and I. Chapple, *J. Dent. Res.*, 2004, **83**, 823–831.
142. Y.-T. Tsai, J. Zhou, H. Weng, J. Shen, L. Tang and W.-J. Hu, *Adv. Healthcare Mater.*, 2014, **3**, 221–229.
143. G. Mahendra, H. Pandit, K. Kliskey, D. Murray, H. S. Gill and N. Athanasou, *Acta Orthop.*, 2009, **80**, 653–659.
144. P. Thomsen, L. Ericson and J. Davies, in *The Bone-Biomaterial Interface*, ed. J. E. Davies, University of Toronto Press, 1991, pp. 153–160.
145. J. M. C. Gutteridge and B. Halliwell, *Biochem. Biophys. Res. Commun.*, 2018, **502**, 183–186.
146. P. Tengvall, I. Lundström, L. Sjöqvist, H. Elwing and L. M. Bjursten, *Biomaterials*, 1989, **10**, 166–175.
147. J. Pan, D. Thierry and C. Leygraf, *J. Biomed. Mater. Res.*, 1994, **28**, 113–122.
148. I. C. Lavos-Valereto, S. Wolynec, I. Ramires, A. C. Guastaldi and I. Costa, *J. Mater. Sci.: Mater. Med.*, 2004, **15**, 55–59.
149. J. Pan, D. Thierry and C. Leygraf, *Electrochim. Acta*, 1996, **41**, 1143–1153.
150. J. J. Muyco, J. J. Gray, T. V. Ratto, C. A. Orme, J. McKittrick and J. Frangos, *Mater. Sci. Eng.: C*, 2006, **26**, 1408–1411.

151. T. Ueno, T. Ikeda, N. Tsukimura, M. Ishijima, H. Minamikawa, Y. Sugita, M. Yamada, N. Wakabayashi and T. Ogawa, *Biomaterials*, 2016, **108**, 177–186.
152. A. Zareidoost, M. Yousefpour, B. Ghaseme and A. Amanzadeh, *J. Mater. Sci.: Mater. Med.*, 2012, **23**, 1479–1488.
153. C. D. R. Kalpidis and R. M. Setayesh, *J. Periodontol.*, 2004, **75**, 631–645.
154. R. A. Gittens, R. Olivares-Navarrete, R. Tannenbaum, B. D. Boyan and Z. Schwartz, *J. Dent. Res.*, 2011, **90**, 1389–1397.
155. H. Zitter and H. Plenk Jr., *J. Biomed. Mater. Res.*, 1987, **21**, 881–896.
156. S. Mittal, J. Rogers, S. Sarkar, J. Koehler and R. S. Passman, *Am. J. Cardiol.*, 2019, **123**, 1967–1971.
157. S. Saba, T. McLaughlin, M. He, A. Althouse, S. Mulukutla and I. Hernandez, *Heart Rhythm*, 2019, **16**(7), 1065–1071.
158. K. Spiers, T. Cardamone, J. B. Furness, J. C. M. Clark, J. F. Patrick and G. M. Clark, *Cochlear Implants Int.*, 2016, **17**, 129–131.
159. A. R. Harris, C. Newbold, P. Carter, R. Cowan and G. G. Wallace, *J. Neural Eng.*, 2018, **15**, 046015.
160. L. C. Kloth, *Int. J. Lower Extremity Wounds*, 2005, **4**, 23–44.
161. J. Szewczenko and J. Marciniak, *J. Mater. Process. Technol.*, 2006, **175**, 404–410.
162. J. M. Khalifeh, Z. Zohny, M. MacEwan, M. Stephen, W. Johnston, P. Gamble, Y. Zeng, Y. Yan and W. Z. Ray, *IEEE Rev. Biomed. Eng.*, 2018, **11**, 217–232.
163. E. Cottrill, Z. Pennington, A. K. Ahmed, D. Lubelski, M. L. Goodwin, A. Perdomo-Pantoja, E. M. Westbroek, N. Theodore, T. Witham and D. Sciubba, *J. Neurosurg.*, 2019, **32**, 1.
164. K. D. Secinti, M. Ayten, G. Kahilogullari, G. Kaygusuz, H. C. Ugur and A. Attar, *J. Clin. Neurosci.*, 2008, **15**, 434–439.
165. T. Berger, J. Spadaro, S. Chapin and R. Becker, *Antimicrob. Agents Chemother.*, 1976, **9**, 357.
166. J. L. Del Pozo, M. S. Rouse, G. Euba, C.-I. Kang, J. N. Mandrekar, J. M. Steckelberg and R. Patel, *Antimicrob. Agents Chemother.*, 2009, **53**, 4064–4068.
167. A. J. Van der Borden, P. G. Maathuis, E. Engels, G. Rakhorst, H. C. van der Mei, H. J. Busscher and P. K. Sharma, *Biomaterials*, 2007, **28**, 2122–2126.
168. B. Ercan, K. M. Kummer, K. M. Tarquinio and T. J. Webster, *Acta Biomater.*, 2011, **7**, 3003–3012.
169. H. Plenk Jr, *Artif. Organs*, 2011, **35**, 237–241.
170. N. Recek, R. Zhou, R. Zhou, V. S. J. Te'o, R. E. Speight, M. Mozetič, A. Vesel, U. Cvelbar, K. Bazaka and K. Ostrikov, *Sci. Rep.*, 2018, **8**, 8252.
171. X. Dai, K. Bazaka, D. J. Richard, E. W. Thompson and K. Ostrikov, *Trends Biotechnol.*, 2018, **36**, 1183–1198.
172. R. Zhou, R. Zhou, P. Wang, B. Luan, X. Zhang, Z. Fang, Y. Xian, X. Lu, K. K. Ostrikov and K. Bazaka, *ACS Appl. Mater. Interfaces*, 2019, **11**, 20660–20669.

173. M. Ishaq, K. Bazaka and K. Ostrikov, *Phys. Plasmas*, 2015, **22**, 122003.
174. R. Zhou, R. Zhou, X. Zhang, J. Li, X. Wang, Q. Chen, S. Yang, Z. Chen, K. Bazaka and K. Ostrikov, *Sci. Rep.*, 2016, **6**, 39552.
175. R. Zhou, R. Zhou, J. Zhuang, Z. Zong, X. Zhang, D. Liu, K. Bazaka and K. Ostrikov, *PLoS One*, 2016, **11**, e0155584.
176. K. Bazaka, I. Levchenko, J. W. M. Lim, O. Baranov, C. Corbella, S. Xu and M. Keidar, *J. Phys. D: Appl. Phys.*, 2019, **52**, 183001.
177. K. Bazaka, M. V. Jacob, R. J. Crawford and E. P. Ivanova, *Appl. Microbiol. Biotechnol.*, 2012, **95**, 299–311.
178. I. Levchenko, K. Bazaka, O. Baranov, R. M. Sankaran, A. Nomine, T. Belmonte and S. Xu, *Appl. Phys. Rev.*, 2018, **5**, 021103.
179. X. Zhang, R. Zhou, K. Bazaka, Y. Liu, R. Zhou, G. Chen, Z. Chen, Q. Liu, S. Yang and K. Ostrikov, *Plasma Processes Polym.*, 2018, **15**, 1700241.
180. M. Ishaq, K. Bazaka and K. Ostrikov, *J. Phys. D: Appl. Phys.*, 2015, **48**, 464002.
181. R. Zhou, R. Zhou, K. Prasad, Z. Fang, R. Speight, K. Bazaka and K. Ostrikov, *Green Chem.*, 2018, **20**, 5276–5284.
182. M. Ishaq, A. Rowe, K. Bazaka, M. Krockenberger, M. D. M. Evans and K. Ostrikov, *Plasma Med.*, 2016, **6**, 67–83.
183. K. Prasad, N. Recek, R. Zhou, R. Zhou, M. Aramesh, A. Wolff, R. E. Speight, M. Mozetič, K. Bazaka and K. Ostrikov, *Sustainable Mater. Technol.*, 2019, **22**, e00123.
184. L. Ponsonnet, M. Boureanu, N. Jaffrezic, A. Othmane, C. Dorel and P. Lejeune, *Mater. Sci. Eng.: C*, 2008, **28**, 896–900.
185. F. Mansfeld, *Electrochim. Acta*, 2007, **52**, 7670–7680.
186. A. Nagiub and F. Mansfeld, *Electrochim. Acta*, 2002, **47**, 2319–2333.
187. W. Q. Hu, L. D. Chen, Y. J. Wang, D. F. Sun and D. W. Han, *Shanghai J. Stomatol.*, 2008, **17**, 88–91.
188. J. C. Souza, P. Ponthiaux, M. Henriques, R. Oliveira, W. Teughels, J.-P. Celis and L. A. Rocha, *J. Dent.*, 2013, **41**, 528–534.
189. L. Li, S. Li, Q. Qu, L. Zuo, Y. He, B. Zhu and C. Li, *Materials*, 2017, **10**, 255.
190. J. Souza, M. Henriques, R. Oliveira, W. Teughels, J.-P. Celis and L. Rocha, *Biofouling*, 2010, **26**, 471–478.
191. F. Laurent, B. Grosgogeat, L. Reclaru, F. Dalard and M. Lissac, *Biomaterials*, 2001, **22**, 2273–2282.
192. P. Silva-Bermudez and S. Rodil, *Surf. Coat. Technol.*, 2013, **233**, 147–158.
193. Y. S. Hedberg, *npj Mater. Degrad.*, 2018, **2**, 26.
194. Y. Hedberg, M.-E. Karlsson, Z. Wei, M. Žnidaršič, I. O. Wallinder and J. Hedberg, *Corrosion*, 2017, **73**, 1423–1436.
195. Y. Hedberg, X. Wang, J. Hedberg, M. Lundin, E. Blomberg and I. Odnevall Wallinder, *J. Mater. Sci.: Mater. Med.*, 2013, **24**, 1015–1033.
196. Y. Hedberg, J. Hedberg, Y. Liu and I. O. Wallinder, *Biometals*, 2011, **24**, 1099–1114.
197. Y. S. Hedberg, M. Žnidaršič, G. Herting, I. Milošev and I. Odnevall Wallinder, *J. Biomed. Mater. Res., Part B*, 2019, **107**, 858–867.

198. Y. Zhang, O. Addison, F. Yu, B. C. R. Troconis, J. R. Scully and A. J. Davenport, *Sci. Rep.*, 2018, **8**, 3185.
199. K. Merritt and S. A. Brown, *J. Biomed. Mater. Res.*, 1988, **22**, 111–120.
200. Y. S. Hedberg, M. Pettersson, S. Pradhan, I. Odnevall Wallinder, M. W. Rutland and C. Persson, *ACS Biomater. Sci. Eng.*, 2015, **1**, 617–620.
201. A. S. Rosenberg, *AAPS J.*, 2006, **8**, E501–E507.
202. A. Freemont, *Diagn. Histopathol.*, 2017, **23**, 221–228.
203. H. J. Cooper, R. M. Urban, R. L. Wixson, R. M. Meneghini and J. J. Jacobs, *J. Bone Jt. Surg.*, 2013, **95**, 865.
204. L. Lucas, R. Buchanan and J. Lemons, *J. Biomed. Mater. Res.*, 1981, **15**, 731–747.
205. S. Shamshoon, P. Thornley and J. de Beer, *Case Rep. Orthop.*, 2018, **2018**, 4.
206. A. J. Smith, P. Dieppe, K. Vernon, M. Porter and A. W. Blom, *Lancet*, 2012, **379**, 1199–1204.
207. E. Frigerio, P. D. Pigatto, G. Guzzi and G. Altomare, *Contact Dermatitis*, 2011, **64**, 273–279.
208. E. Ingham and J. Fisher, *Proc. Inst. Mech. Eng., Part H*, 2000, **214**, 21–37.
209. S. B. Goodman, J. Pajarinen, Z. Yao and T. Lin, *Front. Bioeng. Biotechnol.*, 2019, **7**, 230.
210. G. Vallés, P. González-Melendi, J. L. González-Carrasco, L. Saldaña, E. Sánchez-Sabaté, L. Munuera and N. Vilaboa, *Biomaterials*, 2006, **27**, 5199–5211.
211. S. B. Goodman, P. Huie, Y. Song, D. Schurman, W. Maloney, S. Woolson and R. Sibley, *J. Bone Jt. Surg.*, 1998, **80-B**, 531–539.
212. O. Addison, A. J. Davenport, R. J. Newport, S. Kalra, M. Monir, J. F. W. Mosselmans, D. Proops and R. A. Martin, *J. R. Soc., Interface*, 2012, **9**, 3161–3164.
213. N. J. Hallab and J. J. Jacobs, *Bull. NYU Hosp. Jt. Dis*, 2009, **67**, 182.
214. M. S. Caicedo, R. Desai, K. McAllister, A. Reddy, J. J. Jacobs and N. J. Hallab, *J. Orthop. Res.*, 2009, **27**, 847–854.
215. A. Vignery, *Int. J. Exp. Pathol.*, 2000, **81**, 291–304.
216. N. J. Hallab, S. Anderson, M. Caicedo, A. Brasher, K. Mikecz and J. J. Jacobs, *J. Biomed. Mater. Res., Part A*, 2005, **74A**, 124–140.
217. M. S. Caicedo, P. H. Pennekamp, K. McAllister, J. J. Jacobs and N. J. Hallab, *J. Biomed. Mater. Res., Part A*, 2010, **93A**, 1312–1321.
218. A. Petit, F. Mwale, C. Tkaczyk, J. Antoniou, D. J. Zukor and O. L. Huk, *J. Biomed. Mater. Res., Part A*, 2006, **79A**, 599–605.
219. A. Soloviev, E. M. Schwarz, M. Darowish and R. J. O'Keefe, *J. Orthop. Res.*, 2005, **23**, 1258–1265.
220. M. Abdelhalim, *J. Cancer Sci. Ther.*, 2012, **4**, 046–050.
221. Z. Xing, L. P. Schwab, C. F. Alley, K. A. Hasty and R. A. Smith, *J. Biomed. Mater. Res., Part B*, 2008, **85B**, 37–41.
222. M. K. Wasko and S. B. Goodman, *J. Orthop. Res.*, 2016, **34**, 1497–1504.
223. Y. Bi, J. M. Seabold, S. G. Kaar, A. A. Ragab, V. M. Goldberg, J. M. Anderson and E. M. Greenfield, *J. Bone Miner. Res.*, 2001, **16**, 2082–2091.

224. E. M. Greenfield, M. A. Beidelschies, J. M. Tatro, V. M. Goldberg and A. G. Hise, *J. Biol. Chem.*, 2010, **285**, 32378–32384.
225. L. Yu, L. Wang and S. Chen, *J. Cell. Mol. Med.*, 2010, **14**, 2592–2603.
226. L. Saldaña and N. Vilaboa, *Acta Biomater.*, 2010, **6**, 1649–1660.
227. M. G. Choi, H. S. Koh, D. Kluess, D. O'Connor, A. Mathur, G. A. Truskey, J. Rubin, D. X. F. Zhou and K. L. P. Sung, *Proc. Natl. Acad. Sci. U. S. A.*, 2005, **102**, 4578–4583.
228. A. Fahlgren, C. Bratengeier, C. M. Semeins, J. Klein-Nulend and A. D. Bakker, *J. Orthop. Res.*, 2018, **36**, 1425–1434.
229. M. Niinomi, Fatigue characteristics of metallic biomaterials, *Int. J. Fatigue*, 2007, **29**(6), 992–1000.
230. W. E. Frazier, Metal Additive Manufacturing: A Review, *J. Mater. Eng. Perform.*, 2014, **23**(6), 1917–1928.
231. M. Bryant and A. Neville, Corrosion and mechanical properties, *Orthop. Trauma*, 2016, **30**(3), 176–191.
232. T. Narushima; K. Ueda and Alfirano, Co–Cr Alloys as Effective Metallic Biomaterials, in *Advances in Metallic Biomaterials: Tissues, Materials and Biological Reactions*, ed. M. Niinomi, T. Narushima and M. Nakai, Springer Berlin Heidelberg, Berlin, Heidelberg, 2015, pp. 157–178.
233. M. Esmaily, J. E. Svensson, S. Fajardo, N. Birbilis, G. S. Frankel, S. Virtanen, R. Arrabal, S. Thomas and L. G. Johansson, Fundamentals and advances in magnesium alloy corrosion, *Prog. Mater. Sci.*, 2017, **89**, 92–193.
234. D. Upadhyay, M. A. Panchal, R. S. Dubey and V. K. Srivastava, Corrosion of alloys used in dentistry: A review, *Mater. Sci. Eng.: A*, 2006, **432**(1–2), 1–11.
235. L. Reclaru, R. Lerf, P.-Y. Eschler, A. Blatter and J. M. Meyer, Pitting, crevice and galvanic corrosion of REX stainless-steel/CoCr orthopedic implant material, *Biomaterials*, 2002, **23**(16), 3479–3485.
236. G. B. Higgs, J. A. Hanzlik, D. W. MacDonald, J. L. Gilbert, C. M. Rimnac, S. M. Kurtz and Implant Research Center Writing Centre, Is increased modularity associated with increased fretting and corrosion damage in metal-on-metal total hip arthroplasty devices?: a retrieval study, *J. Arthroplasty*, 2013, **28**(8 Suppl.), 2–6.
237. R. Pokrowiecki, A. Mielczarek, T. Zaręba and S. Tyski, Oral microbiome and peri-implant diseases: where are we now? *Ther. Clin. Risk Manage.*, 2017, **13**, 1529–1542.
238. I. Polyzois, D. Nikolopoulos, I. Michos, E. Patsouris and S. Theocharis, Local and systemic toxicity of nanoscale debris particles in total hip arthroplasty, *J. Appl. Toxicol.*, 2012, **32**(4), 255–269.

CHAPTER 2

Calcium Phosphate Cements: Structure-related Properties

LE FU,[a] MICHAEL PUJARI-PALMER,[b] CAROLINE ÖHMAN-MAGI,[b] HÅKAN ENGQVIST[b] AND WEI XIA*[b]

[a] School of Material Science and Engineering, Central South University, Changsha, China; [b] Materials in Medicine, Div. Applied Materials Science, Department of Materials Science and Engineering, Uppsala University, Sweden

*Email: wei.xia@angstrom.uu.se

2.1 Introduction

It is estimated that every year nearly two million patients worldwide need a bone grafting procedure to repair bone defects stemming from musculoskeletal diseases and bone fracture due to sports and road traffic injuries.[1,2] The number of people suffering from a bone injury is increasing due to longer life expectancy and ageing. For example, osteoporosis, a common disease in elderly people, was responsible for 2 million fractures in 2005, with the number expected to rise to 3 million by 2025.[2] As a result, the demand for bone grafting materials, and their respective market value, is ever-increasing. Bone grafting materials are a 1 billion dollar market, per year, with a yearly growth rate of approximately 10%.[1]

Materials used for replacing missing or damaged bone can be derived from patients themselves (autograft), donors (allograft), or synthetic bone substitutes. Each type of bone graft has its advantages and disadvantages.

Inorganic Materials Series No. 7
The Chemistry of Inorganic Biomaterials
Edited by Christopher Spicer

Published by the Royal Society of Chemistry, www.rsc.org

For instance, an autograft is considered as the gold standard, because bone harvested from the patient themselves will provide an almost perfect fit with the damaged tissue, resulting in no immunological reaction of the patient to the donor bone. However, the additional operation on a second surgical site (typically from the hip joint) is invasive and painful, and only a limited amount of bone can be obtained *via* autografts.[3] Allografts, typically from pigs, do not have the problem of limited quantity; however, they may cause immune problems and infections when implanted in the human body. Synthetic bone substitutes have attracted extensive attention from materials researchers and surgeons over the last few decades since they show many merits, including limitless supply, being free of diseases, *etc.* To date, a wide range of synthetic materials have been proposed and developed as bone substitutes, including metals, ceramics, polymers, and cements.[4–6] Among them, calcium phosphate cements (CPCs) are of particular importance, and have attracted great attention due to their excellent injectability, which allows minimally invasive surgery, and beneficial biological properties, including biocompatibility, bioresorbability, bioactivity, and osteoconductivity.[3,7–10]

2.2 Calcium Phosphate Family

Calcium phosphates (CaP) are primarily composed of calcium cations (Ca^{2+}), orthophosphate (PO_4^{3-}), metaphosphate (PO_3^{-}), or pyrophosphate ($P_2O_4^{7-}$) anions, and sometimes hydrogen (H^+) or hydroxide (OH^-) ions.[11] CaP are widely used as biomaterials since they share chemical and structural similarities with the most important inorganic substituent of hard tissues in mammals: hydroxyapatite ($Ca_{10}(PO_4)_6(OH)_2$). In mineralised tissues, CaP makes up 45–70% of the dry weight of bone, 45–70% of dentine, and ~95% of enamel.[12] In this chapter, CaP cements are referred to as 'CPCs'.

There are a variety of non-ion-substituted CaPs in the ternary system $CaO–P_2O_5–H_2O$. Calcium orthophosphates share a common characteristic, with phosphate groups making up the framework of CaPs, which provides stability to the entire structure, and the other ions or groups being built up around the phosphate group.[13] Figure 2.1 shows the unit cell of hydroxyapatite (HA), which is used as an example to illustrate the microstructure characteristics of CaP. One of the most important features that distinguishes different CaPs is the Ca/P molar ratio. Table 2.1 lists the members of the CaP family that are of interest for biomedical applications, with a Ca/P molar ratio ranging between 0.5 and 2.0. Table 2.2 lists the space group and lattice parameters for each CaP phase. It is evident from Table 2.1 that the aqueous solubility of CaP is closely related to the Ca/P molar ratio, where the lower the Ca/P molar ratio the higher the solubility. The solubility value ranges from ~18 $g\,L^{-1}$ (highly water-soluble, monocalcium phosphate monohydrate (MCPM)) to ~0.0003 $g\,L^{-1}$ (almost insoluble in water, HA). The huge solubility differences between the different CaP phases are determined by the dissolution–precipitation reactions that are used in the setting process of CPCs,[10,14,15] which will be discussed in the following sections.

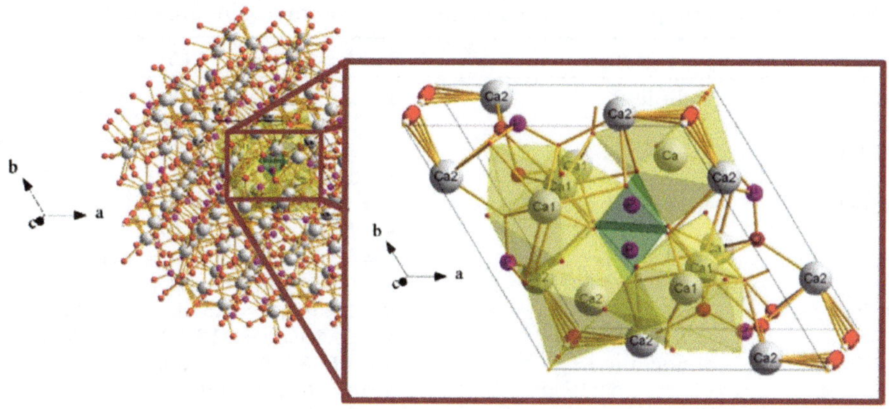

Figure 2.1 The microstructure of HA, showing that the atomic arrangement of HA is built up around a network of orthophosphate (PO_4) groups, which provides stability to the entire structure.
Reprinted from ref. 16 with permission from Elsevier, Copyright 2016.

Table 2.1 Phases in the CaP family of interest for biomedical applications.[11,13,27]

CaPs	Abbreviations	Chemical formula	Ca/P molar ratio	Solubility at 25 °C in water ($g\,L^{-1}$)
Monocalcium phosphate monohydrate	MCPM	$Ca(H_2PO_4)_2 \cdot H_2O$	0.5	~18
Monocalcium phosphate anhydrous	MCPA	$Ca(H_2PO_4)_2$	0.5	~17
Dicalcium phosphate dihydrate	DCPD	$CaHPO_4 \cdot 2H_2O$	1.0	~0.088
Dicalcium phosphate anhydrous	DCPA	$CaHPO_4$	1.0	~0.048
Octacalcium phosphate	OCP	$Ca_8(HPO_4)_2(PO_4)_4 \cdot 5H_2O$	1.33	~0.008
α-Tricalcium phosphate	α-TCP	α-$Ca_3(PO_4)_2$	1.5	~0.025
β-Tricalcium phosphate	β-TCP	α-$Ca_3(PO_4)_2$	1.5	~0.005
Amorphous calcium phosphate	ACP	$Ca_xH_y(PO_4)_z \cdot nH_2O$ $n = 3–4.5$	1.2–2.2	
Calcium-deficient hydroxyapatite	CDHA	$Ca_{10-x}(HPO_4)_x(PO_4)_{6-x}(OH)_{2-x}$ $(0 < x < 1)$	1.5–1.67	~0.094
Hydroxyapatite	HA	$Ca_{10}(PO_4)_6(OH)_2$	1.67	~0.0003
Tetracalcium phosphate	TTCP	$Ca_4(PO_4)_2O$	2	~0.0007

Table 2.2 Space groups and lattice parameters of common CaP phases.[13]

CaPs	Space group	Lattice parameters (Å)
MCPM	Triclinic $P1$	$a=5.6261$, $b=11.889$, $c=6.4731$ $\alpha=98.633°$, $\beta=118.262°$, $\gamma=83.344°$
MCPA	Triclinic $P1$	$a=7.5577$, $b=8.2531$, $c=5.5504$ $\alpha=109.87°$, $\beta=93.68°$, $\gamma=109.15°$
DCPD	Monoclinic Ia	$a=5.812$, $b=5.180$, $c=6.239$ $a=\beta=90°$, $\gamma=116.42°$
DCPA	Triclinic $P1$	$a=6.910$, $b=6.627$, $c=6.998$ $\alpha=96.34°$, $\beta=103.82°$, $\gamma=88.33°$
OCP	Triclinic $P1$	$a=19.692$, $b=9.523$, $c=6.835$ $\alpha=90.15°$, $\beta=92.54(2)°$, $\gamma=108.65°$
α-TCP	Monoclinic $P2_1/a$	$a=12.887$, $b=27.280$, $c=15.219$ $\alpha=\gamma=90°$, $\beta=126.20°$
β-TCP	Rhombohedral $R3cH$	$a=b=10.4183$, $c=37.3464$ $\alpha=\beta=90°$, $\gamma=120.0°$
HA	Monoclinic $P2_1/b$	$a=9.84214$, $b=2a$, $c=6.8814$ $\alpha=\beta=90°$, $\gamma=120.0°$
	Hexagonal $P6_3/m$	$a=b=9.4302$, $c=6.8911$ $\alpha=\beta=90°$, $\gamma=120.0°$
TTCP	Monoclinic $P2_1$	$a=7.023$, $b=11.986$, $c=9.473$ $\alpha=\gamma=90°$, $\beta=90.90°$

MCPM has the highest solubility of any CaP phase, of as high as 18 $g\,L^{-1}$ (Table 2.1), and as a result it does not exist in natural biological calcifications.[17] However, its high solubility makes MCPM a good raw component for CPCs,[9] since when mixed in aqueous solutions it produces large amounts of dissolved Ca^{2+} and PO_4^{3-} ions, which results in the supersaturation and precipitation of other CaP phases with lower solubility, such as HA (Table 2.1). Monocalcium phosphate anhydrous (MCPA) is the anhydrous form of MCPM. Like MCPM, MCPA never appears in calcified tissues due to its acidity and high solubility.[17] It should be noted that the nomenclature of CaPs is referenced to the fixed amount of orthophosphate, P_2O_8 in the materials. Therefore, monocalcium phosphate is CaP with a stoichiometry of Ca_1: $(PO_4)_2$, dicalcium phosphate is Ca_2: $(PO_4)_{2m}$, which can also be written as Ca_1: $(PO_4)_1$, *etc.* Dicalcium phosphate dihydrate (DCPD) can be easily prepared from an aqueous solution containing Ca^{2+} and PO_4^{3-} ions due to its low solubility (~0.088 $g\,L^{-1}$, Table 2.1). DCPD is used as a biomaterial in the form of bone cements in orthopaedic surgery and toothpaste in dentistry for protection against dental caries (cavities).[17] Dicalcium phosphate anhydrous (DCPA) is the anhydrous form of DCPD. DCPA can also be formed by crystallisation from aqueous solution, but this requires heating to 100 °C. Like DCPD, DCPA is currently used as a component of CPCs and is added to toothpaste for protection against caries.[11] Octacalcium phosphate (OCP) is often formed as an intermediate phase during the dissolution/precipitation of other CaP phases that are more thermodynamically stable, such as HA. OCP has a remarkable structural similarity to HA, with both materials consisting of apatitic layers separated by hydrated layers. OCP is widely used in biomaterials, for implantation into bone defects, coatings on metallic implants, and as a component in biocomposites and bone cements.[11]

TCP has two polymorphs, *i.e.*, α- and β-TCP. The two polymorphs consist of tightly bonded PO_4^{3-} tetrahedral units loosely bound to Ca^{2+} ions at various sites.[18] The α-TCP bulk crystal has a very complex monoclinic structure in the space group $P2_1/a$. There are two types of ion columns (indicated as A and B) aligned along the (010) direction in α-TCP. Column A is occupied by Ca^{2+} ions, while column B is occupied by Ca^{2+} and PO_4^{3-} ions. β-TCP has a hexagonal cell in the space group R3c (Figure 2.2b). In β-TCP, the six Ca sites are half occupied, as shown by the dashed circles (I–VI) in Figure. 2.2b. Three Ca vacancies per unit cell are required in order to

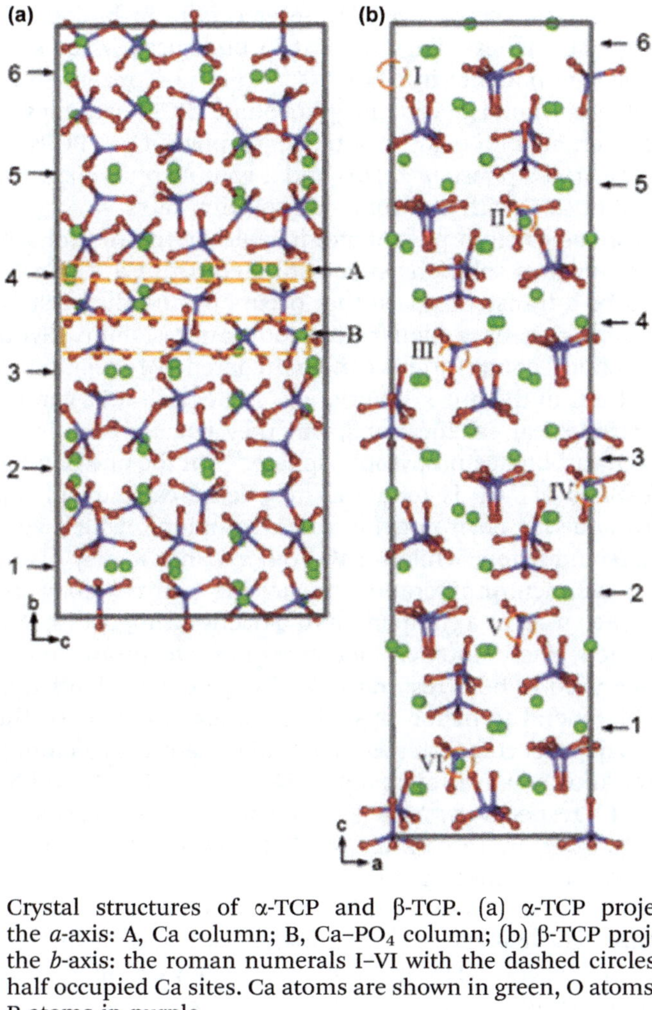

Figure 2.2 Crystal structures of α-TCP and β-TCP. (a) α-TCP projected along the *a*-axis: A, Ca column; B, Ca–PO_4 column; (b) β-TCP projected along the *b*-axis: the roman numerals I–VI with the dashed circles denote six half occupied Ca sites. Ca atoms are shown in green, O atoms in red, and P atoms in purple.
Reprinted from ref. 18 with permission from Elsevier, Copyright 2010.

maintain the charge neutrality of the cell.[18] The α- and β-TCP polymorphs differ significantly in terms of their solubility (Table 2.1) and crystal structure (Table 2.2). It is generally accepted that pure, crystalline TCP cannot be obtained by its direct precipitation from aqueous media, but can only be prepared by calcination of CDHA (calcium-deficient hydroxyapatite) above 800 °C,[19] which results in the occurrence of the following reaction:

$$Ca_9(HPO_4)(PO_4)_5(OH)_1 \rightarrow 3Ca_3(PO_4)_2\ (\beta\text{-TCP}) + H_2O$$

Once β-TCP is heated to temperatures over 1125 °C, β–α transformation occurs. Therefore, α-TCP is typically considered to be the high-temperature phase of β-TCP. β-TCP has seldom been observed in biological systems, excepting magnesium-stabilised β-TCP (Whitlockite), which has been identified during pathological calcification (deposition of calcium salts in sites other than bones and teeth).[11] α-TCP never occurs in biological calcifications. Although TCP is less likely to exist in biological systems, both types can be used as components in CPCs.[7,20,21] The most recent innovation in CaPs and CPCs are biphasic calcium phosphate (BCP) ceramics, which are an attractive alternative to using a pure, single-phase CaP. BCPs are formed by combining a stable phase (*e.g.* HA) and a soluble phase (*e.g.* β-TCP),[15] to produce a CaP or CPC with controlled dissolution rates.

ACP (amorphous calcium phosphate) is typically the first phase that precipitates from aqueous solutions supersaturated with Ca^{2+} and PO_4^{3-}. It is considered to be a transient phase that occurs during the crystallisation of other CaP phases that have higher thermodynamic stability. Synthetic ACP with varied chemical compositions can be prepared by regulating the pH and the concentrations of the mixed solutions.[22] Biological ACPs can be found in soft-tissue pathological calcifications, but they usually contain substituted ions of Na, Mg, carbonate and pyrophosphate.[11] On the one hand, one of the beneficial features of ACP is its ability to release calcium, phosphate and other ions in aqueous environments (high solubility), which contribute to improving osteoinduction, while on the other hand, it may also negatively impact cell attachment/proliferation by causing perturbations in the local pH.[23] ACP is often used as a component in bioactive composites. For instance, ACP-incorporated glass ionomer biocomposites are promising candidates that have been used in both restorative and regenerative dentistry.[23]

Apatite is a general term for crystalline minerals that have the formula $M_{10}(ZO_4)_6X_2$. The most common apatite found in nature is calcium phosphate apatite, better known as hydroxyapatite (HA), where M, ZO_4 and X are Ca^{2+}, PO_4^{3-}, and OH^-, respectively.[24] HA is the most stable and least soluble phase of all the CaPs listed in Table 2.1. Like TCP, HA also has two polymorphs, monoclinic HA (in the space group $P2_1/b$) and hexagonal HA (in the space group $P6_3/m$) (Table 2.2). The monoclinic-to-hexagonal phase transition occurs at temperatures above ~250 °C.[11] Compared with monoclinic HA, the hydroxide ions in hexagonal HA are more disordered within each row, which results in lattice strains in the hexagonal HA that are compensated for by substitutions or ion vacancies. Non-stoichiometric, ion-substituted HA, commonly referred to as 'biological apatite', is the main inorganic component of bones and teeth in mammals.[25] Because of the chemical and structural similarities between synthetic HA and biological HA, once implanted in the human body HA-based biomaterials show excellent osteointegration. In this chapter, osseointegration, osteoinductive and osteoconductive are defined as follows: osseointegration – capable of integrating into surrounding bone; osteoinductive – capable of promoting the differentiation of progenitor cells down an osteoblastic lineage; osteoconductive – support bone growth and encourage the ingrowth of surrounding bone.[26]

2.3 CPCs

Biomaterials based on CaP can be manufactured as nanoparticles,[28,29] coatings,[11,30] scaffolds,[13] or cements.[7,9,21] CPCs can be injected into a defective area to repair missing or damaged bone, and have received extensive attention since their discovery in the 1980s.[3] To date, a variety of CPCs with different compositions have been investigated and are now commercially available. Table 2.3 lists some of the CPCs that are on the market. Generally, CPCs are hydraulic cements (cements in which water has been used as a solvent during their mixing process), formed from a combination of one or more calcium orthophosphates, which upon mixing with the aqueous solution form a paste that sets and hardens after being implanted within the human body. The setting reaction mechanism of CPCs is the same as that of

Table 2.3 Some of the commercially available CPCs.[32]

Producer	Brand name	Formulations	Final product
Stryker–Leibinger Corp	BoneSource	Powder: TTCP, DCP Solution: H_2O, mixture of Na_2HPO_4 and NaH_2PO_4	Apatite
Teknimed	Cementeks	Powder: α-TCP, TTCP Solution: H_2O, $Ca(OH)_2$, H_3PO	Apatite
Mitsubishi materials	Biopexs	Powder: α-TCP, TTCP, DCPD, HA Solution: H_2O, sodium succinate, sodium chondroitin sulfate	Apatite
Shanghai Rebone Biomaterials Co, Ltd	Rebone	Powder: TTCP, DCP Solution: H_2O	Apatite
Synthes–Norian	Norians SRS	Powder: α-TCP, $CaCO_3$, MCPM Solution: H_2O, Na_2HPO_4	Apatite
Synthes–Norian	ChronOSTM Inject	Powder: β-TCP, MCPM, $MgHPO_4 \cdot 3H_2O$, $MgSO_4$, $Na_2H_2P_2O_7$ Solution: H_2O, sodium hyaluronate	Brushite
Kasios	Eurobones	Powder: β-TCP, $Na_4P_2O_7$ Solution: H_2O, H_3PO_4 (3.0M), H_2SO_4 (0.1 M)	Brushite
CalciphOs	VitalOs	Powder: β-TCP, $Na_2H_2P_2O_7$ Solution: H_2O, PBS solution	Brushite

gypsum, which is a widely distributed mineral made up of hydrous calcium sulfate that is used for amending soil and in the making of plaster of Paris. The setting reaction mechanism of CPCs and gypsum occurs via a dissolution–precipitation reaction and entanglement of the precipitated crystals.[31] Taking the formation of apatite cement as an example (Figure 2.3), when α-TCP is mixed with water, Ca^{2+} and PO_4^{3-} ions are released and dissolved into the water until the solution reaches the state of equilibrium with respect to dissolved α-TCP. However, the solution is also in a state of equilibrium with HA, the solubility of which (~0.0003 g L^{-1}) is lower than that of α-TCP (~0.025 g L^{-1}). Thus, the solution is supersaturated with respect to HA, which results in the precipitation of HA crystals. This precipitation makes the solution return to a state of undersaturation, therefore meaning that more α-TCP can be dissolved into the solution. These reactions continue successively, and the precipitated HA crystals become entangled

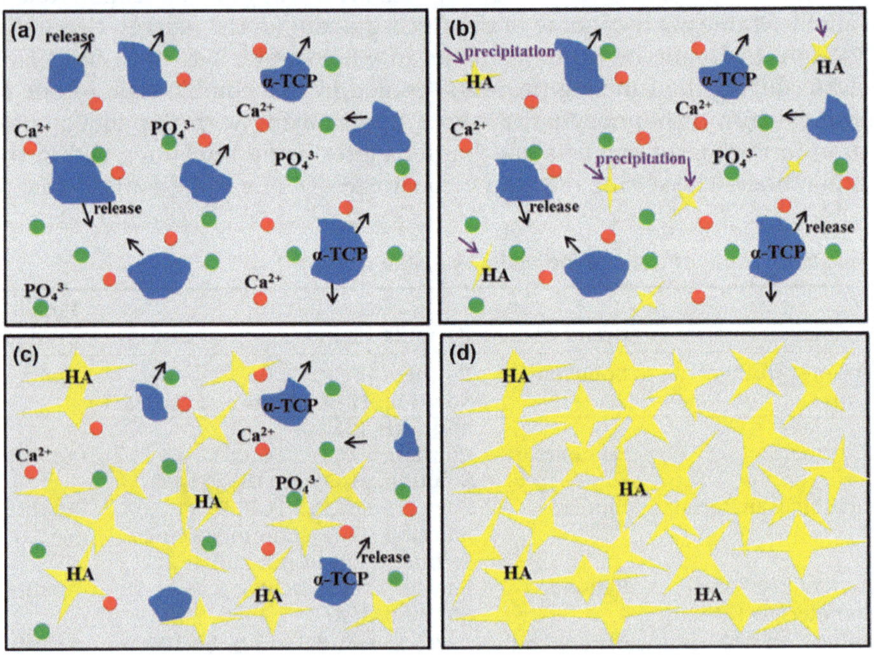

Figure 2.3 A schematic drawing showing the setting mechanism of CPCs. The formation of apatite cement is used as an example. When α-TCP is mixed with water, it dissolves and releases Ca^{2+} and PO_4^{3-} ions into the water (a). The solution is supersaturated with respect to HA, which results in the precipitation of HA crystals (b). The precipitation makes the solution return to a state of undersaturation. More α-TCP can thus be dissolved into the solution and the HA crystals grow larger (c). These reactions continue successively, and the precipitated HA crystals entangle with each other, transforming the initial viscous paste to a set solid (d).

with each other, transforming the initial viscous paste to a set solid. Despite the large number of formulations of CPCs that have been developed, most of the CPCs give as the end-product either precipitated hydroxyapatite (pHA) or brushite (DCPD),[13] commonly referred to as 'apatite cement' and 'brushite cement', respectively.

In 1989, brushite cements were first developed by Mirtchi and Lemaître by mixing water with an acidic calcium phosphate (MCPM) and a basic calcium phosphate (β-TCP). The mixture eventually solidified *via* an exothermic reaction, and the final product was mainly composed of dicalcium phosphate dihydrate.[14] Like HA cement, brushite cement is also biocompatible. Also, it has a unique advantage over HA cement in that brushite cement has higher solubility/dissolution rates and is more easily resorbed under physiological conditions, which results in the faster replacement of the biomaterial with newly formed bone tissue.[9] This is because brushite is stable (least soluble) at an acidic pH of 2.0–4.2, but metastable at physiological pH, and thus dissolves upon exposure to neutral body fluid.[31] Brushite cement resorption involves both biological resorption, by acid and enzymatic containing osteoclastic cells, and physicochemical dissolution.

When designing CPCs for use in orthopaedic applications, the properties of both the viscous paste and set-cement require careful consideration to ensure clinical success. Prior to setting, the cement needs a simple preparation and handling process, to ensure successful delivery and setting reaction during the surgical procedure. After setting *in vivo*, or as a pre-fabricated CPC, the hardened CPC should have adequate porosity to enable bone ingrowth. An implanted CPC also needs to show an adequate dissolution rate and be replaced by host tissue. Finally, the hardened cement must also have sufficient mechanical integrity to stabilise the fracture or implant site, and to resist fragmentation or erosion. The above properties or requirements of CPCs will be briefly discussed in the following sections.

2.3.1 Setting Time

The setting reaction is considered to be the most important feature of a CPC because it not only directly determines the cement hardening time and other setting properties, but also determines the structure of the final products, which in turn affects most of the physical and biological properties of the hardened cement.[3] Generally, the setting time of CPCs is defined as the time needed for the CPC to become strong enough (cohesion) to resist an applied force, which is usually measured using the Gillmore needles method. An optimal setting time (of a few minutes) is a critical parameter for a successful CPC. A CPC paste with a long setting time (of hours or even days) may cause many problems, such as inducing a severe inflammatory response caused by washed-out, or leakage, of CaP particles into tissues.[3] In contrast, a very short setting time (rapid setting reaction, *e.g.* a few seconds) may not provide enough time for surgeons to mix and apply the material during surgery.

CPCs without any additives (*i.e.* accelerants such as phosphoric acid) typically have a long setting time. Many strategies have been used to reduce the setting time of CPCs by kinetically or thermodynamically accelerating the dissolution and precipitation processes,[31] including the use of: (i) smaller particle sizes with higher specific surface area, which results in faster dissolution rates; (ii) raw components with lower crystallinity (high amount of disorder in the crystalline lattice arrangement), which also results in faster dissolution; (iii) accelerators in the liquid and solid compositions, for instance, the addition of the PO_4^{3-} into the liquid phase can lead to immediate supersaturation and precipitation of HA; (iv) a higher setting temperature, which increases the reaction rate; and (v) a lower liquid-to-powder ratio, so that supersaturation occurs rapidly. In summary, it is critical to prepare a CPC with a suitable setting time.

2.3.2 Injectability

The injectability of a paste is defined as its ability to stay homogeneous during injection (resist washout), independent of the injection force.[3] One of the significant advantages of CPCs over other forms of CaP biomaterials (*e.g.* granules, sintered bulk HA) is their injectability. Ideally, the unreacted CPC (paste) should be easy to load into a syringe and injectable into the body at the surgical site through a cannulated needle, which is a vital prerequisite for minimally invasive surgical procedures. However, liquid–solid phase separation (so-called filter-pressing) has often been observed in commercial CPCs during injection, leading to a deviation in the true composition of the extruded paste from that of the initial composition.[33] The underlying mechanism of liquid–solid phase separation is that the extrusion pressure is higher than the liquid filtration pressure. Thus, improved methods of extrusion have mainly focused on the following two aspects:[10] (i) reducing the extrusion pressure, *i.e.* increasing the flowability of the paste so that it flows more easily, and (ii) increasing the pressure required to force the liquid flow past CaP particles. The following improvement methods have been established: increasing the viscosity of the aqueous solution by adding viscous binders, increasing the maximum packing solid volume fraction (SVF_{max}) and reducing the SVF by optimising the particle size, particle size distribution, and particle shape, controlling the setting reactions, and optimising the extrusion parameters by adjusting the needle geometry, syringe size, extrusion rate, *etc.*[10,32,33]

2.3.3 Porosity

Porosity is an important aspect of the microstructure of CPCs, and it not only affects the physical properties of CPCs (*e.g.* mechanical properties) but also influences their biological properties, including bioresorption and bone ingrowth. Porosity in CPCs can be generally divided into intrinsic microporosity and artificial macroporosity. During setting reactions, little water is actually consumed. The majority of the aqueous solution serves instead as a dispersant medium to produce a workable paste.[10] After the setting reaction is completed, the liquid phase remains trapped, as voids between crystals, resulting in intrinsic microporosity. The microporosity of CPCs, which usually varies between 30 and 55%, heavily depends upon the liquid–powder (L/P) ratio and moulding pressure of the cement. The lower the L/P ratio and the higher the moulding pressure, the lower the microporosity.[31] The intrinsic microporosity typically lies in the nano to sub-micron range, which enables fluid flow through the microporous structure and increases surface reactivity, but the micropores are too small to facilitate bone ingrowth.[10] To enable bone ingrowth into CPCs, and concomitantly improve the bioresorption and accelerate replacement by new bone, artificial macroporosity (within the range of hundreds of micrometres) is occasionally introduced into CPCs. Several techniques[3,10,31] have been proposed to produce macropores in CPCs, including the addition of porogenic agents, such as mannitol or sucrose, and the use of gas generating compounds, such as sodium bicarbonate. However, it remains a challenge to reliably control the macropore structure of CPCs with the above methods. The formation of macroporosity also results in a sharp decrease in the mechanical strength of CPCs. Consequently, high macroporosity and lower mechanical strength should be offset by commensurate advantages (*i.e.* accelerated dissolution and tissue formation), which are appropriate for each specific clinical application.

2.3.4 Bioresorbability

Ideally, CPCs should resorb, *i.e.* they should gradually dissolve in bodily fluids and/or be degraded by cells, at a controlled rate that is aligned with the patient healing response, to maintain osseous integrity as the CPC bone graft is gradually replaced by new bone. Here, 'resorb' means that the CPCs should dissolve in the human body and the released ions be reused by bone forming cells, rather than them immediately crystallising as an inorganic calcium phosphate phase, to form new bone. However, the formation rate of new bone tissue is strongly dependent upon the patient (*i.e.* young, healthy patients synthesise new tissue faster than infirm patients), anatomical location (*i.e.* cancellous *versus* cortical bone, in non-load bearing regions of maxillofacial or load-bearing in the spinal vertebra), and the volume of new bone that needs to be synthesised. CPC formulations with an ideal resorption rate are not yet available, due to the complexity of matching the resorption rate with the needs of each specific patient. Generally, brushite CPCs resorb at a faster rate than apatite CPCs, because brushite can be resorbed not only by osteoclastic cells, but also *via* physicochemical dissolution due to the metastability of brushite cements at physiological pH.[10] Apatite is the most thermodynamically stable phase under physiological conditions, and replacement of apatite cement with bone is caused by the resorption by osteoclasts.[31] The concept of biphasic calcium phosphate (BCP) cements is determined by an optimal balance of a stable phase of HA and soluble TCP to regulate the resorption rates of BCP cements.[15] In addition to the composition of the cement, the degree of porosity, pore sizes, and interconnectivity of the pores also exerts significant influence on the resorption rates of CPCs.[1,7,32]

2.3.5 Anti-washout Properties

Anti-washout properties are defined as the ability of a CPC to harden in a dynamic aqueous environment without disintegrating into small particles.[3] Washed out particles released from unset CPCs may cause an inflammatory response in the body. Thus, anti-washout properties are important for the success of a CPC bone graft. Once CPCs are injected into bone defect areas, a struggle between two processes begins. One response is the setting and hardening reaction of CPCs, which results in the entanglement of the precipitated crystals.[3] The other response is the penetration of bodily fluids into the paste, which may result in the washout of small particles from the cement paste. The competition of these two processes determines the overall washout behaviour of CPCs.[31] Accelerating the setting reaction of CPCs is the most straightforward method by which to achieve anti-washout properties. However, as mentioned before, too short a setting time may cause difficulty for surgeons handling CPCs. Thus, a balance should be maintained.

2.4 CaP Nanoparticles

2.4.1 Preparation and Morphologies

It is well established that crystalline or amorphous CaP nanoparticles generally offer improved bioactivity and biocompatibility compared with conventional-sized CaPs (micrometre range) because of their large specific surface area, leading to extensive applications of CaP nanoparticles in biomedical fields. To date, a number of strategies have been explored to synthesise CaP nanoparticles, and these can be principally categorised as either wet-state or solid-state methods. Table 2.4 summarises some of the commonly used methods for preparing CaP nanoparticles. Each method has its advantages and disadvantages. The wet-state techniques are relatively easy and straightforward routes for the synthesis of CaP nanoparticles. The raw materials are solutions containing various phosphate and calcium sources. Usually, only mild conditions are needed in wet-state methods, except in the hydrothermal method. Meanwhile, wet-state methods possess advantages in that their reaction parameters can be adjusted to allow easy control of the size and shape of CaP nanoparticles.[34] Doping elements can also be easily incorporated into CaP nanoparticles to functionalise them. For instance, silver-doped CaP nanoparticles have been prepared by directly adding silver nitrate into an aqueous solution of the raw materials.[35] However, the sol–gel method has drawbacks, such as the possible hydrolysis of the phosphates

Table 2.4 Some of the methods commonly used to prepare CaP nanoparticles.

Methodology	Description	Size and morphology	Crystallinity	Ref.
Co-precipitation (wet-state)	Calcium nitrate and ammonium phosphate solution were mixed. Carboxymethyl cellulose solution was added to stabilise the NPs.	Spherical, 50–60 nm.	Poor	35
Hydrothermal method (wet-state)	Precursors of Ca^{2+} and PO_4^{3-} ions were mixed, moved into autoclave and heated for 3 h at 85 °C.	Spherical, 17–33 nm.	Amorphous	37
Sol–gel method (wet-state)	Precursors of Ca^{2+} and PO_4^{3-} ions were mixed, the pH value of the mixture was adjusted to 3 using nitric acid, followed by drying and calcination.	Spherical and rod-like, 20–70 nm.	High (after calcination)	38
Micelle-templating method (wet-state)	Calcium nitrate and phosphoric acid were used as calcium and phosphorus sources. Mixed with template agent and cosurfactant to form micelles. Drying and calcination to remove the template.	Spherical hollow NPs or nanotubes. Diameter of 60 nm and a hollow diameter of 36 nm.	High	39
Mechanochemical method (solid-state)	The solid-state reaction between Ca-containing raw materials occurs during planetary ball milling.	10–30 nm.	High	40
Plasma decomposition (solid-state)	Spray-dried HA powder was used as feedstock. Inductively coupled plasma torch was used for the nano-powder.	Agglomerated spherical NPs, 10–100 nm.	High	41
Solid-state reaction (solid-state)	Solid-state reaction between diammonium phosphate, calcium nitrate and sodium bicarbonate occur at room temperature, followed by calcination.	Agglomerated, near-spherical, 16–58 nm	High (calcination temperature over 600 °C)	36

and the high cost of the raw materials.[36] Similarly, the hydrothermal method results in a bottle-neck in terms of poor morphology control.

In the co-precipitation technique, high pH values are needed to avoid the formation of a Ca-deficient HA, and a high calcining temperature is needed to form crystalline HA, which may result in agglomeration. Solid-state methods, including solid-state synthesis and mechanochemical methods, have the convenience of being able to produce CaP nanoparticles that exhibit high crystallinity on a large scale using low-cost raw materials. However, solid-state methods usually result in the formation of significantly aggregated products with large particle sizes during calcination. It is also difficult to obtain high phase purity and to regulate crystal sizes using such methods.[34]

CaP crystals with various morphologies and sizes have been synthesised using many methods, including zero-dimensional shapes (atoms, clusters, and similar sphere-like particles); one-dimensional shapes (needle/rod/wire/whisker/fibre-like shapes), the lengths if which are significantly greater than the cross-sectional dimensions; two-dimensional shapes (sheets, plates, disks, flakes, platelets, strips and belts); and three-dimensional shapes (porous and mesoporous spheres, hollow and tube, enamel-like structures, and porous scaffolds). A review by Lin *et al.*[34] details the synthesis of CaP crystals with numerous shapes and sizes. Some of those CaP crystals can be classified as CaP nanoparticles if one of the dimensional sizes is below 100 nm.

2.4.2 Applications

2.4.2.1 Components for Biocomposites

Due to their small size, excellent bioactivity and biocompatibility, CaP nanoparticles can be easily incorporated into other types of biomaterials to give multifunctionality or overcome intrinsic drawbacks of the biomaterials. Owing to their long-term release of fluoride ions, glass ionomer cements have been reliable candidates to promote the mineralisation of enamel and dentin; however, fluoride ions alone are inadequate for preventing secondary caries. Amorphous CaP nanoparticles (5 wt%) have been incorporated into glass ionomer cements to form biocomposites, which have a stimulating effect on ALP (alkaline phosphatase) activity (an early marker of osteoblastic differentiation), biomineralisation, and osteogenic differentiation of hMSCs (human mesenchymal stem cells), due to the balanced and prolonged release of Ca^{2+} and PO_4^{3-} ions.[23] Another example of a biocomposite is injectable methylcellulose (MC) hydrogels that contain CaP nanoparticles.[42] During the hydrogelation of MC, CaP nanoparticles (40–50 nm) are *in situ* synthesised and are homogenously distributed in the gel network. The biocomposite combines the merits of MC hydrogels and CaP nanoparticles, *i.e.*, CaP nanoparticles support the bone healing process *via* osteoconduction, while MC hydrogels provide inter-particular space for bone ingrowth.[42] Biocomposites thus have great potential for use in bone tissue regeneration.

2.4.2.2 Carriers for Drugs, Protein, and Gene Delivery

As carriers for drugs, proteins, and gene delivery, CaP nanoparticles have the following advantages.[25,29,37,43] (1) They are highly biocompatible, can be easily functionalised, and have a predictable fate in the body (*i.e.* dissolution); (2) they are able to penetrate the outer membrane of cells, releasing the incorporated drugs or biomolecules locally; (3) their size can be easily controlled by stabilising agents, such as polymers or nucleic acids;

(4) they can be made to be fluorescent by the incorporation of lanthanide ions, which makes tracking of CaP nanoparticles possible. Three representative examples are described below.

Ye *et al.* developed HA hollow nanoparticles and HA nanotubes for drug delivery.[39] They showed that owing to their hollow microstructure, HA hollow nanoparticles can incorporate a much higher drug payload than traditional HA nanoparticles. Meanwhile, functionalised HA nanotubes show excellent pH-responsivity, *i.e.*, the loaded drug can be stably stored inside the hollow HA nanorods at neutral pH and released when the pH is weakly acidic, making the material a smart drug delivery system.[39]

It has been shown in many studies that scaffolds loaded with bone morphogenetic protein (BMP)-2 can be fabricated by simply immersing the scaffold in BMP-2-containing solutions. BMP-2 is an isolated protein that induces specific cells in the human body to form new cartilage and bone. However, it is difficult to release an adequate quantity of loaded BMP-2 during the later stages of tissue regeneration, and BMP-2 levels rapidly decline. To achieve the gradual release of BMP-2, researchers have developed a gene transfection technique, in which scaffolds are loaded with BMP-2-encoding plasmid DNA-functionalised CaP nanotubes (CPNTs).[43] The multi-shell CPNTs effectively protect plasmid DNA from degradation by DNase or inflammatory reactions, which prolongs the release and increases the yield of the produced BMP-2.[43]

Both DNA and small interfering RNA (siRNA) alone cannot permeate cell membranes due to their high negative charge, which makes a delivery vehicle for DNA and siRNA necessary for gene silencing applications. CaP nanoparticles are an excellent choice for non-viral gene delivery. Epple *et al.* applied siRNA-functionalised CaP nanoparticles to human cervix epithelial cells, which express an enhanced green fluorescent protein (eGFP).[29] The gene silencing efficiency reached $27 \pm 12\%$, and the siRNA-functionalised CaP nanoparticles kept their biological activity for several weeks of storage, enabling convenient handling procedures.[28] Meanwhile, transfection efficiencies of as high as $48 \pm 16\%$ were achieved in triple shell nanoparticles (CaP–DNA–CaP–DNA) with a diameter of around 100 nm, with the triple shell serving to protect DNA against intracellular attack by nucleases, thereby increasing transfection efficiency.[29]

2.4.2.3 *Reparative Materials for Enamel Remineralisation*

The application of CaP nanoparticles, especially HA nanoparticles, in repairing damaged enamel has recently attracted attention in dentistry because of the chemical and structural similarity of HA to tooth minerals. Ling *et al.* developed a rechargeable CaP adhesive that can be repeatedly recharged to re-release calcium and phosphate ions to provide long-term capability for remineralisation and inhibition of caries,[44] where the CaP nanoparticles act as a source of calcium and phosphate ions. It was found that after recharging by simply immersing the adhesive into a recharge solution, the CaP nanoparticle-containing resins showed continuous release of ions for at least 2–3 weeks. Self-healing microcapsules containing amorphous CaP nanoparticles have been prepared, which combine the properties of being self-healing (owing to polymerisation of the healing liquid), antibacterial (owing to an antibacterial agent), and remineralising.[45] Incorporated CaP nanoparticles contribute towards lesion remineralisation and inhibition of secondary caries.

2.5 Setting Reactions

CPCs form *via* the dissolution of precursor phases and reprecipitation into more stable, less soluble products, namely dicalcium phosphate (DCP) in acidic environments, or HA in neutral to basic environments.[3,46] During the setting reaction, disordered and crystalline metastable phases (*i.e.* amorphous CaP, OCP, *etc.*) precipitate rapidly, due to their low solubility, low critical nucleation energy, or the presence of surfaces that provide the free energy needed to overcome the critical nucleation energy. There is a competition between crystal nucleation and growth, since both processes draw upon the limited reserve of soluble ions. Over time the metastable phases dissolve and are replaced by more thermodynamically stable (*i.e.* lower Gibbs free energy) crystalline phases. The processes that occur during a cement setting reaction occur in stages, as described by Bohner *et al.*[46] A schematic drawing of the setting reactions is shown in Figure 2.3.

- Stage 1: ion release leading to supersaturation
- Stage 2: crystal nucleation/precipitation
- Stage 3: crystal growth

In the first stage, precursor powders, or ionic solutions (*e.g.* $CaCl_2 + NaH_2PO_4$) are combined with a liquid (Figure 2.3a). For powders, the initial dissolution step is surface driven, as molecules on the surface are solvated first, forming a dissolution layer at the particle surface. Dissolution then proceeds from the surface inwards, *via* diffusion.[47,48] CPCs are typically formed by combining a soluble acidic phase with a sparingly soluble, neutral or alkaline phase (*i.e.* MCPM $K_{sp} =$ from 1.183×10^{-2}, to 22.38, and β-TCP $K_{sp} = 10^{-28.9}$).[49] The energy needed to solvate the ions during dissolution and to reprecipitate new phases, known as the hydration process, is often provided by the difference between the pH of the precursor phases (exothermic reaction).[50] Consequently, most cementitious reactions are also acid–base reactions. The self-setting CaP cement reactions are summarised in eqn (2.1)–(2.3), reproduced from work by Bohner *et al.*[46] Sequential steps take place during the setting reaction. The first reaction (1) produces DCPD *via* the dissolution of a soluble acidic phase (MCPM) and dissolution/reprecipitation of the surface of β-TCP particles. Since β-TCP is slightly basic and sparingly soluble, surface dissolution is accelerated by low pH liquid containing dissolved MCPM,[51,52] and any additives such as inorganic or organic acids.[53]

$$\beta\text{-TCP} + \text{MCPM} + 7H_2O = 4\ \text{DCPD} \quad (2.1)$$

$$2Ca_4(PO_4)_2O + 2\ \text{DCPA} = Ca_{10}(PO_4)_6(OH)_2 \quad (2.2)$$

$$3\alpha\text{-TCP} + H_2O = Ca_9(HPO_4)(PO_4)_5OH \quad (2.3)$$

In eqn (2.2), DCPA or DCPD is dissolved in the liquid, lowering the pH and accelerating the hydrolysis of metastable, alkaline tetracalcium phosphate.[54] In eqn (2.3) a metastable CaP (α-TCP) undergoes a hydrolysis and transformation reaction, which involves the direct dissolution and reprecipitation of the final phase without the formation of any intermediate phases.[55] The distinctions between the setting reactions in eqn (2.1)–(2.3) lie in the dissolution rate of each precursor, the solubility of the intermediate phases, and the calcium to phosphate ratios of the precursors and products (DCPD = 1.0, TTCP = 2.0, and α-TCP = 1.5).

During the dissolution stage of all three reactions shown in eqn (2.1)–(2.3), supersaturation is reached by dissolving a metastable CaP phase (MCPM, DCPD, or α-TCP), and/or directly supplementing ions in various forms (*i.e.* $Ca_xH_y(PO_4)_z$). In the nucleation stage, the local concentration of ions determines when and where nucleation occurs. As the local ion concentration approaches and exceeds the solubility limit, pre-nucleation clusters appear from the clustering of calcium and orthophosphate ions ($Ca(HPO_4)_x$).[56,57] The exact, stepwise mechanism remains uncertain, as multiple molecular pathways may explain the pre-nucleation process.[57] When the pre-nucleation cluster grows to a critical size, above or below which changes in cluster size lead to a decrease in Gibbs free energy, nucleation occurs. The energy needed for CaP nucleation to occur is defined by the critical nucleation size. The tipping point appears to occur *via* the coordination of a single additional calcium ion to the nucleation centre, once the nucleation centre is stable, thereby allowing amorphous calcium to precipitate.[58] As the nucleation cluster exceeds the critical size, stable nuclei grow into microscopic crystals or particles. The size scale depends upon the CaP phase, although the smallest possible metastable nuclei are amorphous CaP (Posner cluster, 1 nm), which subsequently transform into metastable phases (*e.g.* octacalcium phosphate) or directly to HA.[59–61] Temperature and pH are the second most important factors, after solubility, underlying the nucleation of CaP.[62] An increase in temperature increases the contribution of entropic factors and decreases the free energy needed for nucleation to occur. Changes in pH alter the protonation state of orthophosphate in solution, thereby changing the charge and attractive force between ions, and increasing or decreasing the critical nucleation energy. Crystallisation requires the deprotonation of orthophosphate, although the pH must remain low enough to ensure the high solubility of each ion species, and continued dissolution of any solid precursor powder.[62]

The order, and speed, in which phases nucleate/precipitate is determined by the relative supersaturation of each phase, the local pH and ionic concentration, the composition of the liquid (*i.e.* surface tension and presence of inhibitors or stabilisers), and the surface energy of any available nucleating surfaces (*i.e.* seed crystals) or interfacial energy of nucleating phases.[13,63,64] Crystal nucleation and growth can be controlled in cements by using surfactants to lower the interfacial energy,[65] modifying the surface chemistry of seed crystals, or using polymeric additives to lower the interfacial energy.[66,67] Organic acids are commonly used in cements to slow the setting reaction and lengthen the setting time.[68,69] Organic acids act by chelating calcium to increase the effective solubility, binding to growing crystal surfaces to inhibit growth, or by stabilising nucleating clusters, and also lower the pH to increase the dissolution of precursor powders. Finally, simple ions such as magnesium, strontium and silicon, and some organic compounds like adenosine triphosphate, can stabilise specific phases during the precipitation/nucleation step.[63,70]

2.6 Influence of the CPC Microstructure on the Dissolution Rate

On the macroscale, dissolution is a surface driven process. An increase in the relative surface area of precursor powders: i) increases the dissolution rate during the cement setting reaction, thereby favouring the nucleation of many small crystals over crystal growth; ii) creates a more disordered microstructure because of the dampened crystal growth stage and rapid setting; and iii) leads to an increase in the total porosity, due to the disordered microstructure. Since set cements do not contain ‘particles’, and are composed of a random aggregation of interlocked, precipitated crystals, particle size only plays a role in the initial setting and nucleation reaction.[46]

During the setting reaction, the liquid becomes entrapped between crystals or within grains. These pockets of liquid collectively lead to cement porosity. Individual pores can range in size from micropores (>5 nm) to mesopores (5–50 nm), to macropores (>50 nm). Higher liquid-to-powder ratios produce cements that have higher porosity. Pores increase the surface area of the bulk cement in contact with liquid, and consequently, the total porosity, average pore size, and the liquid-to-powder ratio of a cement are directly related to the dissolution rate.[71] The surface area of a cement that is in contact with liquid is determined by the packing density, which is determined by the pores and crystal morphology. Nanoscale plate-like crystals can pack together with a higher packing density (how much of a given volume of cement is filled with non-cement) than comparably sized amorphous spheres. Therefore, the aggregate microstructure that forms during the cement setting reaction determines the packing density and amount of porosity, which, in turn, determines the relative surface area in contact with the liquid.

In summary, the solubility of the CaP phase that precipitates/nucleates will determine the bulk dissolution rate of the cement. After solubility, the next most important macrostructural features that determine the dissolution rate of the cement are the relative surface area, porosity and pore size, crystallinity, crystal morphology and microstructure.

The macroscale properties of CPCs are also determined by the atomistic and nanoscale events that occur during the setting reaction. During the nucleation stage, excess heat or an increase in ion concentration favours nucleation over crystal growth, leading to a smaller average crystal size in the cement. The setting reaction will also be accelerated, and the resulting micro- and atomic structure will be more disordered.[72,73] Alternatively, if seed crystals are added to precursors, the presence of a hetero-nucleating surface will favour the deposition of additional minerals on the seed crystals, at the expense of spontaneous nucleation. The average crystal size and crystallinity increase, and the growth between individual crystals as they become entangled leads to lower porosity and greater mechanical strength. A decrease in the nucleated crystal size, above a critical size, increases the effective surface area and dissolution rate.

On the atomic to nanoscale range, it is surface energy, rather than surface area, that determines the dissolution rate.[48,74] CaP surfaces with higher free energy dissolve more readily. Dissolution represents the balance between the loss of ions from the surface, and the attraction of ions to the surface of a crystal or particle. If more ions attach to the surface than are released, crystal growth occurs. The probability that an ion will detach from the surface is determined by how strongly it is bonded with its nearest atomic neighbours, and the change in free energy for the free *versus* bound ion.[48] Therefore, dissolution begins with ions that lack neighbours (at the edge of steps, or kinks, and defects). Nanoscale roughness, due to chemical etching or defects in the crystal surface, increases the dissolution rate independent of the surface area. Interestingly, the precipitation of new 'steps' on a crystal surface is length-dependent. Each crystalline CaP phase has a critical length, above and below which the free energy for crystal growth, or dissolution, respectively, is lower (energetically favourable). As the crystal size decreases, approaching this critical length, the length of surface steps is more likely to be shorter than the critical length, thereby favouring dissolution. In this way, even poorly soluble CaP phases (*e.g.* HA) eventually dissolve in liquid.

Nanoscale and microscale CaP crystals are three-dimensional objects, with distinct crystallographic planes and differing energy needed for growth in the direction of a particular plane. The surface energy of each plane depends upon the number of surface atoms exposed, per unit area, and the energy required to 'create' each surface atom (*i.e.* cleavage energy to remove bonds that would have existed if the surface remained within a solid). Planes with a higher surface energy tend towards dissolution, while planes with a lower surface energy tend towards growth.[73,75] Consequently, if the morphology of nucleated crystals that comprise a CPC has a high surface energy, the bulk dissolution rate increases. For example, the morphology of CaP crystals can be modified by the addition of amino acids, ions, or by modifying the temperature and pH during the setting reaction.[76–79] Common variations in crystal morphology include plates, rods, needles, and whiskers. The inclusion of ions can distort the lattice parameters, thereby changing the energy needed to grow along a crystal plane, while amino acids and organic acids tend to bind directly to the surface and stabilise, or inhibit plane growth.[80]

2.7 Influence of the Microstructure on Bioactivity

The bioactivity of a cement involves many of the physical concepts underlying precipitation and setting reactions, though under different conditions. An early definition of a bioactive material defined by Larry Hench was 'one that elicits a specific biological response at the interface of a material, which results in the formation of a bond between the tissues and the material'.[81] More recently, the definition of bioactivity has grown to include both biological and abiotic (acellular) components: in the context of bone regeneration, bioactivity involves the nucleation of mineral on the surface of the implanted material, with the ultimate result being the formation of a chemical bonding to tissue, and replacement of the bond with new tissue.[82] Therefore, the bioactivity of a CPC reflects the intrinsic chemical bonding potential, either with tissue surfaces or as a source of heterogeneous nucleation, and the biological bonding ability, represented by the likelihood that cells and proteins can attach to the material surface, transform any inorganic mineral that forms into biomineral (a mineral containing biologics such as proteins, most commonly collagen) and tissue, and/or directly secrete/catalyse biomineralisation of the materials surface.[83,84]

Early pioneering work in this area includes studies on glasses containing calcium sodium silicate and phosphate (bioglass) by Larry Hench, who identified strong chemical bonding between implanted bioglasses and bone tissue[81,83] and was able to explain the stepwise process of mineral formation on bioglass,[85] and by Tadashi Kokubo, who identified bioactivity in implanted glasses and later developed the simulated body fluid (SBF) that is now routinely used to predict the relative bioactivity of materials surfaces[86] (ISO 23317). Bioactivity in the body and in the laboratory both require supersaturated fluids. Human blood and SBF are supersaturated with respect to calcium and phosphate ions, and specifically supersaturated with respect to the HA phase.[82] The predictive power of SBF is not definitive; false positive and false negative results occur (*e.g.* a material may mineralise in SBF, but does not integrate with mineralised tissue *in vivo*), and newer iterations of SBF that contain more physiological buffering systems, proteins, or metabolites are continually being developed.[82,87,88] Unless otherwise indicated, all subsequent discussion of bioactivity involves studies using either cell culture media, which are rich sources of ions and proteins, or SBF.

Throughout the following discussion, we will distinguish between the effects that arise from physiochemical phenomena, where cells are not required, and/or from biological phenomena. The bioactivity of a CPC is determined by: a) the rate of ion release and dissolution; b) the surface chemistry (*i.e.* charge density, surface energy); and c) cement topography and microstructure (roughness, surface area, presence of defects or kinks, crystal size and morphology, *etc.*).[56,87] Rapid ion release causes mineral to nucleate on the surface of a CPC when the concentration of the ions exceeds the solubility limit. In parallel, ion release also stimulates biological processes such as the differentiation of stem cells and osteoblasts (cells that produce new bone), to produce biomineral directly *via* alkaline phosphatase expression, and indirectly *via* the deposition of collagen.[89–91] Acting synergistically with other proteins that coat the CPC surface *in vivo* (*e.g.* osteopontin, bone sialoprotein), collagen will directly bind ions in solution and lower the free energy needed for the nucleation of CaP. It should be noted that osteoblasts mediate bioactivity in two ways. After attaching to the material surface, osteoblasts secrete an extracellular matrix composed primarily of collagen,[92] which templates the formation of nanoscale biomineral.[42,93] They also secrete alkaline phosphatase, an enzyme that catalyses the cleavage of organic phosphates to rapidly accumulate phosphate in the extracellular environment, leading to local supersaturation, changes in gene expression, and crystal nucleation onto matrix proteins.[94,95]

The most common approach for controlling the ion release rate is by controlling the cement dissolution rate, *via* the porosity and liquid-to-powder ratio, the relative surface area, and precursor phases, as described above. Using an excess of a moderately soluble precursor phase leads to pockets of unreacted precursor, which subsequently increase the release rate of ions as it dissolves. Alternatively, combining cements with different intermediate or final phases (*e.g.* biphasic cement containing β-TCP and α-TCP, leading to HA cement) increases the rate of ion release compared to single-phase cements.

The second most effective way to increase CPC bioactivity is to provide nano- and micro-structures that create high energy surfaces, thereby lowering the energy needed for crystal nucleation and protein/cell adsorption, or that increase the surface roughness and relative surface area, thereby improving cell and protein attachment, and increasing the dissolution rate.[96] Generally, as the surface area increases, cells attach and differentiate with higher efficiency, due in part to better engagement of the cell cytoskeleton and engagement of surface-bound proteins that activate integrin receptors.[3,97] This is only possible in pre-fabricated (*e.g.* scaffolds) or pre-set cements, since it is difficult to control or predict the resulting surface nanotopography of a self-setting cement *in situ*.

Porosity is the most studied property of cements, with respect to bioactivity, because it has a disproportionately large effect on dissolution and bioactivity compared to other factors, and because it is the most easily controlled physical property of cements.[98,99] Cement bioactivity is affected by porosity in the following ways: first, an increase in porosity increases the relative surface area in contact with liquid, with smaller pores ($<$100 nm) having a greater effect than larger pores ($>$1 μm).[99] Second, smaller pores act as areas of concentrated dissolution; rather than diffusion of ions from a two-dimensional flat surface, at the centre of small pores ions that are diffusing from the enclosing surfaces in all directions, in three dimensions, leading to a much higher local concentration of ions and, therefore, increased bioactivity. Third, true bioactivity requires the involvement of cells and proteins/enzymes. A significant number of cells will not fit inside pores smaller than 50–100 μm in diameter (cell diameter is typically 2–20 μm). Pores in the range of 50–400 μm allow for cells to infiltrate, and activate osteogenic genes *via* the stretching of the cell cytoskeleton as cells attempt to attach to the pore surface, and the concentrated release of ions during dissolution.[100] Nevertheless, to favour bone ingrowth, macroporous CPCs are of interest. Macropores are generally introduced through foaming[101,102] or porogenic agents that dissolve or degrade after setting. Materials such as gelatine, poly(lactic-co-glycolic acid) (PLGA) and glucose are often used as porogenic agents within CPCs.[103,104]

As an example of bioactivity in CaPs, consider the case of a CPC that has been implanted into a bone defect *in vivo*. The CPC can directly bond (abiotic) to the organic portion of nearby bone tissue by precipitation of a thin layer of bridging CaP mineral between the two surfaces, deposited from supersaturated bodily fluids. Physical bonding occurs when the precipitated surface mineral dissolves and reprecipitates as a more stable phase (HA, *via* Ostwald ripening), thereby cementing the CaP surface to the collagen-rich tissue surface.[105,106] The CPC can also directly bind to nearby tissue surfaces *via* biologically-mediated effects. The charged CPC surface attracts circulating blood or saliva (dental materials), and proteins and metabolites adsorb which, in turn, increase and control the nucleation of mineral on the implant surface.[107,108] Blood proteins are strongly attracted to the cement surface due to direct ionic interactions, increased free energy resulting from exchange of the protein hydration shell of water for attachment to a solid surface, and from increases in entropy as the protein unfolds on the surface. A thin layer of protein rapidly covers CPC surfaces *in vivo*, and strongly affects which ions are subsequently attracted to the cement surface, which phases nucleate on the surface, and the overall bioactivity of the cement.[109] Bone cell progenitors (mesenchymal stem cells) are also attracted to the material surface, where they attach to the adsorbed blood proteins, secrete an extracellular matrix (osteoid), and differentiate into mature osteoblasts. Osteoblast attachment and differentiation depends upon the profile and conformation of proteins adsorbed on the surface and in the extracellular matrix,[96] and the mechanical properties of the material surface (*e.g.* stiffness, porosity, nano- and micro-topography and morphology).[110,111]

Finally, the concepts of osteoinduction and osteoconduction should also be reviewed briefly. If new biomineral readily forms on the surface of a cement, the cement is considered osteoconductive (it facilitates the deposition of new mineral or mineralised tissue growth). If new biomineral forms within the cement (*i.e.* inside of pores, or clefts in the material, particularly when the pores are far below the surface), or the cement causes new mineral or mineralised tissue to form in regions of the body that do not normally mineralise (*i.e.* implantation of a cement into soft tissue, like skin, causes soft tissue mineralisation and the formation of mature bone-like mineral) the cement is considered osteoinductive. These terms refer more to mineralisation that involves biological processes, rather than purely bioactivity arising from inorganic mineralisation. For further discussion, in particular on the distinction between true osteoinductive stimulation of mineralised tissue *versus* ectopic mineralisation, the reader is directed to ref. 112, 113 and 114. Bioactive mineralisation can also lead to pathological processes. Pathological mineralisation occurs when ion concentrations are artificially elevated (*i.e.* tissue injury leading to the release of ions) or when environmental conditions change (*i.e.* low pH on the surface of teeth from bacteria), leading to an adherent type of bioactivity. As a result, phases that do not normally precipitate, or persist, *in vivo* (brushite, whitlockite and octacalcium phosphate) may form.

2.8 Structure-related Mechanical Properties

An ideal CPC for bone regeneration should have mechanical properties that are similar to the missing or damaged bone that is being replaced. A critical problem that prevents the more extensive clinical application of CPCs is their relatively weak mechanical properties, which limits the use of CPCs in load-bearing bone grafts. The most commonly investigated mechanical property of CPCs is their compressive strength. The brittle nature of these materials, and the more complex specimen geometry and set-up of other types of mechanical tests, makes compressive testing preferable for CPCs. Nevertheless, other properties, such as tensile strength, flexural strength, shear strength, elastic modulus, Poisson's ratio, fatigue strength and life, have also been studied.[115]

There are two types of osseous tissue in the human body: cortical bone and cancellous bone. As its name implies, cortical bone forms the cortex, or outer shell of most bone and it is much denser, harder, stronger and stiffer than cancellous bone. Cancellous bone is porous and much lighter than cortical bone. In the longitudinal direction, cortical bone exhibits a Young's modulus of elasticity $E = 7–30$ GPa, a compressive strength of 167–193 MPa, and fracture toughness of 2–12 MPa m$^{1/2}$.[11] It has been estimated that hip joints are subjected to an average load of up to three times the body weight (approximately 3000 N); peak loads during jumping can even be as high as 10 times the body weight.[11] In contrast, the compressive strength of CPCs typically lies in the range of ∼5–65 MPa,[3] and the fracture toughness of apatite cements lies in the range of 0.05–0.5 MPa m$^{1/2}$ for samples with porosities of 31–77%. Thus, the mechanical properties of CPCs are far lower than those of cortical bone and hip joints. Still, they are comparable to those of cancellous bone, which is characterised by a compressive strength of 1.9–10 MPa and fracture toughness of roughly 0.1 MPa m$^{1/2}$.[11] The medical applications of CPCs are therefore currently focused on the production of non-load-bearing implants, such as pieces for middle ear surgery, and the filling of cancellous bone defects in oral or orthopaedic surgery. Many strategies have been developed to improve the mechanical properties of CPCs,[3] such as reducing the porosity, changing the morphology of entangled apatite crystals, and adding reinforcing agents.

Several factors have an effect on the mechanical properties of CPCs, *e.g.* composition, porosity, filler content, particle size of precursors, the resulting crystal size and shape, and the liquid component.[3,13,27] The porosity is an important factor since it has a great impact not only on the mechanical properties but also resorbability and bioactivity. It is, therefore, one of the most extensively investigated properties of cements. The porosity can be assessed through a variety of different methods, including theoretical calculations, mercury intrusion porosimetry, helium pycnometry, water evaporation and isopropanol resaturation.[115]

As described above, there are many factors that affect the porosity of a CPC. The porosity of both apatite and brushite cements increases with an increased liquid-to-powder ratio (L/P).[116,117] The increased particle size of the precursor powder can both increase,[52] decrease,[116] or not contribute towards the porosity, depending on the cement composition and particle size distribution. The preparation method can also influence the porosity, *e.g.* by introducing air bubbles in the cement. In general, the aim is to have as low porosity as possible in order to maximise the mechanical properties.

As mentioned above, the mechanical properties of CPCs may differ widely – for brushite cements a variation in compressive strength of 1–74 MPa has been reported[115] – due to factors such as chemical components, particle sizes, mixing method, sample preparation and testing method. However, a clear dependency on porosity can be discerned. An increased porosity has a negative effect on the strength, *e.g.* hardness, compressive strength and diametral tensile strength.[101,117] Attempts have therefore been made to predict the strength of CPCs based on porosity *via* a number of different calculations.

Duckworth[31] proposed the following equation based on the Ryshkewitch relationship and previous studies:

$$S_\mathrm{p} = S_0 \exp(-kP)$$

where S_p is the ultimate tensile strength of the porous body, S_0 is the corresponding strength of the non-porous material, P is the porosity, and k is a materials-dependent constant. The modified empirical relationship[118] has been used to describe compressive strength as a function of porosity for both HA[119] and brushite cements,[116] as well as other types of cements:[120]

$$\sigma_\mathrm{C} = \sigma_\mathrm{C0} \exp(-q\Phi)$$

where σ_C is the compressive strength, σ_C0 the compressive strength of a fully dense cement (a cement that has zero porosity), Φ is the porosity and q is a dimensionless constant.

Apatite cements, one typical CPC, generally have a low mechanical strength. Researchers have found that many factors influence their mechanical strength, including the crystal size and phase compositions in the final set samples and powder size, powder-to-liquid ratio, and additives in the liquid and powder phases before mixing.[121] Ishikawa and Asaoka studied the mechanical strength of TTCP–DCPA type apatite cements in terms of their porosity. The ideal tensile strength value was found to be 103 MPa.[122] Ishikawa *et al.* also reported that the mechanical strength of apatite cements increases when the particle size of the raw powder increases. Their results show that large particle sizes are better than small particle sizes for maximising the mechanical strength of cements.[122]

Gbureck *et al.*[123] used 8–10 μm in size metal oxide particles as additives to enhance the strength of apatite cements derived from TTCP (tetra-calcium phosphates). SiO_2 and TiO_2, but not ZrO_2, showed significantly improved compressive strength by approximately 80–100 MPa when the starting particle size of TTCP was around 1–20 μm. However, the improved compressive strength is not high enough to match the strength of cortical bone. The reinforcing mechanism is thought to be due to the changes in the thermodynamic or kinetic solubilities of doped TTCP, and the resultant slower HA crystal formation and a more cross-linked cement structure.

Bioactive glass and carbon fibres have also been used to reinforce CPCs.[124] Different lengths (3–200 mm) *versus* the volume fraction (1.9–9.5%) of the above two types of fibre have been mixed into CaP cements. Their results showed that fibres (diameter of 8 mm) with a length of 75 mm, at 5.7% volume fraction achieved the highest flexural strength, which was four times higher than those of non-reinforced cements. The length and volume fraction of fibres also influence the crack-bridging and pullout effect. Canal *et al.* summarised how fibres, including polymer and ceramic fibers, reinforce CPCs.[125] The deposition/blending of fibres in apatite cements influences the microstructure, and further influences mechanical strength. The structure and orientation of the incorporated fibres, whether regular, such as yarns, nonwovens, woven or knitted fabrics (meshes), or random, appears to be crucial to the resulting flexural strength and modulus of elasticity. If the fibres are biodegradable, the reinforcement is considered to be a temporary effect. One thing that should be mentioned, is that the interaction between fibres and cements is important, independent of the textures, diameter, length and volume.

The crystal size of CDHA in the final α-TCP formed apatite cement can be considered to have a significant influence not only on degradation and biological properties but on mechanical strength. This area has been understudied by the community. Camire *et al.* found a clear reinforcement of mechanical properties when the size of CDHA in cements was <23 nm, rather than >43 nm.[126] The size and shape of CDHA crystals could influence the crystal entanglement in the cement and further affects the final mechanical strength. However, no direct relationship has been proven between the crystal size and shape and the mechanical strength in apatite cements.

2.9 Summary

CPCs were developed decades ago and ever since there has been substantial growth in research into improving their properties in order to satisfy the requirements needed for their clinical applications. Nowadays, CPCs have become a large family, and many of them have been successfully developed as commercial products. Many factors exert great influences on the application of CPCs, including setting time, injectability, bioresorbability and anti-washout properties, *etc.* The above aspects need to be balanced for specific applications. Like other materials, it is the microstructure of CPCs that determines their properties. Specifically, the microstructure of CPCs (grain size, phase composition, porosity, pore size, *etc.*) significantly influences the dissolution rate, bioactivity, and mechanical properties of CPCs. CaP can also be manufactured in the form of nanoparticles. By integrating them with other materials to form a biocomposite, CaP nanoparticles can be used as carriers for drugs, proteins, and gene delivery, reparative materials for enamel remineralisation, *etc.*

References

1. M. Bohner, Resorbable biomaterials as bone graft substitutes, *Mater. Today*, 2010, **13**, 24–30.
2. S. Bose and S. Tarafder, Calcium phosphate ceramic systems in growth factor and drug delivery for bone tissue engineering: A review, *Acta Biomater.*, 2012, **8**, 1401–1421.
3. J. Zhang, W. Liu, V. Schnitzler, F. Tancret and J. M. Bouler, Calcium phosphate cements for bone substitution: Chemistry, handling and mechanical properties, *Acta Biomater.*, 2014, **10**, 1035–1049.
4. S. Lu, M. A. P. McGough, S. M. Shiels, K. J. Zienkiewicz, A. R. Merkel, J. P. Vanderburgh, J. S. Nyman, J. A. Sterling, D. J. Tennent, J. C. Wenke and S. A. Guelcher, Settable polymer/ceramic composite bone grafts stabilize weight-bearing tibial plateau slot defects and integrate with host bone in an ovine model, *Biomaterials*, 2018, **179**, 29–45.
5. S. Shadanbaz and G. J. Dias, Calcium phosphate coatings on magnesium alloys for biomedical applications: A review, *Acta Biomater.*, 2012, **8**, 20–30.
6. M. P. Ginebra, M. Espanol, E. B. Montufar, R. A. Perez and G. Mestres, New processing approaches in calcium phosphate cements and their applications in regenerative medicine, *Acta Biomater.*, 2010, **6**, 2863–2873.
7. A. J. Ambard and L. Mueninghoff, Calcium phosphate cement: Review of mechanical and biological properties, *J. Prosthodontics*, 2006, **15**, 321–328.
8. J. Vecstaudza and J. Locs, Novel preparation route of stable amorphous calcium phosphate nanoparticles with high specific surface area, *J. Alloys Compd.*, 2017, **700**, 215–222.

9. J. Luo, H. Engqvist and C. Persson, A ready-to-use acidic, brushite-forming calcium phosphate cement, *Acta Biomater.*, 2018, **81**, 304–314.
10. R. O'Neill, H. O. McCarthy, E. B. Montufar, M. P. Ginebra, D. I. Wilson, A. Lennon and N. Dunne, Critical review: Injectability of calcium phosphate pastes and cements, *Acta Biomater.*, 2017, **50**, 1–19.
11. N. Eliaz and N. Metoki, Calcium phosphate bioceramics: A review of their history, structure, properties, coating technologies and biomedical applications, *Materials*, 2017, **10**, 334.
12. G. Renaudin, S. Gomes and J. M. Nedelec, First-row transition metal doping in calcium phosphate bioceramics: A detailed crystallographic study, *Materials*, 2017, **10**, 1–22.
13. S. V. Dorozhkin, Calcium orthophosphates in nature, biology and medicine, *Materials*, 2009, **2**, 399–498.
14. F. Tamimi, Z. Sheikh and J. Barralet, Dicalcium phosphate cements: Brushite and monetite, *Acta Biomater.*, 2012, **8**, 474–487.
15. G. Daculsi, O. Laboux, O. Malard and P. Weiss, Current state of the art of biphasic calcium phosphate bioceramics, *J. Mater. Sci.: Mater. Med.*, 2003, **4**, 195–200.
16. M. F. Cipreste, A. M. Peres, A. A. C. Cotta, F. H. Aragón, A. de, M. Antunes, A. S. Leal, W. A. A. Macedo and E. M. B. de Sousa, Synthesis and characterization of 159Gd-doped hydroxyapatite nanorods for bioapplications as theranostic systems, *Mater. Chem. Phys.*, 2016, **181**, 301–311.
17. S. V. Dorozhkin and M. Epple, Biological and medical significance of calcium phosphates, *Angew. Chem., Int. Ed.*, 2002, **41**, 3130–3146.
18. L. Liang, P. Rulis and W. Y. Ching, Mechanical properties, electronic structure and bonding of α-and β-tricalcium phosphates with surface characterization, *Acta Biomater.*, 2010, **6**, 3763–3771.
19. L. Zhou, X. Ge, C. Ren and G. Chen, In-situ high temperature XRD and TEM study of the thermal stability and sintering behavior of octacalcium phosphate, *J. Alloys Compd.*, 2019, **778**, 72–76.
20. M. Roozbahani, M. Alehosseini, M. Kharaziha and R. Emadi, Nano-calcium phosphate bone cement based on Si-stabilized α-tricalcium phosphate with improved mechanical properties, *Mater. Sci. Eng., C*, 2017, **81**, 532–541.
21. M. P. Ginebra, C. Canal, M. Espanol, D. Pastorino and E. B. Montufar, Calcium phosphate cements as drug delivery materials, *Adv. Drug Delivery Rev.*, 2012, **64**, 1090–1110.
22. M. Karimi, S. Hesaraki, M. Alizadeh and A. Kazemzadeh, A facile and sustainable method based on deep eutectic solvents toward synthesis of amorphous calcium phosphate nanoparticles: The effect of using various solvents and precursors on physical characteristics, *J. Non-Cryst. Solids*, 2016, **443**, 59–64.
23. M. Karimi, S. Hesaraki, M. Alizadeh and A. Kazemzadeh, Effect of synthetic amorphous calcium phosphate nanoparticles on the physicochemical and biological properties of resin-modified glass ionomer cements, *Mater. Sci. Eng., C*, 2019, **98**, 227–240.

24. M. Okada and T. Matsumoto, Synthesis and modification of apatite nanoparticles for use in dental and medical applications, *Jpn Dent Sci Rev.*, 2015, **51**, 85–95.
25. S. V. Dorozhkin, Nanosized and nanocrystalline calcium orthophosphates, *Acta Biomater.*, 2010, **6**, 715–734.
26. M. M. Stevens, Biomaterials for bone tissue engineering, *Mater. Today*, 2008, **11**, 18–25.
27. S. V. Dorozhkin, Bioceramics of calcium orthophosphates, *Biomaterials*, 2010, **31**, 1465–1485.
28. V. Sokolova, L. Rojas-Sánchez, N. Białas, N. Schulze and M. Epple, Calcium phosphate nanoparticle-mediated transfection in 2D and 3D mono- and co-culture cell models, *Acta Biomater.*, 2019, **84**, 391–401.
29. M. Epple, K. Ganesan, R. Heumann, J. Klesing, A. Kovtun, S. Neumann and V. Sokolova, Application of calcium phosphate nanoparticles in biomedicine, *J. Mater. Chem.*, 2010, **20**, 18–23.
30. R. V. Chernozem, M. A. Surmeneva, B. Krause, T. Baumbach, V. P. Ignatov, O. Prymak, K. Loza, M. Epple, F. Ennen-Roth, A. Wittmar, M. Ulbricht, E. A. Chudinova, T. Rijavec, A. Lapanje and R. A. Surmenev, Functionalization of titania nanotubes with electrophoretically deposited silver and calcium phosphate nanoparticles: Structure, composition and antibacterial assay, *Mater. Sci. Eng., C*, 2019, **97**, 420–430.
31. K. Ishikawa, Calcium Phosphate Cement, in *Advances in Calcium Phosphate Biomaterials*, 2014, pp. 0–39. ch. 7 https://doi.org/10.1007/978-3-642-53980-0.
32. M. Bohner, U. Gbureck and J. E. Barralet, Technological issues for the development of more efficient calcium phosphate bone cements: A critical assessment, *Biomaterials*, 2005, **26**, 6423–6429.
33. M. Bohner and G. Baroud, Injectability of calcium phosphate pastes, *Biomaterials*, 2005, **26**, 1553–1563.
34. K. Lin, C. Wu and J. Chang, Advances in synthesis of calcium phosphate crystals with controlled size and shape, *Acta Biomater.*, 2014, **10**, 4071–4102.
35. A. Peetsch, C. Greulich, D. Braun, C. Stroetges, H. Rehage, B. Siebers, M. Köller and M. Epple, Silver-doped calcium phosphate nanoparticles: Synthesis, characterization, and toxic effects toward mammalian and prokaryotic cells, *Colloids Surf., B*, 2013, **102**, 724–729.
36. X. Guo, H. Yan, S. Zhao, Z. Li, Y. Li and X. Liang, Effect of calcining temperature on particle size of hydroxyapatite synthesized by solid-state reaction at room temperature, *Adv. Powder Technol.*, 2013, **24**, 1034–1038.
37. W. Chu, Y. Huang, C. Yang, Y. Liao, X. Zhang, M. Yan, S. Cui and C. Zhao, Calcium phosphate nanoparticles functionalized with alendronate-conjugated polyethylene glycol (PEG) for the treatment of bone metastasis, *Int. J. Pharm.*, 2017, **516**, 352–363.

38. W. He, Y. Xie, Q. Xing, P. Ni, Y. Han and H. Dai, Sol-gel synthesis of biocompatible Eu^{3+}/Gd^{3+} co-doped calcium phosphate nanocrystals for cell bioimaging, *J. Lumin.*, 2017, **192**, 902–909.
39. F. Ye, H. Guo, H. Zhang and X. He, Polymeric micelle-templated synthesis of hydroxyapatite hollow nanoparticles for a drug delivery system, *Acta Biomater.*, 2010, **6**, 2212–2218.
40. M. V. Chaikina, N. V. Bulina, O. B. Vinokurova, I. Y. Prosanov and D. V. Dudina, Interaction of calcium phosphates with calcium oxide or calcium hydroxide during the "soft" mechanochemical synthesis of hydroxyapatite, *Ceram. Int.*, 2019, **45**, 16927–16933.
41. J. L. Xu, K. A. Khor, Z. L. Dong, Y. W. Gu, R. Kumar and P. Cheang, Preparation and characterization of nano-sized hydroxyapatite powders produced in a radio frequency (rf) thermal plasma, *Mater. Sci. Eng., A*, 2004, **374**, 101–108.
42. M. H. Kim, B. S. Kim, H. Park, J. Lee and W. H. Park, Injectable methylcellulose hydrogel containing calcium phosphate nanoparticles for bone regeneration, *Int. J. Biol. Macromol.*, 2018, **109**, 57–64.
43. T. Tenkumo, J. R. Vanegas Sáenz, K. Nakamura, Y. Shimizu, V. Sokolova, M. Epple, Y. Kamano, H. Egusa, T. Sugaya and K. Sasaki, Prolonged release of bone morphogenetic protein-2 in vivo by gene transfection with DNA-functionalized calcium phosphate nanoparticle-loaded collagen scaffolds, *Mater. Sci. Eng., C*, 2018, **92**, 172–183.
44. L. Zhang, M. D. Weir, G. Hack, A. F. Fouad and H. H. K. Xu, Rechargeable dental adhesive with calcium phosphate nanoparticles for long-term ion release, *J. Dent.*, 2015, **43**, 1587–1595.
45. J. Wu, M. D. Weir, M. A. S. Melo and H. H. K. Xu, Development of novel self-healing and antibacterial dental composite containing calcium phosphate nanoparticles, *J. Dent.*, 2015, **43**, 317–326.
46. M. Bohner, Reactivity of calcium phosphate cements, *J. Mater. Chem.*, 2007, **17**, 3980–3986.
47. U. Gbureck, J. Probst and R. Thull, Surface properties of calcium phosphate particles for self setting bone cements, *Biomol. Eng.*, 2002, **19**, 51–55.
48. L. Wang, J. Lu, F. Xu and F. Zhang, Dynamics of crystallization and dissolution of calcium orthophosphates at the near-molecular level, *Chin. Sci. Bull.*, 2011, **56**, 713–721.
49. S. V. Dorozhkin, Dissolution mechanism of calcium apatites in acids: A review of literature, *World J. Methodol.*, 2012, **2**, 1–17.
50. Y. Zhang, Q. Li, X. Li, Y. Li, C. Wang, Y. Zhao, Y. Song and Y. Liu, Preparation and Evaluation of Two Apatites with Spherical Nanocrystal Morphology, *J. Nanosci. Nanotechnol.*, 2003, **16**, 2384–2389.
51. D. L. Alge, W. S. Goebel and T.-M. G. Chu, Effects of DCPD cement chemistry on degradation properties and cytocompatibility: comparison of MCPM/β-TCP and MCPM/HA formulations, *Biomed. Mater.*, 2013, **8**, 25010.

52. J. Engstrand, C. Persson and H. Engqvist, The effect of composition on mechanical properties of brushite cements, *J. Mech. Behav. Biomed. Mater.*, 2014, **29**, 81–90.
53. E. Şahin and M. Çiftçioğlu, Monetite promoting effect of citric acid on brushite cement setting kinetics, *Mater. Res. Innovations*, 2014, **18**, 138–145.
54. L. C. Chow, M. Markovic, S. A. Frukhtbeyn and S. Takagi, Hydrolysis of tetracalcium phosphate under a near-constant-composition condition—effects of pH and particle size, *Biomaterials*, 2005, **26**, 393–401.
55. E. Ferńndez, F. J. Gil, M. P. Ginebra, F. C. M. Driessens, J. A. Planell and S. M. Best, Calcium phosphate bone cements for clinical applications. Part I: Solution chemistry, *J. Mater. Sci.: Mater. Med.*, 1999, **10**, 169–176.
56. P. Zhu, Y. Masuda and K. Koumoto, The effect of surface charge on hydroxyapatite nucleation, *Biomaterials*, 2004, **25**, 3915–3921.
57. N. A. Garcia, R. I. Malini, C. L. Freeman, R. Demichelis, P. Raiteri, N. A. J. M. Sommerdijk, J. H. Harding and J. D. Gale, Simulation of Calcium Phosphate Prenucleation Clusters in Aqueous Solution: Association beyond Ion Pairing, *Cryst. Growth Des.*, 2019, **19**, 6422–6430.
58. D. Gebauer, M. Kellermeier, J. D. Gale, L. Bergström and H. Cölfen, Pre-nucleation clusters as solute precursors in crystallisation, *Chem. Soc. Rev.*, 2014, **43**, 2348–2371.
59. A. Lotsari, A. K. Rajasekharan, M. Halvarsson and M. Andersson, Transformation of amorphous calcium phosphate to bone-like apatite, *Nat. Commun.*, 2018, **9**, 4170.
60. F. Betts and A. S. Posner, An X-ray radial distribution study of amorphous calcium phosphate, *Mater. Res. Bull.*, 1974, **9**, 353–360.
61. S. Jiang, W. Jin, Y.-N. Wang, H. Pan, Z. Sun and R. Tang, Effect of the aggregation state of amorphous calcium phosphate on hydroxyapatite nucleation kinetics, *RSC Adv.*, 2017, **7**, 25497–25503.
62. D. Zahn, Mechanisms of Calcium and Phosphate Ion Association in Aqueous Solution, *Z. Anorg. Allg. Chem.*, 2004, **630**, 1507–1511.
63. J. L. Giocondi, B. S. El-Dasher, G. H. Nancollas and C. A. Orme, Molecular mechanisms of crystallization impacting calcium phosphate cements, *Philos. Trans. R. Soc., A*, 2010, **368**, 1937–1961.
64. C. Drouet, A comprehensive guide to experimental and predicted thermodynamic properties of phosphate apatite minerals in view of applicative purposes, *J. Chem. Thermodyn.*, 2015, **81**, 143–159.
65. A. Selmani, I. Coha, K. Magdić, B. Čolović, V. Jokanović, S. Šegota, S. Gajović, A. Gajović, D. Jurašin and M. Dutour Sikirić, Multiscale study of the influence of cationic surfactants on amorphous calcium phosphate precipitation, *CrystEngComm*, 2015, **17**, 8529–8548.
66. Q. Liu, J. Ding, F. K. Mante, S. L. Wunder and G. R. Baran, The role of surface functional groups in calcium phosphate nucleation on titanium foil: a self-assembled monolayer technique, *Biomaterials*, 2002, **23**, 3103–3111.

67. G. K. Toworfe, R. J. Composto, I. M. Shapiro and P. Ducheyne, Nucleation and growth of calcium phosphate on amine-, carboxyl- and hydroxyl-silane self-assembled monolayers, *Biomaterials*, 2006, **27**, 631–642.
68. M. Bohner, H. P. Merkle, P. Van Landuyt, G. Trophardy and J. Lemaitre, Effect of several additives and their admixtures on the physico-chemical properties of a calcium phosphate cement, *J. Mater. Sci.: Mater. Med.*, 2000, **11**, 111–116.
69. D. Tadic and M. Epple, A thorough physicochemical characterisation of 14 calcium phosphate-based bone substitution materials in comparison to natural bone, *Biomaterials*, 2004, **25**, 987–994.
70. N. C. Blumenthal, F. Betts and A. S. Posner, Stabilization of amorphous calcium phosphate by Mg and ATP, *Calcif. Tissue Res.*, 1977, **23**, 245–250.
71. D. Vazquez, S. Takagi, S. Frukhtbeyn and L. C. Chow, Effects of Addition of Mannitol Crystals on the Porosity and Dissolution Rates of a Calcium Phosphate Cement, *J. Res. Natl. Inst. Stand. Technol.*, 2010, **115**, 225–232.
72. A. D. Rafeek, G. Choi and L. A. Evans, Morphological, spectroscopic and crystallographic studies of calcium phosphate bioceramic powders, *J. Aust. Ceram. Soc.*, 2018, **54**, 161–168.
73. V. Uskoković, The role of hydroxyl channel in defining selected physicochemical peculiarities exhibited by hydroxyapatite, *RSC Adv.*, 2015, **5**, 36614–36633.
74. R. Tang, C. A. Orme, G. H. Nancollas and A. New, Understanding of Demineralization: The Dynamics of Brushite Dissolution, *J. Phys. Chem. B*, 2003, **107**, 10653–10657.
75. H. Kim, R. P. Camata, S. Chowdhury and Y. K. Vohra, In vitro dissolution and mechanical behavior of *c*-axis preferentially oriented hydroxyapatite thin films fabricated by pulsed laser deposition, *Acta Biomater.*, 2010, **6**, 3234–3241.
76. S. Chen, K. Grandfield, S. Yu, H. Engqvist and W. Xia, Synthesis of calcium phosphate crystals with thin nacreous structure, *CrystEngComm*, 2016, **18**, 1064–1069.
77. S. Chen, S. Pujari-Palmer, S. Rubino, V. Westlund, M. Ott, H. Engqvist and W. Xia, Highly repeatable synthesis of nHA with high aspect ratio, *Mater. Lett.*, 2015, **159**, 163–167.
78. J. Zhan, Y.-H. Tseng, J. C. C. Chan and C.-Y. Mou, Biomimetic Formation of Hydroxyapatite Nanorods by a Single-Crystal-to-Single-Crystal Transformation, *Adv. Funct. Mater.*, 2005, **15**, 2005–2010.
79. S. Chen, M. Krumova, H. Cölfen and E. V. Sturm, Synthesis of Fiber-like Monetite without Organic Additives and Its Transformation to Hydroxyapatite, *Chem. Mater.*, 2019, **31**, 1543–1551.
80. H. Moussa, W. Jiang, A. Alsheghri, A. Mansour, A. El Hadad, H. Pan, R. Tang, J. Song, J. Vargas, M. D. McKee and F. Tamimi, High strength brushite bioceramics obtained by selective regulation of crystal growth with chiral biomolecules, *Acta Biomater.*, 2020, **106**, 351–359.

81. L. L. Hench, R. J. Splinter, W. C. Allen and T. K. Greenlee, Bonding mechanisms at the interface of ceramic prosthetic materials, *J. Biomed. Mater. Res.*, 1971, **5**, 117–141.
82. M. Bohner and J. Lemaitre, Can bioactivity be tested in vitro with SBF solution? *Biomaterials*, 2009, **30**, 2175–2179.
83. L. L. Hench and H. A. Paschall, Direct chemical bond of bioactive glass-ceramic materials to bone and muscle, *J. Biomed. Mater. Res.*, 1973, **7**, 25–42.
84. V. J. Shirtliff and L. L. Hench, Bioactive materials for tissue engineering, regeneration and repair, *J. Mater. Sci.*, 2003, **38**, 4697–4707.
85. W. Cao and L. L. Hench, Bioactive materials, *Ceram. Int.*, 1996, **22**, 493–507.
86. T. Kokubo, H. Kushitani, C. Ohtsuki, S. Sakka and T. Yamamuro, Chemical reaction of bioactive glass and glass-ceramics with a simulated body fluid, *J. Mater. Sci.: Mater. Med.*, 1992, **3**, 79–83.
87. T. Kokubo and H. Takadama, How useful is SBF in predicting in vivo bone bioactivity? *Biomaterials*, 2006, **27**, 2907–2915.
88. A. A. Zadpoor, Relationship between in vitro apatite-forming ability measured using simulated body fluid and in vivo bioactivity of biomaterials, *Mater. Sci. Eng., C*, 2014, **35**, 134–143.
89. F. Barrère, C. A. van Blitterswijk and K. de Groot, Bone regeneration: molecular and cellular interactions with calcium phosphate ceramics, *Int. J. Nanomed.*, 2006, **1**, 317–332.
90. S. Maeno, Y. Niki, H. Matsumoto, H. Morioka, T. Yatabe, A. Funayama, Y. Toyama, T. Taguchi and J. Tanaka, The effect of calcium ion concentration on osteoblast viability, proliferation and differentiation in monolayer and 3D culture, *Biomaterials*, 2005, **26**, 4847–4855.
91. A. Obata, T. Ogasawara and T. Kasuga, Combinatorial effects of inorganic ions on adhesion and proliferation of osteoblast-like cells, *J. Biomed. Mater. Res., Part A*, 2019, **107**, 1042–1051.
92. L. Niu, S. E. Jee, K. Jiao, L. Tonggu, M. Li, L. Wang, Y. Yang, J. Bian, L. Breschi, S. S. Jang, J. Chen, D. H. Pashley and F. R. Tay, Collagen intrafibrillar mineralization as a result of the balance between osmotic equilibrium and electroneutrality, *Nat. Mater.*, 2017, **16**, 370–378.
93. Y. Wang, T. Azaïs, M. Robin, A. Vallée, C. Catania, P. Legriel, G. Pehau-Arnaudet, F. Babonneau, M.-M. Giraud-Guille and N. Nassif, The predominant role of collagen in the nucleation, growth, structure and orientation of bone apatite, *Nat. Mater.*, 2012, **11**, 724–733.
94. E. Verné, S. Ferraris, C. Vitale-Brovarone, S. Spriano, C. L. Bianchi, A. Naldoni, M. Morra and C. Cassinelli, Alkaline phosphatase grafting on bioactive glasses and glass ceramics, *Acta Biomater.*, 2010, **6**, 229–240.
95. G. R. Beck, B. Zerler and E. Moran, Phosphate is a specific signal for induction of osteopontin gene expression, *Proc. Natl. Acad. Sci.*, 2000, **97**, 8352 LP–8358357.
96. X. D. Zhu, H. S. Fan, Y. M. Xiao, D. X. Li, H. J. Zhang, T. Luxbacher and X. D. Zhang, Effect of surface structure on protein adsorption to

biphasic calcium-phosphate ceramics in vitro and in vivo, *Acta Biomater.*, 2009, **5**, 1311–1318.

97. K. Lin, L. Xia, J. Gan, Z. Zhang, H. Chen, X. Jiang and J. Chang, Tailoring the Nanostructured Surfaces of Hydroxyapatite Bioceramics to Promote Protein Adsorption, Osteoblast Growth, and Osteogenic Differentiation, *ACS Appl. Mater. Interfaces*, 2013, **5**, 8008–8017.
98. F. F. Bakare, Y.-J. Chou, Y.-H. Huang, H. A. Tesfay, T. Moriga and S.-J. Shih, Correlation of Morphology and In-Vitro Degradation Behavior of Spray Pyrolyzed Bioactive Glasses, *Materials*, 2019, **12**, 3703.
99. S. R. Radin and P. Ducheyne, Effect of bioactive ceramic composition and structure on in vitro behavior. III. Porous versus dense ceramics, *J. Biomed. Mater. Res.*, 1994, **28**, 1303–1309.
100. M. Bohner and R. J. Miron, A proposed mechanism for material-induced heterotopic ossification, *Mater. Today*, 2019, **22**, 132–141.
101. M.-P. Ginebra, J.-A. Delgado, I. Harr, A. Almirall, S. D. Valle and J. A. Planell, Factors affecting the structure and properties of an injectable self-setting calcium phosphate foam, *J. Biomed. Mater. Res., Part A*, 2007, **80A**, 351–361.
102. C. Unosson, J. Montufar, E. Engqvist, H. Ginebra and M. P. Persson, Brushite foams - the effect of Tween® 80 and Pluronic® F-127 on foam porosity and mechanical properties, *J. Biomed. Mater. Res.*, 2016, **104**, 67–77.
103. W. J. E. M. Habraken, L. T. De Jonge, J. G. C. Wolke, L. Yubao, A. G. Mikos and J. A. Jansen, Introduction of gelatin microspheres into an injectable calcium phosphate cement, *J. Biomed. Mater. Res., Part A*, 2008, **87**, 643–655.
104. R. P. Félix Lanao, S. C. G. Leeuwenburgh, J. G. C. Wolke and J. A. Jansen, In vitro degradation rate of apatitic calcium phosphate cement with incorporated PLGA microspheres, *Acta Biomater.*, 2011, **7**, 3459–3468.
105. Y. Okada, M. Kobayashi, H. Fujita, Y. Katsura, H. Matsuoka, H. Takadama, T. Kokubo and T. Nakamura, Transmission electron microscopic study of interface between bioactive bone cement and bone: Comparison of apatite and wollastonite containing glass-ceramic filler with hydroxyapatite and β-tricalcium phosphate fillers, *J. Biomed. Mater. Res.*, 1999, **45**, 277–284.
106. X. Wang, F. A. Shah, A. Palmquist and K. Grandfield, 3D Characterization of Human Nano-osseointegration by On-Axis Electron Tomography without the Missing Wedge, *ACS Biomater. Sci. Eng.*, 2017, **3**, 49–55.
107. A. El-Ghannam, P. Ducheyne and I. M. Shapiro, Effect of serum proteins on osteoblast adhesion to surface-modified bioactive glass and hydroxyapatite, *J. Orthop. Res.*, 1999, **17**, 340–345.
108. C. Combes and C. Rey, Adsorption of proteins and calcium phosphate materials bioactivity, *Biomaterials*, 2002, **23**, 2817–2823.
109. K. Wang, C. Zhou, Y. Hong and X. Zhang, A review of protein adsorption on bioceramics, *Interface Focus*, 2012, **2**, 259–277.

110. K. Chatterjee, S. Lin-Gibson, W. E. Wallace, S. H. Parekh, Y. J. Lee, M. T. Cicerone, M. F. Young and C. G. Simon Jr, The effect of 3D hydrogel scaffold modulus on osteoblast differentiation and mineralization revealed by combinatorial screening, *Biomaterials*, 2010, **31**, 5051–5062.
111. I. Titushkin and M. Cho, Modulation of Cellular Mechanics during Osteogenic Differentiation of Human Mesenchymal Stem Cells, *Biophys. J.*, 2007, **93**, 3693–3702.
112. R. J. Miron and Y. F. Zhang, Osteoinduction: A Review of Old Concepts with New Standards, *J. Dent. Res.*, 2012, **91**, 736–744.
113. M. A. Scott, B. Levi, A. Askarinam, A. Nguyen, T. Rackohn, K. Ting, C. Soo and A. W. James, Brief review of models of ectopic bone formation, *Stem Cells Dev.*, 2012, **21**, 655–667.
114. R. M. Klar, The Induction of Bone Formation: The Translation Enigma, *Front. Bioeng. Biotechnol.*, 2018, **6**, 74.
115. I. Ajaxon and C. Persson, Mechanical Properties of Brushite Calcium Phosphate Cements, in *World Sci. Encycl. Nanomedicine Bioeng. II Bioimplants, Regen. Med. Nano-Cancer Diagnosis Phototherapy–Volume 3 Des. Bioact. Mater. Bone Repair Regen.*, World Scientific Pte Ltd, 2017, pp. 285–300. https://doi.org/10.1142/9789813202573_0008.
116. M. P. Hofmann, A. R. Mohammed, Y. Perrie, U. Gbureck and J. E. Barralet, High-strength resorbable brushite bone cement with controlled drug-releasing capabilities, *Acta Biomater.*, 2009, **5**, 43–49.
117. M. Espanol, R. A. Perez, E. B. Montufar, C. Marichal, A. Sacco and M. P. Ginebra, Intrinsic porosity of calcium phosphate cements and its significance for drug delivery and tissue engineering applications, *Acta Biomater.*, 2009, **5**, 2752–2762.
118. E. Ryshkewitch, Compression Strength of Porous Sintered Alumina and Zirconia, *J. Am. Ceram. Soc.*, 1953, **36**, 65–68.
119. V. S. Tsapanos, L. P. Stathopoulou, V. S. Papathanassopoulou and V. A. Tzingounis, Effect of porosity reduction by compaction on compressive strength and microstructure of calcium phosphate cement, *J. Biomed. Mater. Res.*, 2002, **63**, 1–9.
120. T. D. Tonyan and L. J. Gibson, Structure and mechanics of cement foams, *J. Mater. Sci.*, 1992, **27**, 6371–6378.
121. Y. Fukase, E. D. Eanes, S. Takagp, L. C. Chow and W. E. Brown, Setting Reactions and Compressive Strengths of Calcium Phosphate Cements, *J. Dent. Res.*, 1990, **69**, 1852–1856.
122. K. Ishikawa, 1.17 Bioactive Ceramics: Cements, in *Compr. Biomater. II*, ed. P. Ducheyne, Elsevier, Oxford, 2017, pp. 368–391. https://doi.org/10.1016/B978-0-12-803581-8.10170-5.
123. U. Gbureck, K. Spatz and R. Thull, Improvement of Mechanical Properties of Self Setting Calcium Phosphate Bone Cements Mixed with Different Metal Oxides, *Materialwiss. Werkstofftech.*, 2003, **34**, 1036–1040.

124. H. H. K. Xu, F. C. Eichmiller and A. A. Giuseppetti, Reinforcement of a self-setting calcium phosphate cement with different fibers, *J. Biomed. Mater. Res.*, 2000, **52**, 107–114.
125. C. Canal and M. P. Ginebra, Fibre-reinforced calcium phosphate cements: A review, *J. Mech. Behav. Biomed. Mater.*, 2011, **4**, 1658–1671.
126. I. McCarthy, C. L. Camiré, P. Nevsten and L. Lidgren, The effect of crystallinity on strength development of a-TCP bone substitutes, *J. Biomed. Mater. Res., Part B*, 2006, **79**, 159–165.

CHAPTER 3

Inorganic–Organic Hybrids: Mimicking Native Bone

HYEREE PARK, EHSAN REZABEIGI AND SHOWAN N. NAZHAT*

Department of Mining and Materials Engineering, McGill University, Montreal, Canada
*Email: showan.nazhat@mcgill.ca

3.1 Introduction

Inorganic and organic materials are seen as antagonistic materials, with the former described as a 'hard' and the latter a 'soft' material. However, when combined, inorganic–organic hybrid materials can be fabricated with tunable properties, which can be tailored to the desired specifications. In the human body, bone is the representative inorganic–organic hybrid material. Bone damage, due to trauma or disease, is most commonly treated *via* the transplant of a patient's own tissue in a self-derived autograft to limit graft rejection. Nevertheless, more recently, there has been a shift towards allografts and xenografts due to the limited supply of autografts along with donor-site morbidity.[1] However, allografts and xenografts may cause unwanted immune responses and in severe cases be rejected by the patient.

Tissue engineering is an emergent field that aims to provide synthetic biomaterials that are able to restore, maintain or improve the function of tissues.[2] Current tissue engineered materials for bone repair are mostly single-phase materials, which include metals, polymers, ceramics and glasses. However, bone is a complex hybrid tissue that is mainly composed of hydroxyapatite (HA) and collagen type I, inorganic and organic materials,

Inorganic Materials Series No. 7
The Chemistry of Inorganic Biomaterials
Edited by Christopher Spicer

Published by the Royal Society of Chemistry, www.rsc.org

respectively, and is structured in a highly hierarchical manner. Therefore, single phase materials are not ideal bone tissue replacements, predominantly due to the mismatch in mechanical properties between the materials and tissue. Inorganic–organic hybrid materials may offer synergic characteristics of their individual parts, and therefore could better mimic bone in terms of both hierarchical structure and mechanical properties.

A hybrid material can be fabricated by the combination of two (or more) materials, whereby the properties of these discrete materials are superposed.[3] There are, however, discrepancies when deliberating whether composites are included within this definition. Commonly, the term hybrid is given to systems where the phases are blended at the molecular or nanoscale levels. Nevertheless, composites, sandwiches, segmented structures and foams can also superpose their individual properties;[3,4] therefore, the definition of hybrid used in this chapter will be inclusive of the broader definition. As a biomaterial, inorganic–organic hybrid materials may be of use in a variety of tissue engineering applications, including cardiovascular, musculoskeletal and drug delivery applications, due to their customisable mechanical properties.[5] Nevertheless, due to the inherent hybrid nature of bone, the main application of hybrid scaffolds is in bone tissue engineering (BTE). This chapter provides an overview of the structural hierarchy and native biomineralisation of bone and the inorganic–organic hybrids used for BTE, with a specific focus on fabricating mineralised collagen type I *in vitro*.

3.2 Bone as the Native Inorganic–Organic Hybrid Material

3.2.1 The Hierarchical Structure of Bone

The skeletal network within an adult human is made up of more than 200 bones that enable various bodily functions.[6] At the ultrastructural level, bone is a composite material that is composed of a mineral and an organic phase,[7] which is most commonly (though not exclusively) attributed to providing structural support for organs and enabling biomechanical movement within the body.[8,9] Other functions include the housing of minerals, growth factors, and cytokines, as well as energy within the bone marrow tissue to maintain homeostasis.[10] The bulk composition of bone depends on various factors, including the age, type and location of the bone.[11] In general, wet bone is approximately 65 weight% (wt%) HA ($Ca_{10}(PO_4)_6(OH)_2$) mineral, 23 wt% organic phase and 12 wt% water. The mineral and organic materials are well integrated at the nanoscale and therefore can be described as an inorganic–organic hybrid material.[12]

The HA mineral in bone is an impure calcium phosphate, which can be attributed to ion exchange substitutions that occur with ions present in bodily fluid in order to maintain ionic homeostasis.[13] The most prominent substitution is made by carbonates (CO_3^{2-}), which make up 2–8% of bone mineral,[14,15] thereby the mineral present in bone is commonly referred to as hydroxycarbonated apatite (HCA).[16] Within the organic phase, approximately 90 wt% is collagen type I with the rest made up of non-collagenous proteins and proteoglycans.[17–19] In bone, the fibrillar collagen type I protein provides support and scaffolding for HA crystals. Although proportionally less prominent, the non-collagenous proteins are equally important in bone function as they affect mineral formation, mineral nucleation, growth, cell signalling and maintain the homeostasis of ions.[20]

Bone is a highly ordered tissue that shows hierarchical organisation throughout all its levels, from the nano- to the macroscale (Figure 3.1).[21] The foundations of bone are mineralised collagen fibrils that are 80–120 nm in diameter. The collagenous portion of bone is built from, in descending hierarchical order, collagen microfibrils, collagen triple helices or tropocollagen, individual collagen chains or procollagen, and amino acids. In turn, the HCA is built from thin 1.5–4 nm platelet-shaped crystals[22,23] (Figure 3.1, II–V). The *c*-axis of the crystals is aligned with the collagen fibrillar axis.[21]

The mineralised collagen fibrils self-assemble into fibril arrays that can either have an ordered or perceived disordered motif, where the array orientations are changed in a gradational manner.[24] Lamellar bone, the principal form of bone, is composed of anisotropic and co-orientated arrays that are assembled to form parallel sheets with angular offsets of between 30° and 70° (Figure 3.1, IX).[25] Incorporated into the bone, within the spaces between the fibre bundles, are disordered materials that form alternating layers.[25] The lamellar packets are a series of lamellae that are packed into slightly different orientations and the osteons are lamellar bone formed around vascular vessels, including the Haversian canals (Figure 3.1, X).[25] These repeating units form two types of bone, *i.e.*, cortical (compact) and trabecular (cancellous) bone. Compact bone is generally found on the outer layers of most bones, forming a denser envelope, providing support and making up 80% of bone. Trabecular bone accounts for the other 20% and forms a more honeycomb structure that is highly porous and more metabolically active.[10]

The composition and organisation of bone provide it with good mechanical properties, which vary depending on the type and function. For cortical bone, the Young's modulus range if 6–45 GPa, the tensile strength is approximately 150 MPa, and the compressive strength is 250 MPa.[26] The tibia shows a fracture toughness of approximately 4 MPa $m^{1/2}$.[27,28] The hybrid nature of bone is essential in achieving these mechanical properties, in which inorganic HA reinforcement provides the bone with stiffness and strength, while the organic collagen matrix provides the tissue with toughness and ductility.[29]

At the cellular level, three types of bone-specific cells are present: osteoblasts, osteocytes and osteoclasts. Osteoblasts and osteocytes are formed by osteoprogenitor cells or mesenchymal stem cells (MSCs) that undergo a series of differentiation steps, defined by the stages of proliferation, extracellular matrix maturation and mineralisation.[30,31] These cells are heavily influenced by external stimuli.[32] Osteoblasts produce materials that make

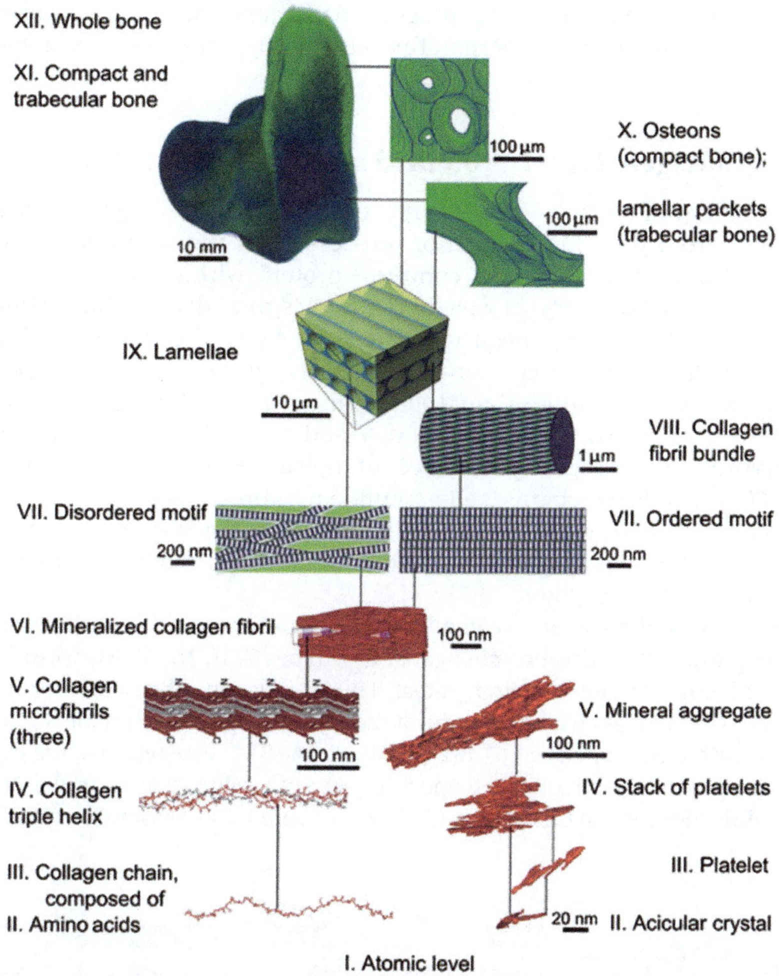

Figure 3.1 Schematic showing the hierarchical organisation of bone.[21] The basis of bone is (VI) mineralised collagen fibrils, which is made up of (II–V, left) the organic collagenous triple-helical structures, and (II–V, right) the inorganic mineral crystal to aggregates. The mineralised collagen fibrils form (IX) lamellae that make up (X) lamellar packets and osteons of (XI) compact and trabecular bones.
Reproduced from ref. 21 with permission from AAAS, Copyright 2018.

up the extracellular matrix, including collagen type I, proteoglycans, while osteocytes maintain homeostasis within bone.[33] Osteoclasts are differentiated from hematopoietic stem cells and function to resorb bone, by acidification of minerals and the enzymatic degradation of the extracellular matrix, which in turn activates osteoblasts in forming new bone.[31] This ongoing remodelling is credited to bone being a structurally supportive and load-bearing tissue,[34] which is imperative for maintaining bone health as imbalances to the remodelling process may cause various bone diseases, including osteoporosis, hyperparathyroidism, hypercalcemia and Paget's disease.[17]

3.2.2 Collagen Type I Protein: The Organic Material

The etymology of collagen comes from 'kolla', the Greek word for 'glue', and is derived from the glue fabrication process from the skin and sinews of animals. Collagen is the most common protein within the human body, making up approximately 25 wt% of the body's total dry weight.[35] Thus far, 29 types of collagen have been identified. All types of collagen are formed of triple-helical structures, the foundations of which are polypeptide chains comprised of more than 1000 amino acids with repeating Gly–X–Y tripeptides, where Gly is glycine, and X and Y are frequently proline and hydroxyproline, respectively.[36] Three of these α-polypeptide chains, also known as procollagen chains, self-assemble into procollagen, a right-handed triple-helical structure centred around glycine molecules with corresponding X and Y amino acids exposed at the surface, with pro-peptides present at the N- and C-termini (Figure 3.2).[35]

The 29 types of collagen are generally separated into either fibrillar or non-fibrillar groups. The fibrillar collagens, *e.g.*, types I, II, III, V, XI, form highly organised supramolecular aggregates. These collagen fibrils are formed of staggered tropocollagens, which exhibit their repeating characteristic d-banding pattern, with a periodicity of 67 nm (Figure 3.2A, B).[40] Pro-peptides are cleaved from the procollagen to form tropocollagen with telopeptides at the N- and C- termini (Figure 3.2A). The tropocollagen is able to self-assemble into fibrils

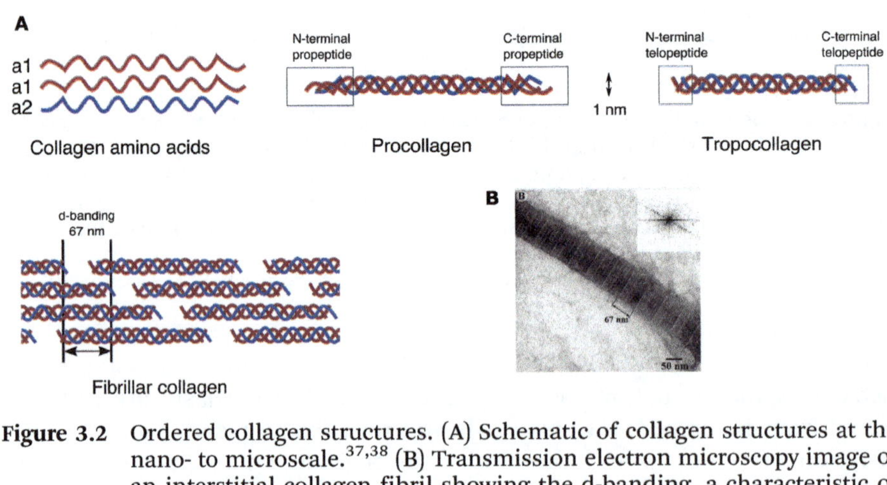

Figure 3.2 Ordered collagen structures. (A) Schematic of collagen structures at the nano- to microscale.[37,38] (B) Transmission electron microscopy image of an interstitial collagen fibril showing the d-banding, a characteristic of fibrillar collagen.[39]
Reproduced from ref. 39 with permission from American Chemical Society, Copyright 2008.

that form building blocks of tissues. Collagen type I is a heterotrimeric collagen, composed of two $\alpha1$ and one $\alpha2$ chains, whereas collagen types II and III are formed from three identical α-chains.[36] Fibrillar collagens are further structured in a hierarchical manner into fibres. In contrast, non-fibrillar collagens, *e.g.*, types IV, IX, XII *etc.*, do not form fibrils and generally act as interactive links between the fibrillar collagens and other molecules present in the extracellular matrix, such as glycosaminoglycans, proteoglycans, elastin and other cell-interactive proteins, *e.g.*, fibronectin and laminin.[41] Amongst the collagens, fibrillar collagens are the most common, aiding the structural and mechanical integrity of various tissues in the native body, and among these, collagen type I is the most common.[42]

3.2.3 Hydroxyapatite Mineral: The Inorganic Material

The mineral phase within bone is HA ($Ca_{10}(PO_4)_6(OH)_2$), which is an impure calcium phosphate that is caused by ionic exchange substitutions with those present within bodily fluid in order to maintain ionic homeostasis.[13] Therefore, the Ca : P molar ratio that defines the pure stoichiometric HA of 1.67 is generally not found within bone due to these naturally occurring impurities. Approximately 20% of Ca^{2+} ions are substituted with other ions present within the body,[14,43] commonly with carbonates, thus forming HCA. The mineral can be found within the ordered lamellar collagen fibres as platelet shaped crystals, in which their *c*-axis is aligned with the long axes of the fibrils.[24] HCA is also observed in the disordered material, between and within the collagen fibrils.[25] The Young's modulus and compressive strength of HCA are 112 GPa and 800 MPa, respectively, but this material is also brittle, with a fracture toughness of 1 MPa $m^{1/2}$.

3.2.4 Development of Collagen Mineralisation in Native Tissues

The formation of HCA occurs through the nucleation and growth of the minerals within collagen fibres. There are two main types of bone formation *in vivo*, endochondral and intrafibrillar ossification. Endochondral ossification occurs during foetal development, where cartilage tissues undergo mineralisation. Due to the looseness and the size of the collagen fibrils (10–20 nm diameter) at this developmental stage,[44,45] HCA clusters or calcification nodules form within the proteoglycan matrix, a process that is generally understood as extrafibrillar mineralisation.[45,46] Osteoclasts resorb the mineralised cartilage and this is replaced by the formation of bone by osteoblasts.[47]

Intrafibrillar or intramembranous ossification occurs within mature bone through osteoblast activity. Collagenous and non-collagenous proteins are produced,[48] and calcium phosphate ions present within the body gather within matrix vesicles that are secreted by osteoblasts in an amorphous state.[21] The main precursor to HCA is amorphous calcium phosphate $Ca_3(PO_4)_2 \cdot nH_2O$,[14] which forms at the nucleation sites at the collagen terminal. Non-collagenous proteins, many of which are acidic and highly charged, play an important role in the mineralisation process.[49,50] Among the non-collagenous proteins, small integrin-binding ligand N-glycosylation (SIBLING) proteins, including osteopontin and bone sialoprotein (BSP), can bind to collagens and promote the interaction and accumulation of calcium ions.[50,51] Furthermore, amorphous calcium phosphates may interact with osteopontin and stabilise into a net negatively charged compound to interact with positively charged sites within collagen fibrils.[52] It has also been proposed that liquid-phase amorphous mineral precursors may be drawn towards the collagen fibrils through capillary action, forming HCA crystals when crystallised.[53] Mineral nucleation typically infiltrates the 40 nm gaps that are created by the staggering of the ordered collagen fibrils on the *c*-axis parallel to the directionality of the collagen fibrils (Figure 3.3).[54] This embedding within the collagen thermodynamically stabilises the crystals. Interfibrillar mineralisation can also occur, where HCA crystals form on the surface of the collagen.[55] It must be noted, however, as with any biological mechanism, that there are numerous other processes and biological agents that are involved within the process of *in vivo* collagen mineralisation.

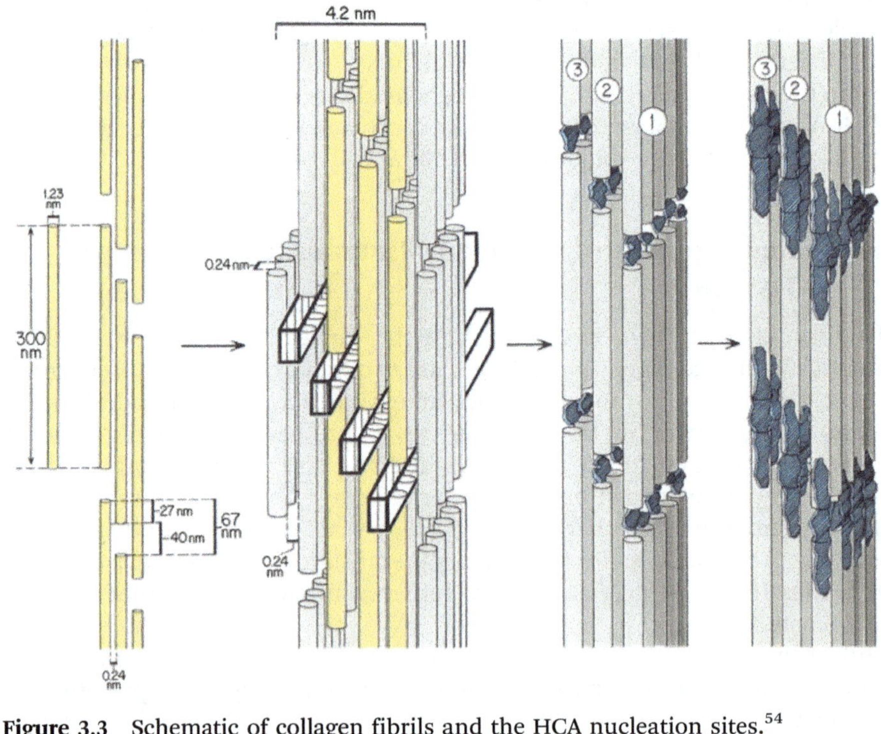

Figure 3.3 Schematic of collagen fibrils and the HCA nucleation sites.[54] Reproduced from ref. 54 with permission from Elsevier, Copyright 1993.

3.2.5 Bone Tissue Engineering (BTE)

Bone is a living and vascularised tissue and therefore able to heal itself. However, if the defects are critical-sized, bone tissues will not be able to heal without supportive aid.[56] Although commonly used, natural bone grafts, including autografts or allografts, are not always readily available or are not ideal alternatives due to potential pathogen transfer leading to rejection by the host.[9,57,58]

BTE and regenerative medicine are emerging as alternative treatment options to natural bone grafts, whereby an engineered tissue is implanted to function similarly to healthy native tissue. In BTE, a biomaterial scaffold is initially created to act as an artificial extracellular matrix, primarily as a template that supports seeded or host cells. With the aid of growth factors and controlled environments, cells are directed to differentiate into mature cells of the desired tissue, produce matrix proteins and in turn remodel the scaffold matrix. Scaffold materials are diverse, consisting of either natural, synthetic materials or of their combinations. The most important criteria for determining material suitability is biocompatibility, or the ability of a material to promote an appropriate host response for a specific application,[59] as well as being non-toxic when implanted into the host.

Moreover, the scaffold should ideally mimic the native microstructure, but the complexity of bone and its hierarchical structures have proven difficult to replicate.[8,9,60,61] For optimal cell adhesion and bone ingrowth, high porosity is desired. In addition, the pores in an ideal BTE scaffold must be interconnected and consist of both meso- and macropores ($>$100 μm).[57,62–64] The mechanical properties of such scaffolds must be comparable to those of the defective bone, which should be maintained throughout the healing process.[57,65] Also, the scaffold should exhibit a degradation rate that is compatible with the rate at which new bone is generated.

3.2.6 Hybrid Materials for BTE

The classification of an inorganic–organic hybrid can be determined by the surface interactions between the inorganic and organic phases within the hybrid material. Specifically, class I hybrids include materials where the inorganic and organic phases interact *via* van der Waals interactions, hydrogen bonding, or electrostatic forces, whereas class II hybrids exhibit stronger covalent or ionocovalent bonding between the inorganic and organic components.[66] Therefore, bone is an inorganic–organic class I hybrid consisting of HCA nanoplatelets and collagen type I fibrils bound by hydrogen bonds and electrostatic forces.[67] Due to its composition and hierarchical nature, it forms a complex structure that is difficult to mimic *in vitro*. It has been widely accepted that hybrid materials have great potential in creating ideal BTE scaffolds. However, in order to successfully design an appropriate hybrid material, the properties of its precursors must be well understood.

3.3 Inorganic and Organic Materials for Bone Tissue Engineering

3.3.1 Inorganic Materials

Calcium phosphate ceramics are widely used inorganic materials for dental and orthopaedic treatments, which can be tailored in composition to mimic that of the HCA mineral in native tissue.[68] Synthetic HA has been shown to be osteoconductive, promoting bone growth when implanted *in vivo*,[69] but its level of bone healing is limited due to high levels of crystallisation and low degradation rates.[70] Tricalcium phosphate (TCP, $Ca_3(PO_4)_2$) degrades more rapidly than HA *in vivo,* and promotes a faster onset of bone healing. In general, calcium phosphate ceramics are influenced by their porosity, chemical composition and crystalline structure.[71] However, calcium phosphate ceramics have poor toughness properties and are therefore unsuitable for high load-bearing applications. Moreover, these ceramics do not have the capacity to promote the differentiation of MSCs into osteogenic lineages.

Bioactive glasses were introduced in the late 1960s, as bone graft materials that would not be rejected by the body.[72] To this end, the formulation of the first bioactive glass was created with the components of native bone in mind, consisting of a silicate network modified with phosphate, calcium and sodium. The main attribute of bioactive glasses is their ability to form a surface HCA layer when in contact with physiological fluids both *in vitro* and *in vivo.* This layer allows for bonding between the newly-formed HCA and native bone,[58,72,73] specifically with the collagen fibrils at the interface.[72] There have been several formulations of bioactive glasses introduced and the level of bioactivity can be classified into either class A, osteoinductive, or B, osteostimulative.[69] Osteoinductive materials exhibit rapid bonding to both soft and hard tissues and promote osteogenesis in non-bone cells, whereas osteostimulative materials bind less rapidly and only to bone itself, promoting bone healing from osteogenic cells.[58,72,74–76] The classes can be predetermined through the design of appropriate compositions of glass, which dictate the levels of ionic dissolution. Through ionic release, bioactive glasses also provide an environment that promotes cell proliferation and osteogenesis leading to enhanced bone formation.[8,74,76] They have also been linked to increased angiogenesis, which prevents cellular necrosis within the scaffold and more importantly the failure of the scaffold.[57,76]

Currently, there are two silicate-based formulations used in clinical applications as alternative bone grafts; Bioglass® (45S5, composition $(45)SiO_2$–$(24.5)CaO$–$(24.5)Na_2O$–$(6)P_2O_5$, in wt%) and Bonalive® (S53P4, composition $(53)SiO_2$–$(23)CaO$–$(20)Na_2O$–$(4)P_2O_5$ in wt%). Glasses are fabricated from network formers, *i.e.*, silicates, phosphates or borates, which are covalently bonded, along with glass modifiers, *i.e.*, calcium and sodium, forming ionic bonds to the main glass structure *via* oxygen bridges.[72,74] These network modifiers promote rapid ionic dissolution, due to the facilitated breakdown of the glass network.[69,73] Although the most common bioactive glasses are silicate-based, phosphate- and borate-based bioactive glasses have recently gathered much interest. Specifically, sol–gel borate-based glasses have been observed to undergo more rapid conversion of HCA, over 25 times as quickly, than their corresponding silicate-based glasses, which has been attributed to their lower chemical durability.[63,77–80] The properties of bioactive glasses depend not only on their composition, but also on microstructure and functionality, which are all affected by the method of synthesis.[69,72,73,81,82]

3.3.2 Organic Materials

Organic polymers have been widely used in tissue engineering applications due to their tunable properties. Polymers can be classified into either natural or synthetic origin, and further into degradable or non-degradable materials. Biodegradable polymers degrade through either hydrolysis or enzymatic routes by chain scission.[58,83] Poly(α-hydroxyacids), including poly(lactide) (PLA), poly(glycolide) (PGA) and poly(lactide-*co*-glycolide) (PLGA) co-polymers, and polycaprolactone (PCL) and their co-polymers are the most widely used synthetic biodegradable polymers, with clinical approval for uses such as sutures, wound dressings and stents. However, due to their low stiffness and strength,[84] most synthetic polymers are not appropriate as standalone bone replacement implants and are most often used as screws and pins in fracture fixation or as surgical sutures.[85] Furthermore, since the process of polyester degradation occurs by the hydrolysis of ester bonds within the polymer backbone chain, this may result in their premature failure. Moreover, these degrade into their monomeric acids, which may cause localised inflammatory responses.[86,87]

Unlike synthetic polymers, naturally derived polymers often exhibit low toxicity levels and may promote cell adhesion. Collagen type I, as the most prominent protein within connective tissues, can be used to form biocompatible and biodegradable materials that promote cell attachment, with low antigenicity.[88–90] Other natural polymers include polysaccharides, such as alginate, chitosan, hyaluronic acid, bacteria-derived cellulose, dextran and other starch-based materials. These natural polymers can be processed into various forms as either porous solids or maintained in a hydrated state as hydrogels, which increases their inherent biocompatibility. Therefore, these materials have been widely investigated for tissue engineering due to their compatibility with cells, but also ease of manipulation.[62] Nevertheless, the application of natural polymers has been limited in BTE mainly due to their low mechanical strength and chemical stability.[62,91,92] To overcome this, the cross-linking of polymers can enhance their mechanical strength and structural integrity. Chemical cross-linkers, such as aldehydes and condensation reactions, and physical cross-linkers, relying on ionic interactions and crystallisation methods, are commonly used to crosslink polymers.[66,93]

3.3.3 The Inorganic–Organic Interface

The creation of inorganic–organic hybrid scaffolds can result in novel complexes with tailored properties that can more closely mimic the inorganic–organic composition of natural bone.[57,94] When used independently, both inorganic and organic materials show less than ideal properties required for BTE. Inorganic materials, although widely used in the clinic, are not mechanically compatible to fully mimic native bone. Synthetic HA and bioactive glasses show good compressive strength, but cannot be used within load-bearing sites due to poor tensile strength and toughness.[91] Furthermore, polymers show insufficient mechanical properties and lack bioactivity as they are generally difficult to mineralise.[9,57,64,91] Therefore, inorganic–organic hybrids have been proposed to combat the misalignment between scaffold materials and native tissues.

However, it is important to consider the interface between the two hybridised phases. Some hybrid materials report a reduction in mechanical properties, which is generally attributed to the surface incompatibility between the two phases.[57,91] These issues arise when there is a lack of interfacial bonding between the precursor materials, where this incompatibility prevents the homogeneous dispersion of inorganic particles throughout the polymeric matrix.[91] A more uniform particle distribution can be achieved by decreasing the particle size of the inorganic phase.[92,95]

3.3.4 Chemical Surface Modification

Inorganic particles are susceptible to agglomeration through hydrogen bonding *via* their surface OH groups (Figure 3.4), where the inorganic micron-sized agglomerates may act as stress concentrators and destabilise the hybrid inorganic–organic interface.[73,95,97,98] Chemical surface modification can prevent particle agglomeration and promote interfacial adhesion to create a homogeneous hybrid system, which can improve the mechanical properties of the material.[64,91]

Therefore, to overcome incompatible inorganic–organic interfaces that may cause weakened mechanical properties, the surface of the inorganic phase can undergo surface modification, commonly through appropriate coupling agents that provide covalent bonding areas, to promote homogenous mixing

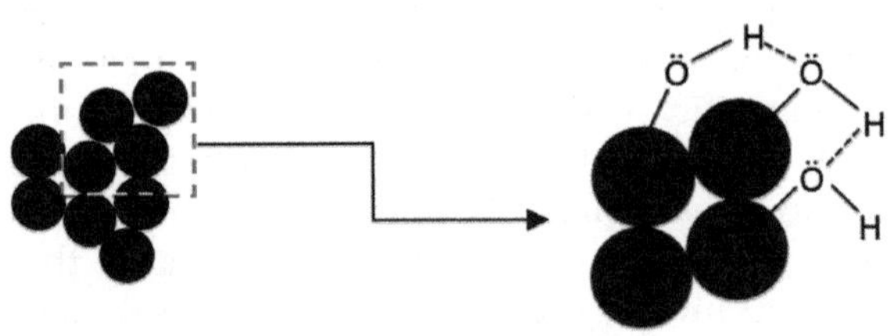

Figure 3.4 Schematic of the agglomeration between inorganic (glass) particles *via* hydrogen bonding.[58,96]
Adapted from ref. 96 with permission from American Chemical Society, Copyright 2008.

between the two phases and improve the mechanical and bioactive properties of the hybrid.[57,64,99] Mechanical methods of hybrid stabilisation, such as stirring or ultrasonication, often have short-term effects in maintaining the separation of agglomerates, whereas the surface modification of inorganic particles using coupling agents provides a long-term solution.[58,64,73,95–97]

Silane coupling agents have been commonly used to modify the surfaces of silicate-based bioactive glasses, which reduces particle agglomeration and provides higher stability within a polymer solution by providing a more hydrophobic inorganic phase (Figure 3.5). The molecular structures of silane coupling agents allows the formation of covalent bonds to both inorganic and organic phases, acting similarly to a cross-linking agent. The chemical composition of a typical silane coupling is $R\text{–}(CH_2)_n\text{–}SiX_3$, where R, $(CH_2)_n$, and X represent an organofunctional, alkyl linker, and hydrolysable groups, respectively.[58,96] The polar end of this molecule is compatible with the hydrolysable groups in the inorganic phase and the non-polar end, including the organofunctional group, is compatible with the organic phase. The length of the alkyl linker group can significantly affect the reactivity and physical properties of the silane coupling agent.[58,96,97,100] Some of the silane agents that have been used to surface modify bioactive glass particles to stabilise them within organic phases are presented in Table 3.1.

The bond between the silane agents and the surface OH groups of the inorganic phase can be formed *via* an anhydrous or a hydrolytic route.[96,101] In the former method, the coupling agent binds directly to the OH groups without water or catalyst through a lengthy process that requires relatively high temperatures. The latter hydrolytic approach is more common, where hydrolysed silane agents form secondary bonds with the inorganic surface OH groups that convert to covalent bonds during a reflux step at ~70–100 °C. To control the rate of the reactions, this approach may include an acidic or basic catalyst.[96,101,102]

The hydrolytic surface modification process is sensitive to the amount of water and silane, the pH of the system and the characteristics of inorganic materials, including their morphology, specific surface area and surface chemistry.[95–97,101–103] All of these parameters have an impact on the reaction processes and should be carefully determined. The water level within the

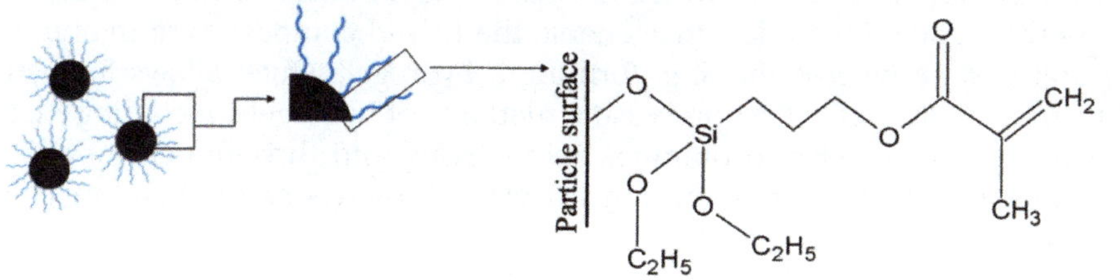

Figure 3.5 Schematic of coupling agent molecules grafted onto inorganic particles to prevent agglomeration. The silane coupling agent used in this figure is 3-methacryloxypropyltriethoxysilane (Table 3.1).[58,96]
Adapted from ref. 96 with permission from American Chemical Society, Copyright 2008.

Table 3.1 Examples of silane coupling agents that have been commonly used for surface modification purposes to improve the interfacial adhesion in inorganic–organic systems.[58,96,99]

	Chemical formula		
Chemical name	*R*	*n*	*X*
Methacryloxymethyltriethoxysilane	$CH_2{=}C(CH_3)COO$	1	OC_2H_5
3-Methacryloxypropyltrimethoxysilane	$CH_2{=}C(CH_3)COO$	3	OCH_3
3-Methacryloxypropyltriethoxysilane (MPTES)	$CH_2{=}C(CH_3)COO$	3	OC_2H_5
Vinyltrimethoxysilane	$CH_2{=}CH$	0	OCH_3
3-Aminopropyltriethoxysilane (APTES)	H_2N	3	OC_2H_5
3-Glycidoxypropyltrimethoxysilane	$CH_2(O)CHCH_2O$	3	OCH_3
Methyltriethoxysilane	CH_3	0	OC_2H_5
3-Acryloxypropyltrimethoxysilane	$CH_2{=}CHCOO$	3	OCH_3
Dimethyldichlorosilane	$(CH_3)_2$	0	Cl
3-Isocyanatopropyltriethoxysilane (ICPTES)	OCN	3	OC_2H_5
3-Glycidyloxypropyltrimethoxysilane (GPTMS)	$CH_2(O)CHCH_2O$	3	OCH_3

system impacts the hydrolysis of the silane agents, where excess water may result in rapid hydrolysis and cause undesirable self-polymerisation or multi-layer deposition that limits interfacial adhesion.[104,105]

For many polymers, the functionalisation of the inorganic phase is essential to form a class II hybrid. After binding to the organic phase, the polar end of the coupling agent is made available as a binding site with the inorganic phase.[57,58,61]

3.4 Methods of Hybridising Inorganic–Organic Hybrids

3.4.1 Composites

Several methods can be utilised in order to fabricate an inorganic–organic hybrid material, which can be optimised for specific applications. Generally, polymers that are melt-mixed or solvent-solubilised can be blended or mixed with inorganic particles within the polymer, which may require cooling, cross-linking or dehydration. To stabilise or further enhance the hybrid bonding, inorganic materials can be surface modified to create a class I or II hybrid material.

However, some materials may enable the hybrid composites to maintain a hydrated state by cross-linking, forming a hydrogel, which allows for better integration with cells. Post-processing methods of hydrogels, including mechanical processing, electrospinning, freeze-drying and three-dimensional (3D) printing, can be used to produce hydrogels with more structural and mechanical integrity.

3.4.2 Hybrids

To prepare materials with strong covalent bonding, the sol–gel method is an effective method of producing inorganic–organic hybrid scaffolds with unique micro- and nanostructures.[57,99,106] In this method, the inorganic phase is formed *in situ* and is exposed to an aqueous-soluble or functionalised organic polymer to form a hybrid.[57,96]

The sol–gel process for the synthesis of the inorganic phase involves the hydrolysis of appropriate precursors to form a sol, which then undergoes condensation and gelation. Depending on the composition of the system and the presence, or lack thereof, of a catalyst, the gelation process may take from a few minutes up to several days. In some cases, a hydrolysis and/or gelation catalyst is incorporated into the systems to control the hydrolysis and gelation rates, respectively, *via* pH control of the environment.[58,100,107,108] Eventually, the aged gel is dried and stabilised in order to generate the final hybrid material.[57] The sol–gel process is a sensitive technique, where the properties of the end material, including particle size, specific surface area and surface functionality, can be affected by the precursor compositions and the processing conditions.[58,99,109]

Metal alkoxides ($M(OR)_n$) are the most common sol–gel precursors, where M is a metal centre of valency n and R is an alkyl or aryl group.[110] Hydrolysis of the alkoxy group attached to the metal centre creates a hydroxy–metal compound, and water or alcohol (X = H or R) is removed by condensation. In some cases, hydroxo-bridges are formed.[111] The three equations below show the general progression of the sol–gel method with metal alkoxides.

$$\text{Hydrolysis: } \text{M–OR} + H_2O \rightarrow \text{M–OH} + \text{ROH}$$

$$\text{Condensation: } \text{M–OH} + \text{M–OX} \rightarrow \text{M–O–M} + \text{X–OH}$$

$$\text{Hydroxo-bridge formation: } \text{M–OH} + \text{M–OX} \rightarrow \text{M–(OH)–M–OX}$$

Under acidic conditions, a higher concentration of X–OH monomers is expected within the sol due to the higher rates of hydrolysis. The monomers are gradually polymerised *via* a 'cluster-cluster growth' mechanism to form a 3D structure, where the inorganic polymeric chains have undergone gelation. However, if the hydrolysis catalyst is alkaline, polymerisation is governed by a 'monomer–cluster growth' mechanism, resulting in the formation of a colloidal gel.[58,112,113]

To form class I inorganic–organic hybrids using the sol–gel method, the organic phase can be added after the hydrolysis of the alkoxides.[61,114] Therefore, the inorganic precursors should be carefully selected in order to remove the need for high-temperature stabilisation and residue removal processes.[73,114] In this hybridisation process, the hydroxyl groups of the inorganic component bind to amide, carboxyl or carbonyl groups present in the organic phase (Figure 3.6A of polyvinylpyrrolidone (PVP)–silica bonding).[61] Another approach of forming a class I inorganic–organic hybrid materials is through the simultaneous gelation of both components (Figure 3.6B).[61,96]

Class II inorganic–organic hybrids can also be formed using the sol–gel method by modifying the alkoxyl precursors with the organic material,

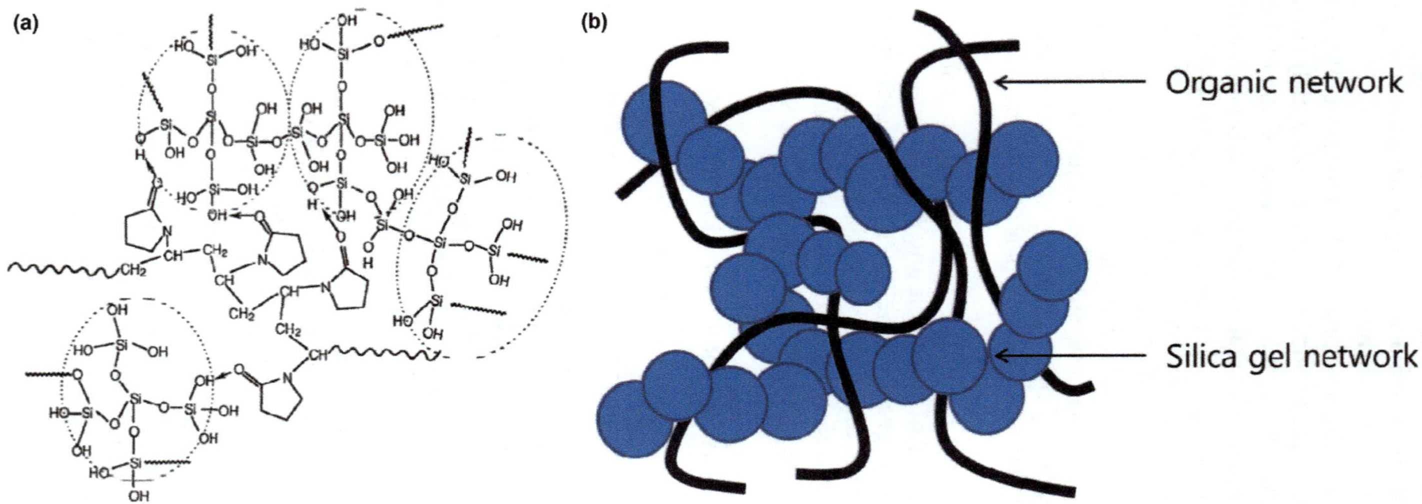

Figure 3.6 Schematic approaches for producing class I hybrids: (a) PVP-silica hybrid system and (b) interpenetrating a silica gel and an organic network.[61]
Reproduced from ref. 61, https://doi.org/10.1016/j.pmatsci.2015.12.001, under the terms of the CC BY 4.0 license, https://creativecommons.org/licenses/by/4.0/.

followed by hydrolysis and condensation.[61] When the organic phase is added at this stage, chemical surface modifications in the form of coupling agents may be necessary to provide a stable hybrid material. Coupling agents can enable covalent bonding between the inorganic and organic phases to produce a strongly bonded hybrid material by modifying the inorganic surface or the chemical bonds within the organic polymer, or by initiating a co-polymerisation process to produce a functionalised polymer.[96]

3.5 Mimicking Native Bone

There are numerous materials that can and have been used as hybrid materials for BTE using various incorporation techniques. Nevertheless, the most biomimetic approach is to provide a collagenous organic phase with a HA inorganic phase. Therefore, in this section, HA and collagen type I and its mineralised hybrids will be discussed. Bioactive glasses have been included within this scope due to their remarkable ability to form HCA *in situ*. It must be recognised, however, that the hierarchical structure of bone found within native tissue is extremely difficult to replicate and that none of the methods reported produce an ideal biomimetic scaffold for BTE.

3.5.1 Synthetic Hydroxyapatite

HA is the most biomimetic inorganic material to be used within a scaffold for BTE. The mineral can be extracted from bovine bone,[115] which is exposed to sodium hypochlorite, sodium hydroxide, and hydrogen peroxide, and sintered at 1100 °C. The particle size of the extracted HA can be controlled through sieving. However, this method is time-consuming and may invoke immunogenic responses, due to its source and potential xeno-contamination. Synthetic HA can also be precipitated *in vitro* through different reactions using several inorganic precursors, as described below:

i) Calcium nitrate ($Ca(NO_3)_2$) and ammonium dihydrogen phosphate ($(NH_4)H_2PO_4$) can be reacted together to form synthetic HA. To keep the solution at pH 11–12, ammonia is added to the reaction. The HA can then be extracted by heating for 10 minutes at 80 °C then incubated for 24 h at 37 °C, or alternatively, filtered and washed and dried at 140 °C, and then sintered at 1000 °C.[116,117]

$$5Ca(NO_3)_2 \cdot H_2O + 3(NH_4)H_2PO_4 \rightarrow Ca_5(PO_4)_3OH + 3NH_4NO_3 + 7NHO_3$$

ii) Calcium acetate ($Ca(CH_3COO)_2 \cdot H_2O$) and $(NH_4)H_2PO_4$ can be heated to 60–80 °C at neutral pH to obtain HA. The mineral is extracted by filtration and subsequently dried.[118]

$$5Ca(CH_3COO)_2 \cdot H_2O + 3(NH_4)H_2PO_4 \rightarrow$$
$$Ca_5(PO_4)_3OH + 3NH_4(CH_3O_2) + 7CH_3CO_2H$$

iii) Calcium ions in calcium chloride ($CaCl_2 \cdot H_2O$) and phosphate ions in potassium dihydrogen orthohosphate (KH_2PO_4) can be used as counterions to form HA.[119] This reaction reduces the risk of impurities compared to the previous reactions, due to the lack of ammonia and calcium nitrate.[120]

$$5CaCl_2 \cdot 2H_2O + 3KH_2PO_4 \rightarrow Ca_5(PO_4)_3OH + 3KCl + 7HCl + H_2O$$

The properties of both the tissue extracted and *in vitro* precipitated HA are determined by various properties of the mineral, including particle size, morphology, chemical composition, crystallinity and sintering temperatures.

3.5.2 Bioactive Glass: Hydroxyapatite Precursor

Bioactive glasses are a class of inorganic materials that are able to form a HA-like surface layer when exposed to physiological fluids either *in vivo* or *in vitro*.[121] Within this context, a bioactive material is one 'that elicits a specific biological response at the interface of the material, which results in the formation of a bond between the tissues and material'.[121] The HA layers that form on bioactive glasses are able to bind with both hard and soft tissues *in vivo*.[122] Furthermore, the inherent homogenous degradation and ionic release of bioactive glasses may stimulate natural healing process within native bone,[72] where the amorphous nature of bioactive glasses has been linked to more rapid biomineralisation than their crystalline counterparts.[123,124]

Ionic release *in vivo* has been, depending on glass composition and classification, observed to have either an osteoinductive or osteostimulative effect.[125–127] Bioactive glasses can be based on either silicate, phosphate or borate, and have been studied extensively for BTE applications.[72] Currently, only silicate-based 45S5 Bioglass® and S53P4 Bonalive® are clinically approved for bone grafts. The bone regeneration levels induced by these glasses are high, but being a ceramic, the materials are brittle and show comparably poor mechanical properties to bone.[128]

3.5.2.1 Silicate-based Bioactive Glass

The most commonly used and available types of glasses, amongst both bioactive and non-bioactive, are silicate-based glasses that form amorphous glass from SiO_4 tetrahedra.[129] For silicate-based bioactive glasses, silica is the glass network former while calcium oxide and sodium oxide are network modifiers, which are added to facilitate processing. The calcium (Ca^{2+}) and sodium (Na^+) ions bind to the silicon oxide ($Si_xO_y^-$), which can modify the glass properties, such as lowering the glass melting temperature from 1700 °C (pure silica) to 1450 °C (45S5).[128–130] The high amounts of calcium and sodium oxides also promote faster ion exchange. Phosphorus pentoxide decreases the chemical durability of the glass, which increases the HCA conversion rates and bonding to bone.[131]

The mechanism of HCA formation on the surface of silicate-based glasses has been widely investigated. The following stages have been proposed following immersion within physiological fluids:[121,128,130]

1. Ion exchange of the Na^+ and Ca^{2+} with H^+ or H_3O^+ in solution (body fluid or simulated body fluid), creating Si–OH bonds on the glass surface and thus forming a silica-rich area. The pH of the solution is increased, and the phosphate component of the glass is released.
2. The increase in pH leads to the break-up of the silica glass network *via* the breaking of Si–O–Si bonds with OH^- ions. Soluble $Si(OH)_4$ is lost in solution, thereby increasing the presence of Si–OH on the glass surface.
3. The Si–OH on the glass surface is condensed into a silica-rich layer.
4. Ca^{2+} and $PO_4{}^{3-}$ migrate *via* the silica-rich layer to the surface forming an amorphous $CaO–P_2O_5$.
5. Hydroxyl and carbonate ions in the solution and $CaO–P_2O_5$ undergo crystallisation to form HCA.

A disadvantage of silicate bioactive glasses may be found within their relatively slow degradation rates, which is determined by their composition.[129] Also, considering that the ratio of calcium to phosphate is an important factor during crystalline apatite formation,[130] the composition of bioactive glass has to be carefully controlled.

3.5.2.2 Phosphate-based Glasses

The network former for phosphate-based glasses is phosphorus pentoxide (P_4O_{10}), with phosphate tetrahedral units. However, phosphate-based glasses have different structures than silicate-based glasses due to the differences in the valences of silicon (4+) and phosphorus (5+). Furthermore, the bone regenerating potential of phosphate-based bioactive glasses does not involve the promotion of apatite formation, but acts instead by influencing cells to promote bone growth through calcium ion release.[10,132]

The dissolution mechanism of phosphate-based glasses has been described as a hydration and hydrolysis reaction, where Na^+ ions from the glass and H^+ within body fluid form a hydrated layer on the glass surface, and the water and H^+ ions present break down the P–O–P bonds and the glass network.[79,133] The rate of breakdown of phosphate-based glasses can be predetermined by their composition,[79,134] where the pH can also control the hydration stage of dissolution.[133] Moreover, phosphate-based glasses can be easily doped with various ions to tailor the material properties.[135,136] The dissolution of ions can provide desirable properties, for example, copper, silver and zinc ions are antibacterial, whereas strontium can increase bioactivity, prevent tooth decay and increase bone mass and density.[136]

3.5.2.3 Borate-based Glasses

There have been recent developments in borate-based glasses for tissue engineering applications.[137,138] The network forming agent is boron oxide (B_2O_3), which forms a combination of trigonal and tetrahedral structures when alkali metals are incorporated as glass modifiers.[139,140] This combination of both structures decreases the stability of the glasses, especially when compared to silicate-based glasses.[139,140] Interestingly, the partial or full substitution of silicate with borate has been shown to accelerate the degradation rates of glasses.[141,142] The dissolution and apatite forming characteristics of borate glasses are similar to those of silicate-based glasses, with the exception that borate bioactive glasses release BO_3^{3-} ions and thus do not create a silica-rich layer.[141] The mechanism for apatite formation has been indicated to occur after the dissolution of sodium and calcium ions, where the calcium reacts immediately with PO_4^{3-} ions leading to the formation of HCA.

3.5.2.4 Fabrication Methods of Bioactive Glass

The traditional melt-quenching process is the most common and well-documented technique for the production of any type of glass.[143] In this method, the oxide precursors are heated to above 1000 °C. Once a low viscosity system is reached, it is rapidly quenched to inhibit glass crystallisation and to maintain its amorphous structure. This is commonly achieved by pouring the melt onto a cool steel plate or into a graphite mold.[144] The glass can be heat treated by annealing, at a temperature lower than the transition temperature, thereby removing residual stresses in order to create a more durable glass that degrades more slowly.[131] It can then be cast into a monolithic shape, or ground to particles of a desired size.[144] Only a limited number of bioactive glass compositions can be produced *via* this method due to the high melting temperatures and occasionally the lack of bioactivity in the final glass. Furthermore, the melting process can be costly as it requires high temperature furnaces and expensive platinum crucibles. With the melt-derived method, the direct synthesis of submicron particles, which are essential for the production of hybrid systems, is not possible.

Ultrasonic spray pyrolysis (USP) is a more recently identified method for fabricating bioactive glasses and glass–ceramic nanoparticles. This method results in spherical particles (Figure 3.7), where aqueous solutions of precursors, commonly salts, are sprayed and atomised in an ultrasonic chamber. The atomised particles are dried, decomposed and melted through a hot reaction column followed by rapid cooling.[58,80] The resulting nanoparticles have a low tendency to agglomerate, due to the limited OH groups present on the surface. Another similar method, flame spray pyrolysis (FSP), can be used to directly synthesise bioactive glass nanoparticles (1–200 nm). In this process, precursors are transferred into a flame using a solvent fuel, where the precursors melt and form droplets that are rapidly cooled to form nanoparticles that are gathered by a filter. The FSP-derived nanoparticles

Figure 3.7 SEM image of typical USP-derived bioactive glass particles (synthesis temperature of 1400 °C).[80]
Reproduced from ref. 80 with permission from Elsevier, Copyright 2009.

maintain their amorphous structure due to the rapid quenching process, potentially promoting their bioactivity. As with the USP-derived particles, the FSP-derived particles also exhibit a low tendency to agglomerate due to the minimal amount of surface OH groups.[58,145,146] This aspect, however, can become undesirable when aiming to improve inorganic–organic interfacial adhesion *via* chemical modification of the surface. Both USP and FSP synthesis methods are expensive and complex processes as specialised equipment is required.

A sol–gel is created when a sol, a colloidal suspension of particles, undergoes hydrolysation and condensation into a gel.[147] The sol–gel method can also be used to synthesise bioactive glasses using a relatively low-temperature approach (Figure 3.8). As previously described, alkoxide precursors are mixed at room temperature with either an acidic or basic catalyst, to enhance hydrolysis and condensation processes.[148] The gel is aged in order to strengthen the gel network, where it forms more bonds, resulting in a slight shrinkage of the gel.[149,150] Then, to remove the excess organic material still present, the gel is dried either at room temperature to form a xerogel, or by critical point drying to form an aerogel that is highly porous with low density. The final stage of the sol–gel method involves a calcination process, typically in the range of 400–600 °C, to become a glass.[78,128] The sol–gel method produces nanoporous glasses, which have higher surface area than melt-derived glasses, significantly increasing their dissolution rates and bioactivity.[78,147]

Using the sol–gel method, a 100% SiO_2 bioactive glass was created and its bioactivity was confirmed,[8,147,151] which was previously unfeasible using the melt-derived method. Furthermore, sol–gel derived bioactive glasses form HCA more rapidly relative to melt-quench derived compositions.[152,153]

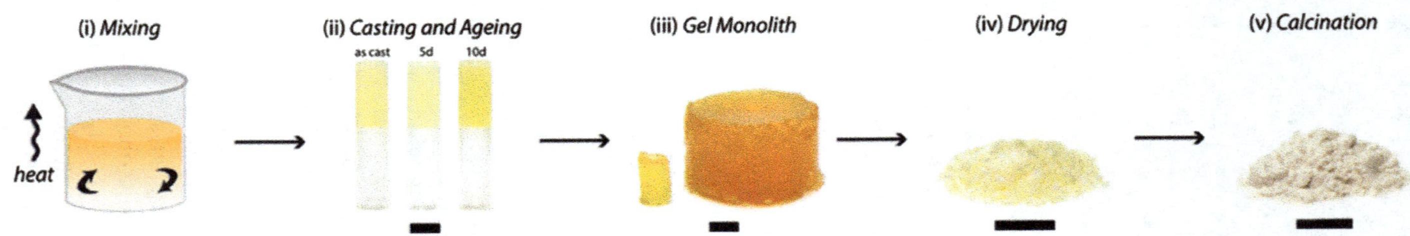

Figure 3.8 Schematic showing a typical sol–gel fabrication method of bioactive glass.[78] Reproduced from ref. 78 with permission from American Chemical Society, Copyright 2015.

This is attributable to the their inherently high specific areas and the concentration of surface OH groups, favouring surface HCA formation, which indicates higher bioactivity.[69,106,154] Indeed, HCA formation in sol–gel derived borate-based glasses occurs after only 3 h of immersion in simulated body fluid (SBF), a fluid formulated to mimic physiological fluids,[122] leading to a ~25 fold increase in mineralisation compared with melt–quench derived equivalents.[78]

3.5.3 Collagen as a Biomaterial

Collagen type I is the most prominent protein within bone and has been recognised as a suitable biomaterial for use in BTE.[38,90,155] Collagens are extensively used in clinically approved medical devices, including bioprosthetic heart valves, arteriovenous shunts, suture threads, haemostatic membranes, wound dressings, tissue augmentations, ophthalmic barriers, and drug delivery vehicles.[90] Collagen type I is a preferred material for use in tissue engineering as it provides an excellent environment for cells and is a versatile material.

The origin of the collagens used within these products is generally mammalian, where the source animal greatly influences the properties of the collagen, especially in the cellular responses induced.[156] Marine sources have been investigated, and collagen has been extracted from sponges, fish and jellyfish, but these proteins have different amino acid compositions and thermal sensitivity than mammalian collagens.[156,157] Human collagen is available, but at high costs and low supply. Therefore, there have been advances in producing recombinant human collagens that are produced by bacteria.[158,159] The benefits of this type of collagen include their reproducibility and reduced immunogenicity compared to animal collagens.[159] However, recombinant collagens suffer from high cost and low yield, and lack enzymes that are critical to forming mammalian-like stable structured collagens.[160]

One of the most important factors to consider when using biological materials for medical applications is the immunogenicity and antigenicity caused by the material.[1] Non-autologous collagens may evoke antigenesis by the host antibodies if the collagens have not been polymerised and have available antigenic amino acid sequences.[161] However, if the collagens are implanted post-fibrillogenesis, these sequences are masked within the procollagen.[162,163] Moreover, the host reactions can be predetermined by a simple diagnostic test.[164]

3.5.3.1 Extraction of Collagen

Collagenous biomaterials can be obtained through decellularisation of the extracellular matrix and solubilisation of collagenous tissues.[165,166] Decellularisation processes require purification and removal of cellular components, including DNA and RNA, to minimise immunogenic responses. Rigorous processing occurs in order to remove cell debris, but this does not negate the necessity of immunosuppressants.[167,168] It also compromises the structural and mechanical properties of the collagen, which significantly influences the cellular responses and its functionality both *in vitro* and *in vivo*.[169] This decellularised structure can be strengthened by chemical cross-linking, which increases the durability of the collagen matrix.

Collagen type I can also be extracted through the solubilisation of whole tissues.[166] Three methods of solubilisation are available: acidic,[165,166] neutral salt[170] or proteolytic solutions.[171,172] Telopeptides, which enable the self-assembly of the tropocollagens, can be maintained in acid-solubilised collagen, but these are generally cleaved or denatured during the latter two processes. Furthermore, acid-solubilisation yields the most material[171] and has generally become the extraction method of choice, with several protocols available.[173]

In vivo, there are several factors that aid the assembly of collagen fibrils to regulate the alignment and cross-linking between the collagen chains.[174] The cross-linking process influences the mechanical properties of collagen and is induced through enzyme catalysis[175] However, *in vitro* fibrillogenesis is induced by neutralising the acidified tropocollagen in a physiological environment, *i.e.*, pH of 7.2–7.4 and temperature of 34–37 °C. The self-assembly of the tropocollagen is initiated through the bonding of peptide chains by hydrogen bonds and van der Waals forces.[176,177] The tropocollagen polymerisation of the acid-solubilised collagens results in highly hydrated collagen hydrogels that hold over 99% water,[178] with the final collagen concentration being below 5 $mg\,mL^{-1}$ or approximately 0.5 wt% collagen fibrillar density. Furthermore, the level of structural representation, as seen in bone and other native tissues, occurs only at the nanoscale where d-banding is exhibited, but not at the microscale, where fibres are randomly orientated.[179,180] Self-assembly does not align the collagen fibres and the mechanical properties exhibited by these highly hydrated collagen gels are poor, with tensile strengths of 4–6 kPa and elastic moduli of 20–30 kPa,[181] and are thus incapable of withstanding the appropriate mechanical loads needed for tissue engineering, particularly for load bearing tissues such as bone.[182] Moreover, cell-mediated remodelling occurs when these gels are seeded with osteoblasts and fibroblasts, which limits cell proliferation.[183] In order to improve the mechanical properties of highly hydrated collagen gels, several methods have been proposed.[184] Cross-linking methods are available, through chemical, physical and enzymatic means,[185] but the *in vivo* effects of these processes are not fully understood. Furthermore, many of these processes involve chemicals that are known to be toxic towards cells.[186,187]

3.5.4 *In Vitro* Mineralisation of Collagen

Reconstituted acid-solubilised collagen hydrogels do not replicate the mineralisation properties seen *in vivo*,[188–190] and therefore several strategies of mineralising collagen *in vitro* have been proposed. While fibrillar collagen networks act as templates for HCA nucleation in native bone, the *in vitro* reproduction of intrafibrillar mineralisation of collagen scaffolds is difficult to achieve. To this end, *in vivo* nucleation pathways have inspired the development of numerous mineralisation solutions and the introduction of functional groups within the collagen fibrils in order to increase the nucleating potentials of the collagen.

3.5.4.1 Supersaturation Method

Direct mineralisation of collagen utilises calcium and phosphate solutions, using similar precursors to synthetic HA, as described above. Calcium and phosphate ions are able to diffuse into the fibrils of a solid collagen film, which enable the deposition of HA on the collagen surface (Figure 3.9).

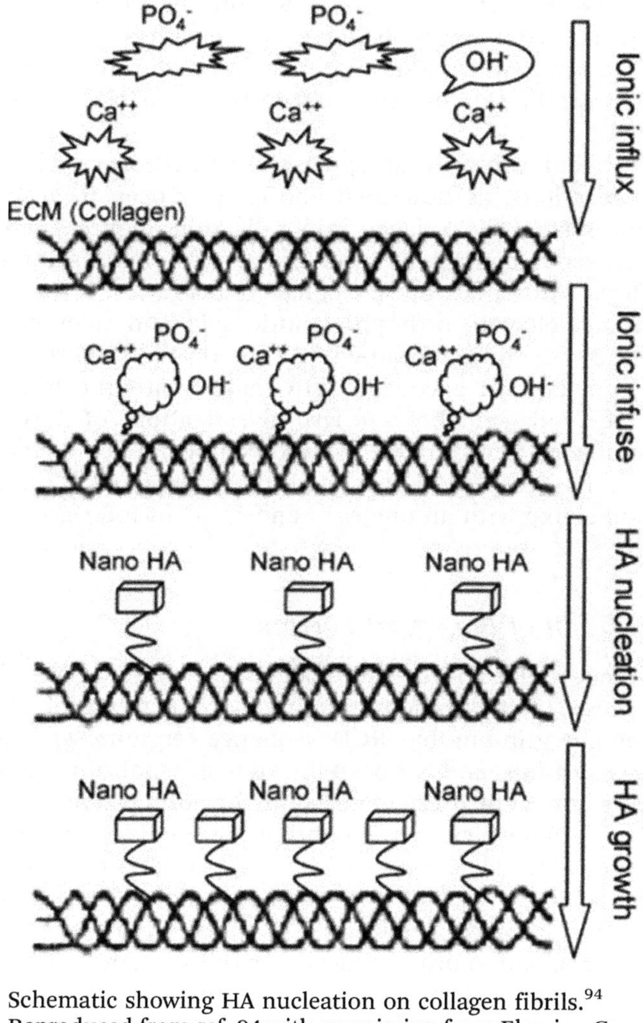

Figure 3.9 Schematic showing HA nucleation on collagen fibrils.[94]
Reproduced from ref. 94 with permission from Elsevier, Copyright 2005.

This method, however, does not significantly increase the mechanical properties of the collagen.[119,191] Several modifications of this soaking approach have been investigated, including submersion in brushite (dicalcium phosphate dihydrate) and the introduction of biological factors, *e.g.*, phosphoserine, a naturally occurring phosphorylated amino acid, to improve nucleation. This method increases the mechanical properties of the scaffold, resulting in strength and Young's moduli values of 45 MPa and 1.5 GPa, respectively.[192] Phosphoserine has been credited to be a bridging component between the organic collagen and inorganic calcium phosphate that can increase bonding strengths as well as provide mineral nucleation sites.[67]

Ultimately, mineralisation of the collagen surface *via* the supersaturation method occurs by the diffusion of calcium and phosphate ions and interactions with the collagen within the solution. Calcium phosphate precipitation and crystallisation occur within the collagen membrane. However, calcium ions diffuse faster than phosphate ions, which may interfere with the mineralisation process,[193] although this discrepancy can be solved by controlling the relative concentrations of the two ions within the system.

In this vein, calcium chloride and potassium dihydrogen orthophosphate precipitation reactions, as described above, have been used in an automated titration system for collagen mineralisation (Figure 3.10A). The solution pH was used to control the input of the two solutions and to achieve a hybrid material with a 60 : 40 wt% collagen–HA ratio.[193] Similarly, aqueous solutions of collagen, orthophosphoric acid and calcium hydroxide were mixed at 37 °C and maintained at a pH of 9[194] (Figure 3.10B). This titration method created a collagen–HA matrix with the mineral being aligned with the collagen fibres. *In vivo* investigations of these materials within subcutaneous mice models showed composite debris encapsulated within fibrotic tissues.[195] However, within critical-sized tibia defects in beagles, bone healing with an unclear bone–implant interface at 12 weeks was observed.[196]

3.5.4.2 Addition of Biological Factors

Factors that are known to influence mineral nucleation in native bone have also been suggested to aid mineralisation *in vitro*. For example, BSP, which consists of an integrin-binding RGD sequence (arginine–glycine–aspartic acid) near its C-terminus and has been linked to mineralisation *in vivo*,[197,198] is regarded as a critical medium in cell signalling and promoting osteoblast differentiation.[197] Nevertheless, the bonding mechanism between BSP protein and collagen type I is not fully understood, yet it has been observed that the collagen is required to be in its triple helical state.[199] Recombinant BSP has been observed to bind to collagen regardless of the cleaving of the telopeptide terminals, and to promote acellular HA formation by 10-fold when compared to BSP-agarose gels in buffer containing $Ca(NO_3)_2$ and Na_2HPO_4, indicating that the mineralisation of collagen may be supported by the

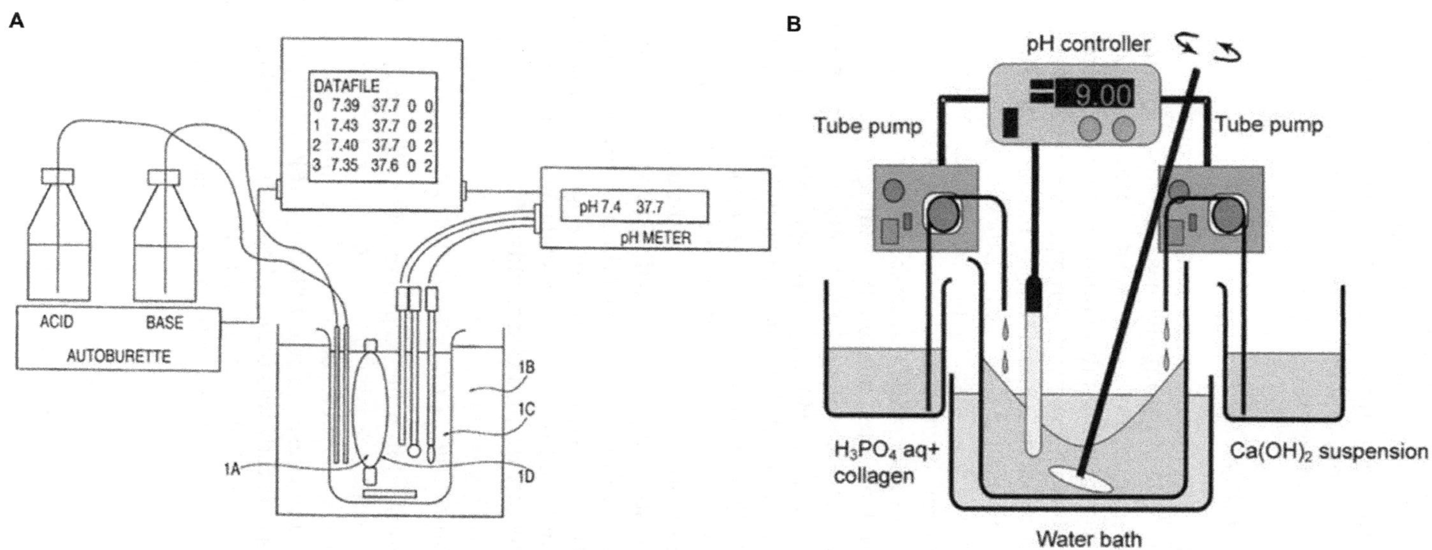

Figure 3.10 Schematic of automated titration–precipitation systems optimising collagen mineralisation by controlling the pH (A) calcium chloride and potassium dihydrogen orthophosphate precipitation[193] (credit: Jan Czernuszka, Alison Lawson and Alasdair Simpson) and (B) orthophosphoric acid and calcium hydroxide titration.[194]
Reproduced from ref. 194 with permission from Elsevier, Copyright 2004.

presence of BSP.[199] Moreover, alkaline phosphatase (ALP) is an enzyme that has been observed to induce mineralisation *in vivo*. Within a collagen/calcium phosphate sheet, ALP increased HA nucleation when exposed to $CaCl_2$ and Na_2HPO_4. The mineralised sheets promoted osteogenic differentiation of MC3T3-E1 murine pre-osteoblasts.[200]

3.5.4.3 Simulated Body Fluid

SBF is a solution that has almost equal ionic concentrations as those of human blood plasma.[122] It is commonly used as an *in vitro* alternative to observe bioactivity, or apatite formation *in vitro*. However, there has been increasing interest in using SBF as a mineralising agent of different scaffolds, including collagen. Acid-solubilised collagen type I scaffolds have been found to form crystalline HA through submersion in 1×SBF with exposure to calcium silicate glass,[201] and also in 1.5×SBF.[202] Another study investigated the mineralisation of collagen and octacalcium phosphate, a HA precursor, in SBF[203] and while mineralisation was observed, HA was not identified. To increase the nucleation of HA on SBF immersed collagens, negatively charged groups were introduced on the collagen by adding different acids, which acted similarly to anionic non-collagenous proteins and provided areas of negative charge with nuclei for HA formation. Acids such as citric acid and poly-L-aspartic acid have both shown nucleating abilities.[39,204–206]

Acid-solubilised collagen type I was also co-precipitated *in situ* with modified SBF for 24 h at 40 °C, followed by freeze-drying to form a porous structure and subsequently cross-linked to strengthen the collagenous matrix.[207] The scaffold showed good interactions with bone marrow derived MSCs and was implanted into mouse calvaria post cell-seeding. Bone was observed to heal after 3 weeks, and the results were comparable to those obtained following implantation of the clinically available bone graft alternative Healos® (DePuy Synthes, USA) scaffolds.[207]

3.5.4.4 Sol–Gel Bioactive Glass Precipitation

Sol–gel processing was used in order to precipitate a silicate-based bioactive glass directly onto a collagen type I disc (Asspress Cirurgica, Brazil).[208] Precursors of bioactive glass were dissolved in solution and the collagen was submerged for 5 minutes, to form $(80)SiO_2$–$(16)CaO$–$(4)P_2O_5$ (in wt%) bioactive glass. An *in vivo* study was undertaken with this collagen–bioactive glass scaffold, where it was implanted into a murine subcutaneous site. After 14 days *in vivo*, superior angiogenesis and mineralisation was observed when compared to collagen alone.[208]

3.5.5 Fabricating Composites

The hybridisation of collagen with either HA or bioactive glasses has also been utilised for BTE. In order to form these composites, the ceramics are mixed during the reconstitution stage of collagen. All of the following methods involve an additional step to an already existing method of collagen scaffold fabrication.

3.5.5.1 Electrospinning

Electrospinning is a widely applied method for obtaining thin polymer fibres by manipulating an electric field created by charge imbalance.[209] To achieve this, an electrical potential is applied to the starting polymer solution or melt, which forms an electrically charged jet by overcoming the surface tension at the critical voltage. Electrospinning collagen can enable fibre alignment and anisotropy[210] and can also be used to form collagen–nano-HA matrices.[211–213] The weight or volume ratios of HA to collagen can be adjusted as well as the porosity within the collagen network.[213] Osteoblast and MSC attachment and proliferation have been observed, indicating the potential of these scaffolds for BTE.[211,214] Electrospinning generally involves a cytotoxic solvent step, commonly 1,1,3,3,3-hexafluoro-2-propanol. Although the solvent is evaporated off during the electrospinning process, its use limits cell-seeding processes.

3.5.5.2 Freeze Drying

The freeze-drying method enables the production of collagen scaffolds, dictating pore structure, size and fibre alignment. An ice crystal network is formed within the collagenous matrix, which is removed by sublimation, creating a porous scaffold.[215] The porosity and pore size of collagen scaffolds have been shown to be directly related to a decrease in freezing temperatures. Due to the ease of manipulation and absence of harsh solvents, this method has been studied extensively as a means to incorporate HA and bioactive glasses within collagenous networks. Freeze-dried collagen hybrid scaffolds typically require cross-linking in order to maintain their structure.[216,217] These scaffolds have been shown to adhere to and proliferate cells, including pre-osteoblasts and MSCs, and to heal bone defects *in vivo* in a rabbit radius defect model.[216]

Freeze-dried collagen–bioactive glass ($(60)SiO_2$–$(36)CaO$–$(4)P_2O_5$ in mol%) scaffolds have also been shown to improve MSC adhesion, compared to unhybridised bioactive glasses, leading to the formation of a homogeneous cell network by day 21 in culture.[218] It was found that modifying the polymer phase with either phosphatidylserine or hyaluronic acid increased the functionality for BTE.[217] *In vivo* healing was also reported to be increased using the modified scaffolds. In another study, a mesoporous silicate-bioactive glass with the composition $(85)SiO_2$–$(15)CaO$ (in mol%) and collagen type I were cast in a plastic mould, where the collagen underwent fibrillogenesis.[219] The highly hydrated collagen and bioactive glass hydrogel was freeze-dried at -60 °C to form porous hybrid scaffolds. The scaffolds after freeze-drying showed poor bone healing in a rat calvaria model, but scaffolds that were submerged in SBF for 7 and 14 days prior to implantation were observed to heal the defect better than the as-made scaffold.

3.5.5.3 Additive Manufacturing

Additive manufacturing, or 3D printing, is a robust tool for fabricating scaffolds.[220] It has garnered interest due to its cost-effectiveness, but also its capability of producing scaffolds that are geometrically complex.[221] Collagen and HA scaffolds have been bioprinted using a 3D bioplotter (Envision Tec, Germany) to form scaffolds of different porosities.[222,223] The collagen–HA scaffolds were stiffer than collagen-only scaffolds and showed increased proliferation and osteogenic differentiation of human bone marrow MSCs *in vitro*.[222] A bioink has also been developed using collagen type I and strontium-containing mesoporous glass particles. Although these scaffolds showed promising rheological and extrusion properties for bioprinting,[224] the actual printing of a collagen–bioactive glass scaffold has yet to be demonstrated. Nevertheless, the use of additive manufacturing in tissue engineering is rapidly increasing due to the increasing availability of hardware and biomaterial inks, and the precise control over the spatial deposition of biomaterials and biological materials.[221]

3.5.5.4 Dense Collagen

Two methods of mechanically stabilising reconstituted, highly hydrated collagen gels from acid-solubilised collagen have been proposed: plastic compression (PC) and gel aspiration–ejection (GAE),[225–228] as illustrated in Figure 3.11A. Both utilise rapid, non-invasive and mechanically-driven approaches in order to remove the casting fluid and increase the collagen fibrillar density.[225–228] PC utilises a load that is placed on top of the self-assembled highly hydrated collagen gel and removes the fluid *via* a load-expulsion approach, forming a thin membrane-like sheet of collagen ranging from 30 to 150 μm in thickness, referred to as dense collagen (DC).[229] This collagenous sheet maintains the random orientation of the collagen fibrils and has frequently been rolled during a spiral assembly stage to achieve a more stable three-dimensional material (Figure 3.12A). It has since been investigated for BTE both after spiral assembly and as a sheet.[188,229–231] The collagenous sheets were observed to show good bone healing in non-load bearing critical-sized calvaria defects with the delivery of dental pulp stem cells.[188,189,232,233]

GAE uses a blunt-tip needle and an angioplasty inflation device to provide a pressure differential that expels the fluid during aspiration.[226] This method also enables the reorientation and alignment of collagen fibres, which is mainly dictated by the combination of the needle gauge and the surface area of the gel precursor used (Figure 3.11B). The use of a needle enables the possibility to inject the scaffold into a desired location

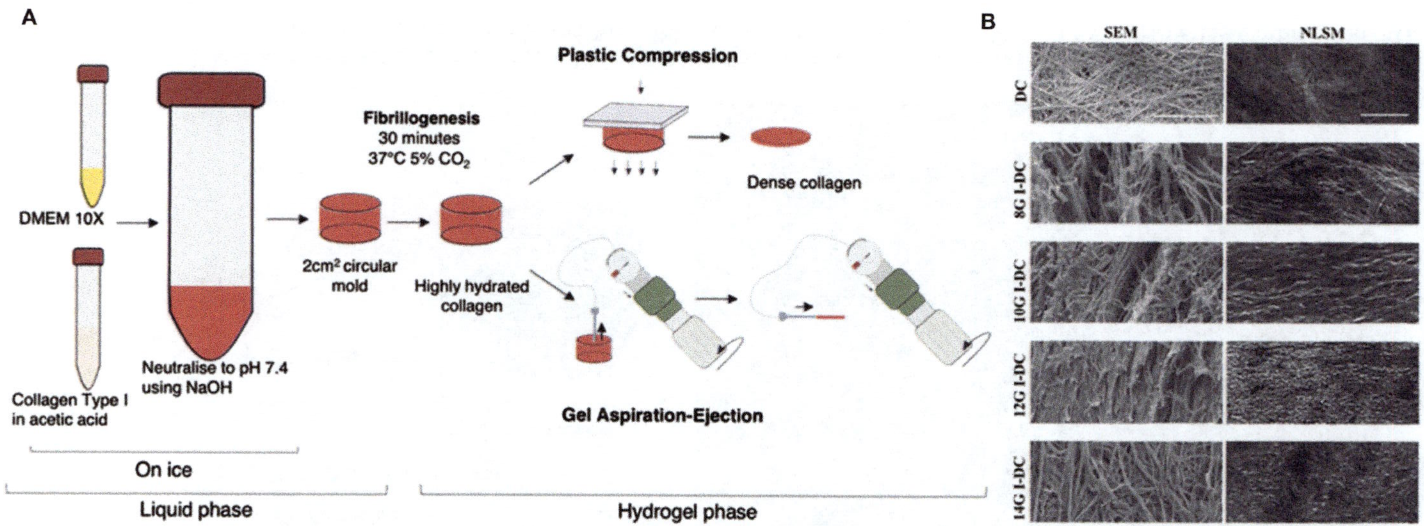

Figure 3.11 (A) Schematic representation of PC and GAE techniques. (B) SEM and non-linear scanning microscopy (NLSM) of DC (PC) and IDC (GAE) fabricated using 8–14 G needles. Scale bars = 10 and 25 μm for SEM and NLSM, respectively.[226] Reproduced from ref. 226 with permission from IOP Publishing, Copyright 2016.

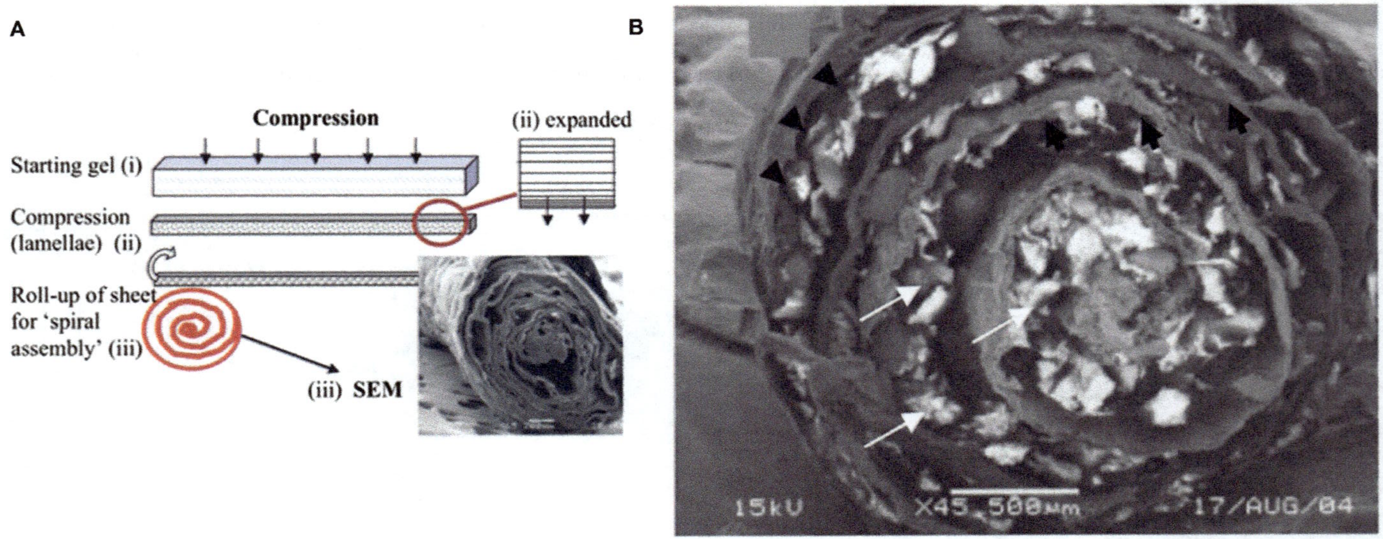

Figure 3.12 (A) Schematic of the PC and spiral assembly of DC. (B) SEM image of spiral assembled DC–HA.[225] Reproduced from ref. 225 with permission from John Wiley & Sons, Copyright 2005 WILEY-VCH Verlag GmbH & Co. KGaA, Weinheim.

within a living tissue, therefore it has frequently been termed injectable DC (IDC).[226,227] The alignment of collagen fibres is beneficial for BTE as it resembles the ordered structures found within the hierarchical architecture of native bone (as shown earlier in Figure 3.1, VII–VIII). Recently, a method of automating the GAE method has been proposed that enables the precise use of multiple syringe pumps and needles.[228] This automation enables the rapid fabrication of highly defined, mechanically stable, mini-tissue building blocks that do not require cross-linking agents.[228] The method has yet to be utilised for hybrid systems, but the mechanism of densification is identical to that used by GAE. Therefore, the GAE-fabricated scaffolds can easily be scaled up using the automated-GAE system.

Both PC and GAE methods provide a rapid and reproducible fabrication method with collagen fibrillar densities up to 17 wt%,[226,234] thus mimicking the mechanical, biological and microstructural properties of the native osteoid.[235–237] The compressive moduli of PC–DC hydrogels are between 20 and 40 kPa, similar to that of an osteoid.[238]

The first demonstration of incorporating ceramics within DC gels was achieved by sprinkling HA along the length of the construct before the compression stage. Post-compression, the collagen–HA membrane was spirally assembled, and the HA could be seen between the collagen layers (Figure 3.12B). This method did not utilise the key benefit of both PC and GAE of being able to incorporate particles and/or cells within the collagen network, by their introduction as part of the processing route within the neutralised solution stage, prior to self-assembly. Bioactive glass particles have also been incorporated within collagenous solutions prior to collagen fibrillogenesis and processing by PC or GAE, resulting in their homogenous distribution within the dense collagenous network, with no effects on the collagen self-assembly.[239]

Melt-derived phosphate glass fibres (PGFs) of 30–40 μm in thickness[240] have also been incorporated into dense collagen gels by first placing the PGF and neutralised collagen solution within a mould used for collagen self-assembly, and then plastically compressing and rolling the compressed sheets (Figure 3.13). The addition of up to 30 wt% PGF increased the Young's modulus of the sheets from 1.3 (0 wt% PGF) to 160 MPa (30 wt% PGF), significantly changing the mechanical profile of the material. Cell viability was confirmed to be over 80% for all conditions, indicating good compatibility with human fibroblasts.[240] During the degradation process, microchannels were created within the construct, which could aid the potential formation of blood vessels.

Several silicate–bioactive glass formulations have been investigated within PC–DC hydrogels. Bioactive glass with the formulation (70)SiO_2–(25)CaO–(5)P_2O_5 mol% was hybridised with collagen in a weight ratio of 30 : 70, which could be mineralised in SBF and promoted osteogenic differentiation of human dental pulp stem cells.[241] Mesoporous glass (85)SiO_2–(15)CaO mol% bioactive glass in PC–DC hydrogels was also shown to improve mechanical

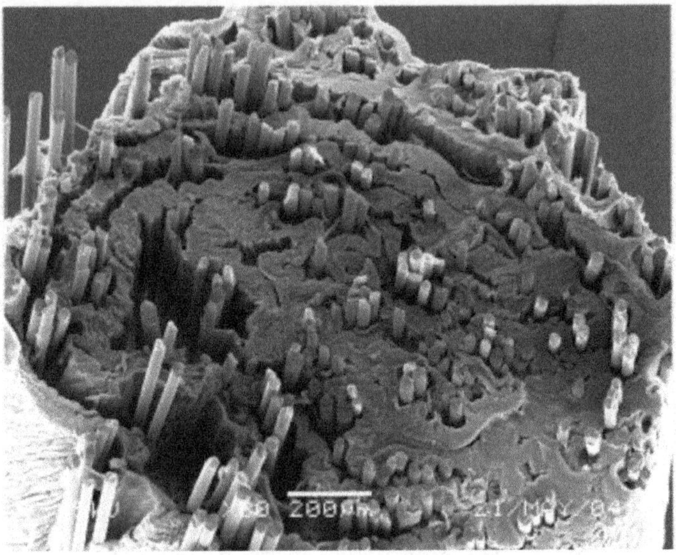

Figure 3.13 SEM image of phosphate glass fibres within a spiral assembled DC matrix.[240]
Reproduced from ref. 240 with permission from American Chemical Society, Copyright 2007.

properties compared to unhybridised collagen hydrogels, and to promote osteogenic differentiation when seeded with rat MSCs.[242] Further studies of hybrid PC–DC hydrogels included micro- and nanosized 45S5 (μ45S5 and n45S5, respectively).[239,243] Both DC-45S5 scaffolds formed HCA upon immersion in SBF, which could be observed after 14 days for the μ45S5 hybridised scaffolds, or 3 days for those with n45S5, with the latter showing 95% collagen mineralisation by day 7 (Figure 3.14A and B).[239,243] In both DC-45S5 scaffolds, the mechanical properties were improved upon exposure to SBF, confirming the increase in mineralisation (Figure 3.14C).[239,243] These studies indicate the potential of silicate-based bioactive glasses within PC–DC hydrogels for use in BTE.

The hybridisation of bioactive glasses and DC fibres is translatable to the GAE method, whereby GAE–DC–45S5 hybrid scaffolds were fabricated.[244] Similarly to the PC–DC–45S5 hydrogels, *in vitro* acellular mineralisation through immersion in SBF was confirmed by HCA formation between days 7 and 14. The efficacy of this acellular hybrid scaffold for bone formation *in vivo* was confirmed in a subcutaneous rat model, where ectopic bone was observed at day 21 (Figure 3.15). Cell infiltration, mineralisation, and angiogenesis, as well as osteoblastic and osteoclastic cell activities, were observed, thereby suggesting the osteoinductive nature of the GAE–DC–45S5 scaffolds and its potential for use in BTE.[244]

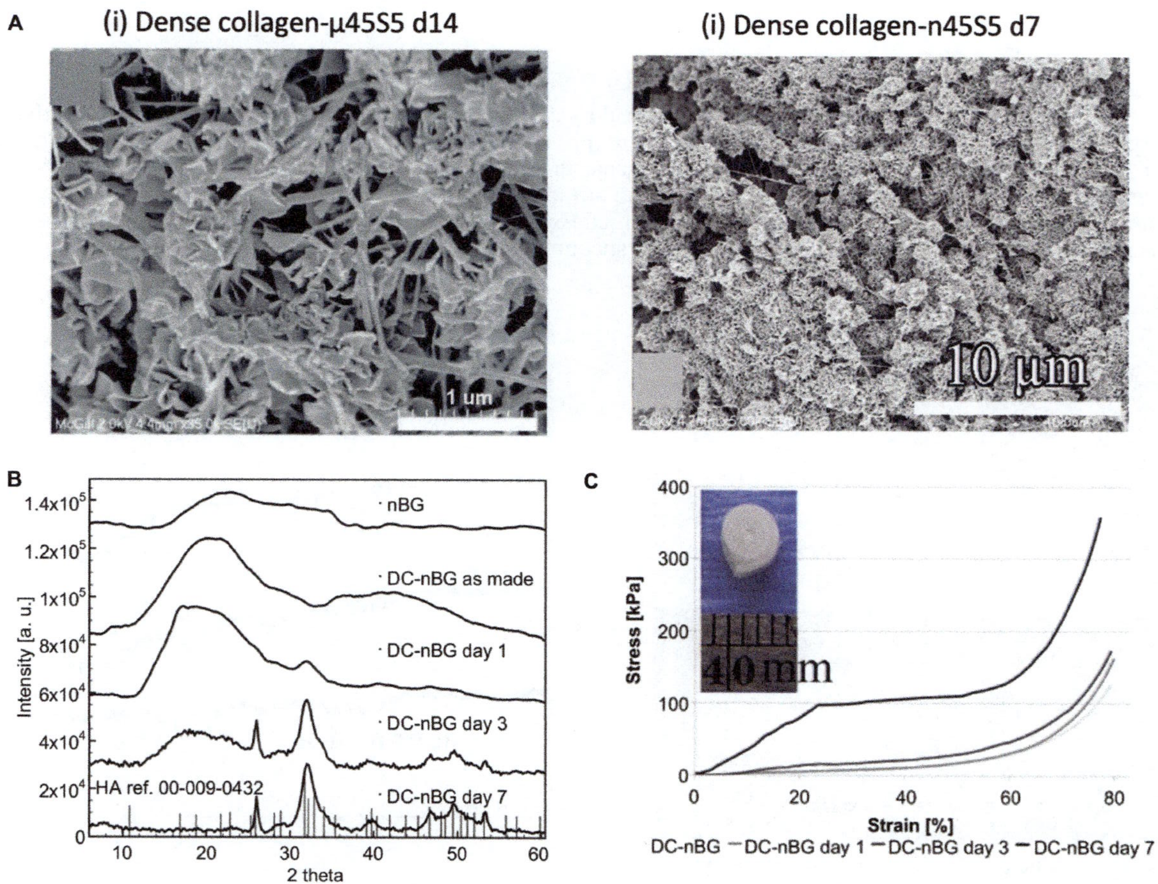

Figure 3.14 Mineralisation of DC-45S5[239,243] (A) SEM micrographs of (i) DC-µ45S5 at day 14 and (ii) DC-n45S5 at day 7. (B) X-ray diffractograms of DC-n45S5 at different immersion stages in SBF with the reference HA diffractogram, showing its increased mineralisation. (C) Representative compression curves for DC-n45S5 over time in SBF.
Adapted from ref. 239 with permission from American Chemical Society, Copyright 2010; and from ref. 243 with permission from Elsevier, Copyright 2011.

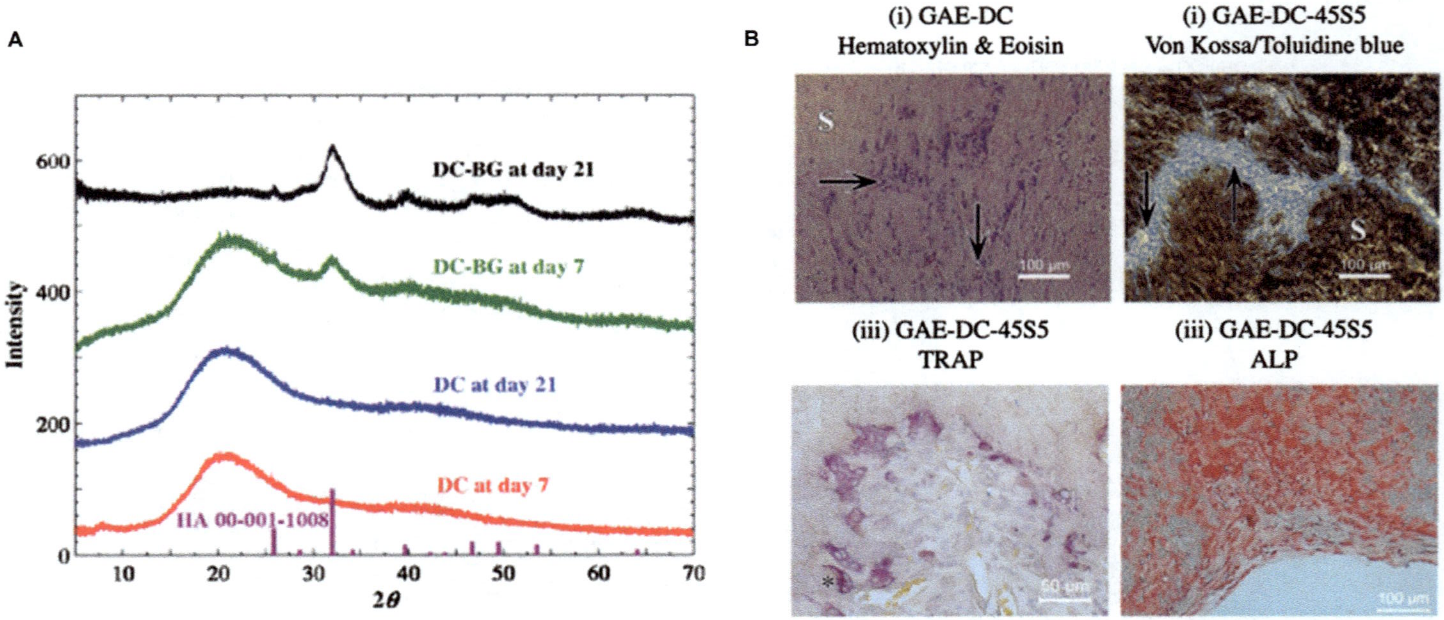

Figure 3.15 GAE–DC and 45S5 hybrid scaffolds *in vivo* in a subcutaneous rat model.[244] (A) X-ray diffractograms of DC and DC–45S5 within the subcutaneous model over 21 days showing hydroxyapatite like peaks for GAE–DC–45S5 at day 21 (B) Histological analysis at day 21, (i) Hematoxylin and eosin of DC only, with the arrows showing neovascularisation, (ii) Von Kossa and Toluidine blue of DC–45S5 scaffolds, showing mineralisation in black and the arrows show neovascularisation, (iii) TRAP markings within DC–45S5, showing osteoclast activity and (iv) ALP markings within DC–45S5 showing osteoblast activity. Adapted from ref. 244 with permission from Elsevier, Copyright 2016.

3.6 Conclusions

Bone can be classed as an inorganic–organic class II material composed of predominantly collagen type I and HCA. Although various inorganic and organic materials exist for BTE, based on the composition of native bone there is scope for further biomimicry. In reproducing the bulk native bone composition, mineralised collagen type I scaffolds may provide appropriate hybrid materials with suitable properties for use in BTE applications. As demonstrated, numerous methods of fabricating collagen–HA materials *in vitro* exist, where some methods have looked to native *in vivo* bone mineralisation for inspiration, and some exploit more direct hybridisation or composite methods. Although these scaffolds mimic the composition of bone and can stimulate healing *in vivo*, they do not reflect the extensive hierarchical organisation of bone, which ultimately limits their mechanical properties and applications in load-bearing sites.

The current challenge for BTE is, therefore, the development of a collagenous–mineral hybrid material that is formed in a highly structured manner. Within the collagen densification approach, GAE can be used to fabricate dense collagen and hybrids, including bioactive glass, with aligned fibres that display a more organised anisotropic structure. Furthermore, this fabrication process has recently been automated and improved to enable high throughput of aligned dense collagens, where their properties, including alignment and density, can be pre-determined. These recent advances may provide a system for fabricating a more bone-like material, in terms of composition and structure, for use in BTE.

References

1. *Biomaterials Science: An Introduction to Materials*, ed. B. D. Ratner, A. S. Hoffman, F. J. Schoen and J. E. Lemons, Academic Press Inc., 3rd edn, 2013.
2. R. Langer and J. P. Vacanti, *Science*, 1993, **260**, 920–926.
3. M. Ashby, H. Shercliff and D. Cebon, *Materials: Engineering, Science, Processing and Design*, Butterworth-Heinemann, Oxford, UK, 1st edn, 2007.
4. M. F. Ashby and Y. J. M. Bréchet, *Acta Mater.*, 2003, **51**, 5801–5821.
5. A. Abalymov, B. Parakhonskiy and A. G. Skirtach, *Polymers*, 2020, **12**, 620.
6. P. E. Neumann and T. R. Gest, *Clin. Anat.*, 2020, **33**, 187–191.
7. M. J. Glimcher, *Rev. Mineral. Geochem.*, 2006, **64**, 223–282.
8. S. M. Best, A. E. Porter, E. S. Thian and J. Huang, *J. Eur. Ceram. Soc.*, 2008, **28**, 1319–1327.
9. G. Stylios, T. Wan and P. Giannoudis, *Injury*, 2007, **38**, S63–S74.
10. B. Clarke, *Clin. J. Am. Soc. Nephrol.*, 2008, **3**, S131–S139.
11. S. Lees, *Connect. Tissue Res.*, 1987, **16**, 281–303.
12. J. D. Currey, *Bones: Structure and Mechanics*, Princeton University Press, 2nd edn., 2013.

13. G. H. Nancollas and M. B. Tomson, *Faraday Discuss. Chem. Soc.*, 1976, **61**, 175–183.
14. E. A. B. Hughes, T. E. Robinson, D. B. Bassett, S. C. Cox and L. M. Grover, *J. Mater. Chem. B*, 2019, **7**, 7460–7470.
15. F. C. M. Driessens, *Bull. Soc. Chim. Belges*, 2010, **89**, 663–689.
16. A. Boskey and N. Pleshko Camacho, *Biomaterials*, 2007, **28**, 2465–2478.
17. G. A. Rodan, *Introduction to Bone Biology*, 1992, vol. 13.
18. S. Weiner and H. D. Wagner, *Annu. Rev. Mater. Sci.*, 1998, **28**, 271–298.
19. M. F. Young, *Osteoporos. Int.*, 2003, **14**, 35–42.
20. A. L. Boskey, *Bone Miner.*, 1989, **6**, 111–123.
21. N. Reznikov, M. Bilton, L. Lari, M. M. Stevens and R. Kröger, *Science*, 2018, **360**, eaao2189.
22. R. A. Robinson, *J. Bone Jt. Surg.*, 1952, **34**, 389–476.
23. S. Weiner and P. A. Price, *Calcif. Tissue Int.*, 1986, **39**, 365–375.
24. N. Reznikov, R. Shahar and S. Weiner, *Acta Biomater*, 2014, **10**, 3815–3826.
25. N. Reznikov, H. Chase, V. Brumfeld, R. Shahar and S. Weiner, *Bone*, 2015, **71**, 189–195.
26. J. Currey, in *Collagen: Structure and Mechanics*, ed. P. Fratzl, Springer, US, 2008, pp. 397–420.
27. T. L. Norman, D. Vashishth and D. B. Burr, *Fracture Toughness of Human Bone under Tension*, 1995, vol. 28.
28. G. De With, H. J. A. Van Dijk, N. Hattu and K. Prijs, *J. Mater. Sci.*, 1981, **16**, 1592–1598.
29. P. Fratzl, H. S. Gupta, E. P. Paschalis and P. Roschger, *J. Mater. Chem.*, 2004, **14**, 2115–2123.
30. U. H. Lerner, *Semin. Orthod.*, 2012, **18**, 237–248.
31. G. S. Stein and J. B. Lian, *Endocr. Rev.*, 1993, **14**, 424–442.
32. I. Owan, D. B. Burr, C. H. Turner, J. Qiu, Y. Tu, J. E. Onyia and R. L. Duncan, *Am. J. Physiol. Physiol.*, 1997, **273**, C810–C815.
33. V. Kartsogiannis and K. W. Ng, *Mol. Cell. Endocrinol.*, 2004, **228**, 79–102.
34. M. D. McKee and A. Nanci, *Cells Mater*, 1993, **3**, 219–243.
35. R. Parenteau-Bareil, R. Gauvin and F. Berthod, *Materials*, 2010, **3**, 1863–1887.
36. S. Ricard-Blum, F. Ruggiero and M. Van Der Rest, in *Collagen*, ed. J. Brinckmann, H. Notbohm and P. K. Müller, Springer, 2005, pp. 35–84.
37. M. J. Buehler, *Proc. Natl. Acad. Sci. U. S. A.*, 2006, **103**, 12285–12290.
38. J. A. M. Ramshaw, J. A. Werkmeister and V. Glattauer, *Biotechnol. Genet. Eng. Rev.*, 1996, **13**, 335–382.
39. A. S. Deshpande and E. Beniash, *Cryst. Growth Des.*, 2008, **8**, 3084–3090.
40. B. R. Williams, R. A. Gelman, D. C. Poppke and K. Piez, *J. Biol. Chem.*, 1978, **253**, 6578–6585.
41. K. Merrett, M. K. Ljunggren, D. Mondal, M. Griffith and M. Rafat, in *Type I Collagen: Biological Functions, Synthesis and Medicinal Applications*, ed. M. E. Henriques and M. Pinto, Nova Science Publishers, 2012, pp. 1–43.

42. S.-T. Li, in *The Biomedical Engineering Handbook*, ed. J. D. Bronzino, CRC Press, 2nd edn, 2000.
43. A. Farzadi, F. Bakhshi, M. Solati-Hashjin, M. Asadi-Eydivand and N. A. A. Osman, *Ceram. Int.*, 2014, **40**, 6021–6029.
44. E. Bonucci, in *Calcification of Biological Systems*, ed. E. Bonucci, CRC Press, 1992, pp. 19–40.
45. D. A. Cameron, *Clin. Orthop. Relat. Res*, 1963, **26**, 199–228.
46. N. Shepard, in *Calcification in Biological Systems*, ed. E. Bonucci, CRC Press, 1992, pp. 41–58.
47. M. J. Seibel, S. P. Robins and J. P. Bilezikian, *Dynamics of Bone and Cartilage Metabolism*, Academic Press, 2nd edn, 2006.
48. M. Murshed and M. D. McKee, *Curr. Opin. Nephrol. Hypertens.*, 2010, **19**, 359–365.
49. M. D. McKee and W. G. Cole, *Pediatr. Bone*, 2012, 9–37.
50. G. K. Hunter and H. A. Goldberg, *Proc. Natl. Acad. Sci. U. S. A.*, 1993, **90**, 8562–8565.
51. C. Jiang, K. Zurick, C. Qin and M. T. Bernards, *Connect. Tissue Res.*, 2018, **59**, 274–286.
52. F. Nudelman, K. Pieterse, A. George, P. H. H. Bomans, H. Friedrich, L. J. Brylka, P. A. J. Hilbers, G. De With and N. A. J. M. Sommerdijk, *Nat. Mater.*, 2010, **9**, 1004–1009.
53. M. J. Olszta, X. Cheng, S. S. Jee, R. Kumar, Y.-Y. Kim, M. J. Kaufman, E. P. Douglas and L. B. Gower, *Mater. Sci. Eng. R Rep.*, 2007, **58**, 77–116.
54. W. J. Landis, M. J. Song, A. Leith, L. McEwen and B. F. McEwen, *J. Struct. Biol.*, 1993, **110**, 39–54.
55. K. Kawasaki, A. V. Buchanan and K. M. Weiss, *Annu. Rev. Genet.*, 2009, **43**, 119–142.
56. F. Loi, L. A. Córdova, J. Pajarinen, T. hua Lin, Z. Yao and S. B. Goodman, *Bone*, 2016, **86**, 119–130.
57. J. R. Jones, *J. Eur. Ceram. Soc.*, 2009, **29**, 1275–1281.
58. E. Rezabeigi, PhD thesis, Concordia University, 2015.
59. D. F. Williams, *Williams Dictonary of Biomaterials*, Liverpool University Press, 1st edn, 1999.
60. E. Gentleman and J. M. Polak, *J. Mater. Sci. Mater. Med.*, 2006, **17**, 1029–1035.
61. G. J. Owens, R. K. Singh, F. Foroutan, M. Alqaysi, C. M. Han, C. Mahapatra, H. W. Kim and J. C. Knowles, *Prog. Mater. Sci.*, 2016, **77**, 1–79.
62. D. Puppi, F. Chiellini, A. M. Piras and E. Chiellini, *Prog. Polym. Sci.*, 2010, **35**, 403–440.
63. M. S. Mohammadi, E. Rezabeigi, J. Bertram, B. Marelli, R. Gendron, S. N. Nazhat and M. N. Bureau, *Polymers*, 2020, **12**, 231.
64. E. Rezabeigi, P. M. Wood-Adams and R. A. L. Drew, *J. Polym. Sci. Part B, Polym. Phys.*, 2017, **55**, 1055–1062.
65. A. Ficai, E. Andronescu, G. Voicu, C. Ghitulica, B. S. Vasile, D. Ficai and V. Trandafir, *Chem. Eng. J.*, 2010, **160**, 794–800.

66. M. S. Saveleva, K. Eftekhari, A. Abalymov, T. E. L. Douglas, D. Volodkin, B. V. Parakhonskiy and A. G. Skirtach, *Front. Chem.*, 2019, **7**, 179.
67. M. J. Glimcher, *Philos. Trans. R. Soc. London, Ser. B*, 1984, **304**, 479–508.
68. M. Bohner, *Injury*, 2000, **31**, D37–D47.
69. L. L. Hench, D. L. Wheeler and D. C. Greenspan, *J. Sol-Gel Sci. Technol.*, 1998, **13**, 245–250.
70. A. Kolk, J. Handschel, W. Drescher, D. Rothamel, F. Kloss, M. Blessmann, M. Heiland, K. D. Wolff and R. Smeets, *J. Cranio-Maxillofacial Surg*, 2012, **40**, 706–718.
71. A. R. Vaccaro, K. Chiba, J. G. Heller, T. C. Patel, J. S. Thalgott, E. Truumees, J. S. Fischgrund, M. R. Craig, S. C. Berta and J. C. Wang, *Spine J*, 2002, **2**, 206–215.
72. L. L. Hench, *J. Mater. Sci. Mater. Med*, 2006, **17**, 967–978.
73. E. Rezabeigi, P. M. Wood-Adams and R. A. L. Drew, *Mater. Sci. Eng. C*, 2014, **40**, 248–252.
74. E. A. Abou Neel, T. Mizoguchi, M. Ito, M. Bitar, V. Salih and J. C. Knowles, *Biomaterials*, 2007, **28**, 2967–2977.
75. J. Zhong and D. C. Greenspan, *J. Biomed. Mater. Res.*, 2000, **53**, 694–701.
76. L. L. Hench, *J. Eur. Ceram. Soc.*, 2009, **29**, 1257–1265.
77. W. C. Lepry, E. Rezabeigi, S. Smith and S. N. Nazhat, *Biomed. Glas*, 2019, **5**, 98–111.
78. W. C. Lepry and S. N. Nazhat, *Chem. Mater.*, 2015, **27**, 4821–4831.
79. H. Gao, T. Tan and D. Wang, *J. Control. Release*, 2004, **96**, 21–28.
80. J. S. Cho and Y. C. Kang, *Ceram. Int.*, 2009, **35**, 2103–2109.
81. A. Zandi Karimi, E. Rezabeigi and R. A. L. Drew, *J. Non. Cryst. Solids*, 2018, **502**, 176–183.
82. A. Zandi Karimi, E. Rezabeigi and R. A. L. Drew, *J. Mech. Behav. Biomed. Mater.*, 2019, **97**, 396–405.
83. K. Madhavan Nampoothiri, N. R. Nair and R. P. John, *Bioresour. Technol*, 2010, **101**, 8493–8501.
84. A. Biswas, I. S. Bayer, H. Zhao, T. Wang, F. Watanabe and A. S. Biris, *Biomacromolecules*, 2010, **11**, 2545–2549.
85. J. C. Middleton and A. J. Tipton, *Biomaterials*, 2000, **21**, 2335–2346.
86. E. J. Bergsma, F. R. Rozema, R. R. M. Bos and W. C. D. Bruijn, *J. Oral Maxillofac. Surg.*, 1993, **51**, 666–670.
87. C. Martin, H. Winet and J. Y. Bao, *Biomaterials*, 1996, **17**, 2373–2380.
88. G. Griffanti, M. James-Bhasin, I. Donelli, G. Freddi and S. N. Nazhat, *Biomed. Mater*, 2019, **14**, 15006.
89. F. Chicatun, E. Rezabeigi, N. Muja, M. T. Kaartinen, M. D. McKee and S. N. Nazhat, *Emergent Mater*, 2019, **2**, 245–262.
90. J. A. M. Ramshaw, Y. Y. Peng, V. Glattauer and J. A. Werkmeister, *J. Mater. Sci. Mater. Med*, 2009, **20**, 3–8.
91. K. Rezwan, Q. Z. Chen, J. J. Blaker and A. R. Boccaccini, *Biomaterials*, 2006, **27**, 3413–3431.

92. S. K. Misra, D. Mohn, T. J. Brunner, W. J. Stark, S. E. Philip, I. Roy, V. Salih, J. C. Knowles and A. R. Boccaccini, *Biomaterials*, 2008, **29**, 1750–1761.
93. J. Maitra and V. K. Shukla, *Am. J. Polym. Sci*, 2014, **4**, 25–31.
94. R. Murugan and S. Ramakrishna, *Compos. Sci. Technol.*, 2005, **65**, 2385–2406.
95. Y. Gao and J. Chang, *J. Biomater. Appl.*, 2009, **24**, 119–138.
96. H. Zou, S. Wu and J. Shen, *Chem. Rev.*, 2008, **108**, 3893–3957.
97. L. Wei, N. Hu and Y. Zhang, *Materials*, 2010, **3**, 4066–4079.
98. E. Rezabeigi, P. M. Wood-Adams and R. A. L. Drew, *J. Biomed. Mater. Res. – Part B Appl. Biomater*, 2017, **105**, 2433–2442.
99. G. Poologasundarampillai and A. L. B. Maçon, in Bioactive Glasses: Fundamentals, *Technology and Applications*, ed. A. R. Boccaccini, D. S. Brauer and L. Hupa, The Royal Society of Chemistry, 2017, pp. 286–302.
100. Q. Z. Chen, I. Ahmed, J. C. Knowles, S. N. Nazhat, A. R. Boccaccini and K. Rezwan, *J. Biomed. Mater. Res. Part A*, 2008, **86A**, 987–995.
101. A. Perro and V. N. Manoharan, *Langmuir*, 2010, **26**, 18669–18675.
102. Y. Xie, C. A. S. Hill, Z. Xiao, H. Militz and C. Mai, *Compos. Part A Appl. Sci. Manuf.*, 2010, **41**, 806–819.
103. R. S. Pryce and L. L. Hench, *J. Mater. Chem.*, 2004, **14**, 2303–2310.
104. A. Perro, S. Reculusa, S. Ravaine, E. Bourgeat-Lami and E. Duguet, *J. Mater. Chem.*, 2005, **15**, 3745–3760.
105. K. Kailasam, S. Mascotto, S. Gross, C. MacCato and K. Müller, *J. Mater. Chem.*, 2010, **20**, 2345–2355.
106. D. Arcos and M. Vallet-Regí, *Acta Biomater.*, 2010, **6**, 2874–2888.
107. Z. Hong, A. Liu, L. Chen, X. Chen and X. Jing, *J. Non. Cryst. Solids*, 2009, **355**, 368–372.
108. M. Catauro, M. G. Raucci, F. De Gaetano and A. Marotta, *J. Mater. Sci. Mater. Med.*, 2004, **15**, 831–837.
109. W. C. Lepry and S. N. Nazhat, in *Bioactive Glasses: Materials, Properties and Applications*, ed. A. R. Boccaccini, D. S. Brauer and L. Hupa, The Royal Society of Chemistry, 2017, pp. 496–521.
110. D. C. Bradley, *Chem. Rev.*, 1989, **89**, 1317–1322.
111. C. Sanchez and Soler-Illia, in *Encyclopedia of Chemical Processing*, ed. S. Lee, Taylor & Francis, 2006, pp. 1267–1280.
112. M. N. Rahaman, *Ceramic Processing and Sintering*, CRC Press, 2nd edn, 2017.
113. D. Stephan and S. Wistuba, *Cem. Concr. Res.*, 2006, **36**, 2011–2020.
114. F. Baino, S. Hamzehlou and S. Kargozar, *J. Funct. Biomater.*, 2018, **9**, 25.
115. C. V. M. Rodrigues, P. Serricella, A. B. R. Linhares, R. M. Guerdes, R. Borojevic, M. A. Rossi, M. E. L. Duarte and M. Farina, *Biomaterials*, 2003, **24**, 4987–4997.
116. L. Sukhodub, C. Moseke, L. Sukhodub, B. Sulkio-Cleff, V. Y. Maleev, M. Semenov, E. Bereznyak and T. Bolbukh, *J. Mol. Struct.*, 2004, **704**, 53–58.

117. D. Bakoš, M. Soldán and I. Hernández-Fuentes, *Biomaterials*, 1999, **20**, 191–195.
118. M. Okazakl, H. Ohmae, J. Takahashi, H. Kimura and M. Sakuda, *Biomaterials*, 1990, **11**, 568–572.
119. A. C. Lawson and J. T. Czernuszka, *Proc. Inst. Mech. Eng. Part H J. Eng. Med.*, 1998, **212**, 413–425.
120. D. A. Wahl and J. T. Czernuszka, *Eur. Cells Mater.*, 2006, **11**, 43–56.
121. L. L. Hench, *J. Am. Ceram. Soc.*, 1991, **74**, 1487–1510.
122. T. Kokubo and H. Takadama, *Biomaterials*, 2006, **27**, 2907–2915.
123. T. K. Greenlee, C. A. Beckham, A. R. Crebo and J. C. Malmorg, *J. Biomed. Mater. Res.*, 1972, **6**, 235–244.
124. O. P. Filho, G. P. Latorre and L. L. Hench, *J. Biomed. Mater. Res.*, 1996, **30**, 509–514.
125. M. Ojansivu, X. Wang, L. Hyväri, M. Kellomäki, L. Hupa, S. Vanhatupa and S. Miettinen, *Eur. Cells Mater.*, 2018, **35**, 53–71.
126. P. V. Phan, M. Grzanna, J. Chu, A. Polotsky, A. El-Ghannam, D. Van Heerden, D. S. Hungerford and C. G. Frondoza, *J. Biomed. Mater. Res., Part A*, 2003, **67**, 1001–1008.
127. L. L. Hench and J. R. Jones, *Front. Bioeng. Biotechnol.*, 2015, **3**, 194.
128. J. R. Jones, *Acta Biomater.*, 2015, **23**, S53–S82.
129. M. N. Rahaman, D. E. Day, B. Sonny Bal, Q. Fu, S. B. Jung, L. F. Bonewald and A. P. Tomsia, *Acta Biomater.*, 2011, **7**, 2355–2373.
130. L. L. Hench, R. J. Splinter, W. C. Allen and T. K. Greenlee, *J. Biomed. Mater. Res.*, 1971, **5**, 117–141.
131. N. J. Lakhkar, I. H. Lee, H. W. Kim, V. Salih, I. B. Wall and J. C. Knowles, *Adv. Drug Delivery Rev.*, 2013, **65**, 405–420.
132. V. Salih, K. Franks, M. James, G. W. Hastings, J. C. Knowles and I. Olsen, *J. Mater. Sci. Mater. Med.*, 2000, **11**, 615–620.
133. B. C. Bunker, G. W. Arnold and J. A. Wilder, *J. Non. Cryst. Solids*, 1984, **64**, 291–316.
134. K. Franks, I. Abrahams and J. C. Knowles, *J. Mater. Sci. Mater. Med.*, 2000, **11**, 609–614.
135. J. C. Knowles, *J. Mater. Chem.*, 2003, **13**, 2395.
136. E. A. Abou Neel, V. Salih and J. C. Knowles, in *Comprehensive Biomaterials*, ed. P. Ducheyne, Elsevier, 2011, vol. 1, pp. 285–297.
137. Q. Fu, M. N. Rahaman, H. Fu and X. Liu, *J. Biomed. Mater. Res., Part A*, 2010, **95**, 164–171.
138. Q. Fu, M. N. Rahaman, B. S. Bal, L. F. Bonewald, K. Kuroki and R. F. Brown, *J. Biomed. Mater. Res., Part A*, 2010, **95**, 172–179.
139. M. Schuch, C. Trott and P. Maass, *RSC Adv.*, 2011, **1**, 1370–1382.
140. E. I. Kamitsos, *Modifying Role of Alkali-metal Cations in Borate Glass Networks*, 1989, vol. 93.
141. W. Huang, D. E. Day, K. Kittiratanapiboon and M. N. Rahaman, *J. Mater. Sci. Mater. Med.*, 2006, **17**, 583–596.
142. A. Yao, D. Wang, W. Huang, Q. Fu, M. N. Rahaman and D. E. Day, *J. Am. Ceram. Soc.*, 2007, **90**, 303–306.

143. D. Arcos, D. C. Greenspan and M. Vallet-Regí, *J. Biomed. Mater. Res., Part A*, 2003, **65**, 344–351.
144. J. E. Shelby, *Introduction to Glass Science and Technology*, The Royal Society of Chemistry, 2007.
145. L. Mädler, H. K. Kammler, R. Mueller and S. E. Pratsinis, *J. Aerosol Sci.*, 2002, **33**, 369–389.
146. T. J. Brunner, R. N. Grass and W. J. Stark, *Chem. Commun.*, 2006, 1384.
147. R. Li, A. E. Clark and L. L. Hench, *J. Appl. Biomater.*, 1991, **2**, 231–239.
148. J. Livage, in *Sol-Gel Technologies for Glass Producers and Users*, ed. M. A. Aegerter and M. Mennig, Springer, 2004, pp. 3–14.
149. L. L. Hench and J. K. West, *Chem. Rev.*, 1990, **90**, 33–72.
150. C. J. Brinker and G. W. Scherer, *Sol-Gel Science: The Physics and Chemistry of Sol-Gel Processing*, Elsevier, 2013.
151. P. Sepulveda, J. R. Jones and L. L. Hench, *J. Biomed. Mater. Res.*, 2001, **58**, 734–740.
152. I. A. Silver, J. Deas and M. Erecińska, *Biomaterials*, 2001, **22**, 175–185.
153. P. Sepulveda, J. R. Jones and L. L. Hench, *J. Biomed. Mater. Res.*, 2002, **59**, 340–348.
154. V. Maquet, A. R. Boccaccini, L. Pravata, I. Notingher and R. Jérôme, *Biomaterials*, 2004, **25**, 4185–4194.
155. A. Mandal, S. Panigrahi and C. Zhang, *Biol. Eng.*, 2010, **2**, 63–88.
156. Y. K. Lin and D. C. Liu, *Food Chem.*, 2006, **99**, 244–251.
157. C. Meena, S. A. Mengi and S. G. Deshpande, *Proc. Indian Acad. Sci. Chem. Sci.*, 1999, **111**, 319–329.
158. H. Stein, M. Wilensky, Y. Tsafrir, M. Rosenthal, R. Amir, T. Avraham, K. Ofir, O. Dgany, A. Yayon and O. Shoseyov, in *Biomacromolecules*, American Chemical Society, 2009, vol. 10, pp. 2640–2645.
159. C. Yang, P. J. Hillas, J. A. Báez, M. Nokelainen, J. Balan, J. Tang, R. Spiro and J. W. Polarek, *BioDrugs*, 2004, **18**, 103–119.
160. S. Browne, D. I. Zeugolis and A. Pandit, *Tissue Eng., Part A*, 2013, **19**, 1491–1494.
161. F. O. Schmitt, L. Levine, M. P. Drake, A. L. Rubin, D. Pfahl and P. F. Davison, *Proc. Natl. Acad. Sci. U. S. A.*, 1964, **51**, 493–497.
162. D. Michaeli, G. R. Martin, J. Kettman, E. Benjamini, D. Y. K. Leung and B. A. Blatt, *Science*, 1969, **166**, 1522–1523.
163. H. Lindsley, M. Mannik and P. Bornstein, *J. Exp. Med.*, 1971, **133**, 1309–1324.
164. H. Furthmayr and R. Timpl, in *International Review of Connective Tissue Research*, Elsevier, 1976, vol. 7, pp. 61–99.
165. F. S. Steven and D. S. Jackson, *Biochem. J.*, 1967, **104**, 534–536.
166. N. Rajan, J. Habermehl, M. F. Coté, C. J. Doillon and D. Mantovani, *Nat. Protoc.*, 2007, **1**, 2753–2758.
167. S. F. Badylak, *Transpl. Immunol.*, 2004, **12**, 367–377.
168. A. J. Allman, T. B. McPherson, L. C. Merrill, S. F. Badylak and D. W. Metzger, *Tissue Eng.*, 2002, **8**, 53–62.

169. T. W. Gilbert, T. L. Sellaro and S. F. Badylak, *Biomaterials*, 2006, **27**, 3675–3683.
170. J. Gross, J. H. Highberger and F. O. Schmitt, *Proc. Natl. Acad. Sci. U. S. A.*, 1955, **41**, 1–7.
171. E. J. Miller and R. K. Rhodes, *Methods Enzymol.*, 1982, **82**, 33–64.
172. M. P. Drake, P. F. Davison, S. Bump and F. O. Schmitt, *Biochemistry*, 1966, **5**, 301–312.
173. G. Chandrakasan, D. A. Torchia and K. A. Piez, *J. Biol. Chem.*, 1976, **251**, 6062–6067.
174. D. Silver, J. Miller, R. Harrison and D. J. Prockop, *Proc. Natl. Acad. Sci. U. S. A.*, 1992, **89**, 9860–9864.
175. K. Gelse, E. Pöschl and T. Aigner, *Adv. Drug Delivery Rev.*, 2003, **55**, 1531–1546.
176. J. A. Fallas, L. E. R. O'Leary and J. D. Hartgerink, *Chem. Soc. Rev.*, 2010, **39**, 3510–3527.
177. C. Dong and Y. Lv, *Polymers*, 2016, **8**, 42.
178. V. Serpooshan, T. M. Quinn, N. Muja and S. N. Nazhat, *Soft Matter*, 2011, **7**, 2918–2926.
179. N. Nassif, F. Gobeaux, J. Seto, E. Belamie, P. Davidson, P. Panine, G. Mosser, P. Fratzl and M. M. Giraud Guille, *Chem. Mater.*, 2010, **22**, 3307–3309.
180. B. V. Slaughter, S. S. Khurshid, O. Z. Fisher, A. Khademhosseini and N. A. Peppas, *Adv. Mater.*, 2009, **21**, 3307–3329.
181. B. A. Roeder, K. Kokini, J. E. Sturgis, J. P. Robinson and S. L. Voytik-Harbin, *J. Biomech. Eng.*, 2002, **124**, 214–222.
182. V. Mouriño and A. R. Boccaccini, *J. R. Soc. Interface*, 2010, **7**, 209–227.
183. R. Sarber, B. Hull, C. Merrill, T. Soranno and E. Bell, *Mech. Ageing Dev.*, 1981, **17**, 107–117.
184. S. J. Kew, J. H. Gwynne, D. Enea, M. Abu-Rub, A. Pandit, D. Zeugolis, R. A. Brooks, N. Rushton, S. M. Best and R. E. Cameron, *Acta Biomater.*, 2011, **7**, 3237–3247.
185. D. I. Zeugolis, G. R. Paul and G. Attenburrow, *J. Biomed. Mater. Res., Part A*, 2009, **89**, 895–908.
186. Y. Di and R. J. Heath, *Polym. Degrad. Stab.*, 2009, **94**, 1684–1692.
187. J. M. Lee, C. A. Pereira and L. W. K. Kan, *J. Biomed. Mater. Res.*, 1994, **28**, 981–992.
188. F. Chamieh, A. M. Collignon, B. R. Coyac, J. Lesieur, S. Ribes, J. Sadoine, A. Llorens, A. Nicoletti, D. Letourneur, M. L. Colombier, S. N. Nazhat, P. Bouchard, C. Chaussain and G. Y. Rochefort, *Sci. Rep.*, 2016, **6**, 38814.
189. A. M. Collignon, G. Castillo-Dali, E. Gomez, T. Guilbert, J. Lesieur, A. Nicoletti, S. Acuna-Mendoza, D. Letourneur, C. Chaussain, G. Y. Rochefort and A. Poliard, *Stem Cells*, 2019, **37**, 701–711.
190. P. Habibovic, D. C. Bassett, C. J. Doillon, C. Gerard, M. D. McKee and J. E. Barralet, *Adv. Mater.*, 2010, **22**, 1858–1862.

191. N. J. Mathers and J. T. Czernuszka, *Growth of Hydroxyapatite on Type 1 Collagen*, 1991, vol. 10.
192. K. I. Clarke, S. E. Graves, A. T. C. Wong, J. T. Triffitt, M. J. O. Francis and J. T. Czernuszka, *J. Mater. Sci. Mater. Med*, 1993, **4**, 107–110.
193. J. T. Czernuszka, A. C. Lawson, A. Hamish and R. W. Simpson, Composite material and methods of making the same, US6395036B1, 2002.
194. M. Kikuchi, T. Ikoma, S. Itoh, H. N. Matsumoto, Y. Koyama, K. Takakuda, K. Shinomiya and J. Tanaka, *Compos. Sci. Technol.*, 2004, **64**, 819–825.
195. S. Itoh, M. Kikuchi, K. Takakuda, Y. Koyama, H. N. Matsumoto, S. Ichinose, J. Tanaka, T. Kawauchi and K. Shinomiya, *J. Biomed. Mater. Res.*, 2001, **54**, 445–453.
196. M. Kikuchi, S. Itoh, S. Ichinose, K. Shinomiya and J. Tanaka, *Biomaterials*, 2001, **22**, 1705–1711.
197. J. A. R. Gordon, C. E. Tye, A. V. Sampaio, T. M. Underhill, G. K. Hunter and H. A. Goldberg, *Bone*, 2007, **41**, 462–473.
198. L. W. Fisher and N. S. Fedarko, *Connect. Tissue Res.*, 2003, **44**, 33–40.
199. G. S. Baht, G. K. Hunter and H. A. Goldberg, *Matrix Biol.*, 2008, **27**, 600–608.
200. O. Tomomatsu, A. Tachibana, K. Yamauchi and T. Tanabe, *J. Ceram. Soc. Japan*, 2008, **116**, 10–13.
201. D. Lickorish, J. A. M. Ramshaw, J. A. Werkmeister, V. Glattauer and C. R. Howlett, *J. Biomed. Mater. Res., Part A*, 2004, **68**, 19–27.
202. L. J. Zhang, X. S. Feng, H. G. Liu, D. J. Qian, L. Zhang, X. L. Yu and F. Z. Cui, *Mater. Lett.*, 2004, **58**, 719–722.
203. Y. Honda, S. Kamakura, K. Sasaki and O. Suzuki, *J. Biomed. Mater. Res. Part B Appl. Biomater.*, 2007, **80B**, 281–289.
204. J. H. Bradt, M. Mertig, A. Teresiak and W. Pompe, *Chem. Mater.*, 1999, **11**, 2694–2701.
205. S. H. Rhee and J. Tanaka, *Biomaterials*, 1999, **20**, 2155–2160.
206. W. Jiang, G. Griffanti, F. Tamimi, M. D. McKee and S. N. Nazhat, *J. Struct. Biol.*, 2020, **212**, 107592.
207. M. M. Villa, L. Wang, J. Huang, D. W. Rowe and M. Wei, *J. Biomed. Mater. Res., Part B Appl. Biomater.*, 2015, **103**, 243–253.
208. Â. L. Andrade, S. P. Andrade and R. Z. Domingues, *J. Biomed. Mater. Res., Part B Appl. Biomater.*, 2006, **79B**, 122–128.
209. N. G. Rim, C. S. Shin and H. Shin, *Biomed Mater.*, 2013, **8**, 14102.
210. J. A. Matthews, G. E. Wnek, D. G. Simpson and G. L. Bowlin, *Biomacromolecules*, 2002, **3**, 232–238.
211. J. Venugopal, S. Low, A. T. Choon, T. S. Sampath Kumar and S. Ramakrishna, *J. Mater. Sci. Mater. Med.*, 2008, **19**, 2039–2046.
212. S. H. Teng, E. J. Lee, P. Wang and H. E. Kim, *Mater. Lett.*, 2008, **62**, 3055–3058.
213. V. Thomas, D. R. Dean, M. V. Jose, B. Mathew, S. Chowdhury and Y. K. Vohra, *Biomacromolecules*, 2007, **8**, 631–637.
214. M. C. Phipps, W. C. Clem, S. A. Catledge, Y. Xu, K. M. Hennessy, V. Thomas, M. J. Jablonsky, S. Chowdhury, A. V. Stanishevsky, Y. K. Vohra and S. L. Bellis, *PLoS One*, 2011, **6**, e16813.

215. M. Madaghiele, A. Sannino, I. V. Yannas and M. Spector, *J. Biomed. Mater. Res., Part A*, 2008, **85**, 757–767.
216. Y. Liu, J. Gu and D. Fan, *Polymers*, 2020, **12**, 61.
217. B. Sarker, J. Hum, S. N. Nazhat and A. R. Boccaccini, *Adv. Healthcare Mater.*, 2015, **4**, 176–194.
218. C. Xu, P. Su, X. Chen, Y. Meng, W. Yu, A. P. Xiang and Y. Wang, *Biomaterials*, 2011, **32**, 1051–1058.
219. A. El-Fiqi, J. H. Kim and H. W. Kim, *Mater. Sci. Eng. C*, 2020, **110**, 110660.
220. S. V. Murphy and A. Atala, *Nat. Biotechnol.*, 2014, **32**, 773–785.
221. P. Ahangar, M. E. Cooke, M. H. Weber and D. H. Rosenzweig, *Appl. Sci.*, 2019, **9**, 1713.
222. Q. Li, X. Lei, X. Wang, Z. Cai, P. Lyu and G. Zhang, *Tissue Eng., Part A*, 2019, **25**, 1261–1271.
223. I. L. Ardelean, D. Gudovan, D. Ficai, A. Ficai, E. Andronescu, M. G. Albu-Kaya, P. Neacsu, R. N. Ion, A. Cimpean and V. Mitran, *Mater. Lett.*, 2018, **231**, 179–182.
224. G. Montalbano, S. Fiorilli, A. Caneschi and C. Vitale-Brovarone, *Materials*, 2018, **11**, 700.
225. R. A. Brown, M. Wiseman, C. B. Chuo, U. Cheema and S. N. Nazhat, *Adv. Funct. Mater.*, 2005, **15**, 1762–1770.
226. N. O. Kamranpour, A. K. Miri, M. James-Bhasin and S. N. Nazhat, *Biofabrication*, 2016, **8**, 15018.
227. B. Marelli, C. E. Ghezzi, M. James-Bhasin and S. N. Nazhat, *Biomaterials*, 2015, **37**, 183–193.
228. G. Griffanti, E. Rezabeigi, J. Li, M. Murshed and S. N. Nazhat, *Adv. Funct. Mater.*, 2020, **30**, 1903874.
229. V. Mudera, M. Morgan, U. Cheema, S. N. Nazhat and R. Brown, *J. Tissue Eng. Regen. Med.*, 2007, **1**, 192–198.
230. B. R. Coyac, F. Chicatun, B. Hoac, V. Nelea, C. Chaussain, S. N. Nazhat and M. D. McKee, *J. Dent. Res.*, 2013, **92**, 648–654.
231. P. G. Buxton, M. Bitar, K. Gellynck, M. Parkar, R. A. Brown, A. M. Young, J. C. Knowles and S. N. Nazhat, *Bone*, 2008, **43**, 377–385.
232. A. M. Collignon, J. Lesieur, N. Anizan, R. Ben Azzouna, A. Poliard, C. Gorin, D. Letourneur, C. Chaussain, F. Rouzet and G. Y. Rochefort, *Acta Biomater*, 2018, **82**, 111–121.
233. A. Novais, J. Lesieur, J. Sadoine, L. Slimani, B. Baroukh, B. Saubaméa, A. Schmitt, S. Vital, A. Poliard, C. Hélary, G. Y. Rochefort, C. Chaussain and C. Gorin, *Stem Cells Transl. Med.*, 2019, **8**, 844–857.
234. C. E. Pedraza, B. Marelli, F. Chicatun, M. D. McKee and S. N. Nazhat, *Tissue Eng., Part A*, 2010, **16**, 781–793.
235. B. Marelli, C. E. Ghezzi, J. E. Barralet and S. N. Nazhat, *Soft Matter*, 2011, **7**, 9898–9907.
236. F. Chicatun, C. E. Pedraza, C. E. Ghezzi, B. Marelli, M. T. Kaartinen, M. D. McKee and S. N. Nazhat, *Biomacromolecules*, 2011, **12**, 2946–2956.

237. B. Marelli, C. E. Ghezzi, Y. L. Zhang, I. Rouiller, J. E. Barralet and S. N. Nazhat, *Biomaterials*, 2015, **37**, 252–259.
238. A. J. Engler, S. Sen, H. L. Sweeney and D. E. Discher, *Cell*, 2006, **126**, 677–689.
239. B. Marelli, C. E. Ghezzi, J. E. Barralet, A. R. Boccaccini and S. N. Nazhat, *Biomacromolecules*, 2010, **11**, 1470–1479.
240. S. N. Nazhat, E. A. Abou Neel, A. Kidane, I. Ahmed, C. Hope, M. Kershaw, P. D. Lee, E. Stride, N. Saffari, J. C. Knowles and R. A. Brown, *Biomacromolecules*, 2007, **8**, 543–551.
241. W. J. Bae, K. S. Min, J. J. Kim, J. J. Kim, H. W. Kim and E. C. Kim, *Dent. Mater.*, 2012, **28**, 1271–1279.
242. J. H. Lee, A. El-Fiqi, C. M. Han and H. W. Kim, *Tissue Eng. Regen. Med.*, 2015, **12**, 90–97.
243. B. Marelli, C. E. Ghezzi, D. Mohn, W. J. Stark, J. E. Barralet, A. R. Boccaccini and S. N. Nazhat, *Biomaterials*, 2011, **32**, 8915–8926.
244. A. K. Miri, N. Muja, N. O. Kamranpour, W. C. Lepry, A. R. Boccaccini, S. A. Clarke and S. N. Nazhat, *Biomaterials*, 2016, **85**, 128–141.

CHAPTER 4

Bioelectronics and Neural Interfaces

W. TONG,[a,b,c] U. AREGUETA ROBLES[d] AND A. GELMI*[e]

[a] National Vision Research Institute, Australian College of Optometry, VIC, Australia; [b] Department of Optometry and Vision Sciences, Melbourne School of Health Sciences, The University of Melbourne, VIC, Australia; [c] School of Physics, The University of Melbourne, VIC, Australia; [d] Graduate School of Biomedical Engineering, University of New South Wales, NSW, Australia; [e] Applied Chemistry and Environmental Science, School of Science, RMIT University, Melbourne, Australia
*Email: amy.gelmi@rmit.edu.au

4.1 Introduction

Brain disease and injuries place a huge burden on society, both in terms of economic and social, and humanitarian aspects. The breadth and range of brain diseases and injuries includes those caused by sudden trauma or strokes, neurological disorders such as epilepsy or Parkinson's, disorders common in the ageing population, such as dementia and Alzheimer's, and increasingly, mood and anxiety disorders. In 2010, the total European cost of brain disorders, including direct healthcare (37%), non-medical costs (23%), and indirect costs (40%) was €780 billion.[1] Emerging biomedical approaches use neural prostheses and therapeutic modulation to deliver electrical signals that can alleviate symptoms and even restore function to patients, through both stimulating and recording electrically-active nerve and neural cells. When interfacing electronics with living tissue, we enter the field of bioelectronics; here we can benefit from the common properties of inorganic

Inorganic Materials Series No. 7
The Chemistry of Inorganic Biomaterials
Edited by Christopher Spicer

Published by the Royal Society of Chemistry, www.rsc.org

biomaterials to facilitate the formation of neural–electrode interfaces, which are the principal component of these new biomedical therapies. Figure 4.1 illustrates the many clinically developed neural interfaces and the different diseases and conditions they aim to treat.

Neural and nerve tissues are physiologically electroactive and transmit and receive electrical signals, much like an electronic circuit or network. The cochlear implant is arguably one of the most successful examples of a bioelectronic medical device; the cochlear implant has over 600 000 users worldwide, and on average 700 adults and 550 children are implanted with one every year in the United Kingdom.[3] The device allows users to 'hear', by converting sound into electrical signals and directly stimulating auditory nerves in the cochlear *via* an electrode array. New materials and fabrication methods are continuously being researched to improve the electrode component of the cochlear implant; the smaller the electrode size, the more targeted and specific the stimulation can be, and new materials can be used to better modulate the interface between electrode and tissue for improved long-term performance.

For brain diseases and tissue injuries, the cellular 'circuitry' can misfire or be damaged. Interfacing bioelectronics with the damaged tissue of interest, to provide electrical stimulation for neural modulation, can aid treatment and therapies in patients suffering from these conditions. As the electrical stimulation can be applied to a specific region of interest *via* the bioelectronic interface, this type of therapy can be effectively targeted, as opposed to

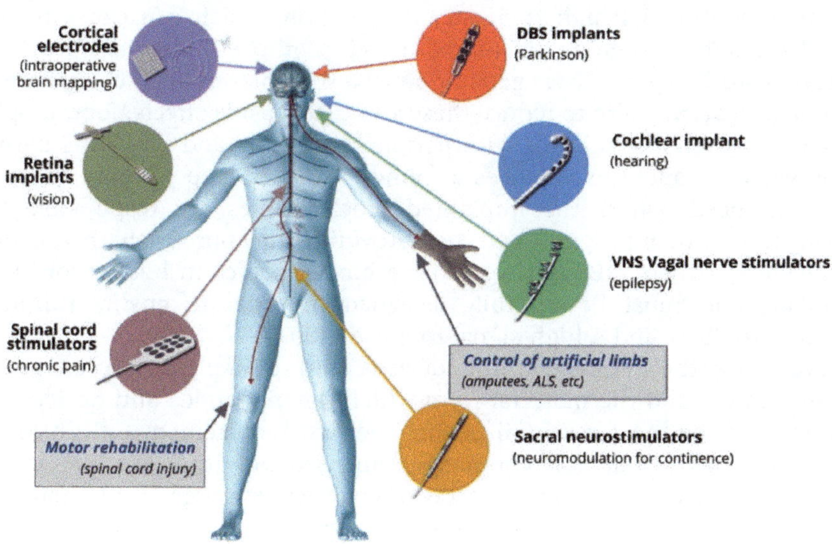

Figure 4.1 Clinically developed neural interfaces. Schematics of the actual devices are shown (not to scale) along with each disease indication (in parentheses) that they are mainly used or developed for.
Reproduced from ref. 2 with permission from John Wiley & Sons, Copyright 2017 WILEY-VCH Verlag GmbH & Co. KGaA, Weinheim.

more conventional drug therapies, which suffer from typically systemic effects. This approach has proven to be beneficial for many brain related diseases and injuries. Epilepsy therapy has greatly benefitted from bioelectronics; up to one third of epilepsy patients have a drug-resistant version of the disease[4] and not all patients can undergo surgery. Hence, therapeutic electrical stimulation can greatly improve the prognosis and treatment of these patients. Vagus nerve stimulation (VNS) is an approved therapy for epilepsy, with efficacy comparable to antiepileptic drugs.[5] VNS uses a bioelectronic device to deliver electrical pulse signals to reduce the frequency and severity of epileptic seizures. More recent advances in epilepsy research involve neural prostheses capable of both recording and transmitting an electrical signal. In these systems, the oncoming seizure can be detected, and an appropriate signal generated to treat the seizure before it occurs.[6,7] This approach of creating neural prostheses that record electrical signals from the neural tissue and then deliver therapeutic electrical stimulation is the forward focus of the field.

Bioelectronics are also paving the way for new electrical stimulation therapies to help alleviate the debilitating symptoms of Parkinson's disease. Deep brain stimulation (DBS) has been demonstrated to improve mobility and reduce dyskinesias (impairment of voluntary movement), and significantly reduce the accompanying medication dose.[8] As a chronic disease, quality of life and efficacy of treatment are paramount for a patient. The medications for Parkinson's disease can often lose efficacy over time, and so DBS is highly beneficial to help manage the symptoms long-term.[9] However, this type of DBS uses continuous high frequency stimulation, which can cause neurological side effects such as disordered mood, cognitive problems, speech issues, or disturbed gait.[10] Next generation DBS therapies tailored to a patient's neurological activity aim to reduce these stimulation side effects; for example, tracking the tremors in limbs with peripheral sensors and providing stimulation when a patient experiences a tremor.[11] Closed-loop systems are even more advanced, where the implanted bioelectronics are responsible for sensing tremor neural signals and then providing the appropriate therapeutic stimulation.[12] These DBS devices require bioelectronics to both record and stimulate, and must be carefully designed in order to ensure minimal stimulus artefacts and a high signal-to-noise ratio.[13]

Developing these bioelectronic devices and neural prostheses needs careful design, and the material from which the electrodes and devices are fabricated is the most essential factor. The first requirement for a bioelectronic material is that it is conductive, and secondly that the material is biocompatible. There are a wide array of conductive inorganic biomaterials which are used in bioelectronics; silicon, metals, metal alloys, metal oxides, as well as carbon-based materials such as diamond and graphene. These materials are used to create a variety of bioelectronic devices, each of which require careful and precise fabrication. In this chapter we discuss the use of inorganic biomaterials in bioelectronics, and the complex relationships at the interface between the living tissue and implanted device.

4.2 The Inorganic Biomaterial Electrode Interface

The interface between the living tissue and the electrode will dictate the efficacy of the device, hence understanding this interface for the selected material is vital when discussing bioelectronics. Individual cells must maintain close physical coupling with the electrode for good fidelity when recording, and for efficacious charge transfer when stimulating. For the cells to couple effectively, the material needs to have good biocompatibility and support neuronal integration. Inorganic biomaterials can deliver electrical signals to cells through several different mechanisms, which need to be carefully considered when designing the neural device. To promote efficacious and biocompatible electrical signalling, understanding, and optimising the interface of the tissue and electrode is vital.

4.2.1 Inorganic Biomaterial Electrodes

Inorganic biomaterials commonly used in bioelectronics include metals, metal alloys, metal compounds, and inorganic carbon allotropes (Table 4.1).

Metallic electrodes have long been regarded as the standard in bioelectronics as they do not produce substantial tissue response, facilitate electrical charge transfer, and can be observed radiographically following implantation.[49] The most common substrate electrode materials encompass platinum (Pt), Pt alloys with iridium (PtIr), gold (Au), and to a lesser extent titanium (Ti) and stainless steel (SS). Historically, Pt has been regarded as the material of choice for electrode fabrication due to its electrochemical stability, resistance to corrosion, and limited reactivity in biological environments compared to other metals.[50] Other noble metals such as gold, although regarded as biocompatible, due to its biological inertness, will readily dissolve

Table 4.1 Inorganic materials considered as candidates for neural interfaces.

Material		References
Metal	Platinum (Pt)	14–16
	Platinum–iridium alloy (PtIr)	17,18
	Iridium (Ir)	18,19
	Gold (Au)	20–22
	Stainless steel (SS)	23,24
	Titanium (Ti)	25
	Tantalum (Ta)	26,27
Metal compounds	Iridium oxide	28–33
	Tantalum oxide	26,34
	Titanium nitride	17,30,35,36
Carbon allotropes	Carbon nanotubes (CNTs)	37–41
	Glassy carbon	42,43
	Graphene	44–46
	Diamond	47,48

under electrical stimulation conditions that increase the electrode potential by a few hundreds of millivolts with respect to its resting potential. Pt, Ti and Ir are the only metals reported to date that resist dissolution when their electrode potential is significantly increased from their resting potential.

Metal compounds, including commonly used iridium oxide (IrOx) and titanium nitride (TiN) and to a lesser extent tantalum oxide (Ta_2O_5), have also been applied in bioelectronics. IrOx can be fabricated and prepared using different approaches depending on the electrode design, including electro-deposited, activated, or sputter-coated IrOx. IrOx has been used in stimulating microelectrodes,[51] demonstrating enhanced charge injection capacity compared to traditional noble metals (Pt and PtIr). TiN is a chemically stable and biocompatible metallic conductor, and has been used in medical devices such as cardiac pacemakers, cardiac valves, and orthopaedic prostheses.[51,52] The application of this metal compound in neural bioelectronics has been demonstrated through the fabrication of TiN-coated Pt microelectrodes for cortical stimulation.[17] Ta_2O_5 has been explored for electrode fabrication for neural interfaces, but has not seen much application in the field.[26,27]

Inorganic carbon allotropes include graphene, diamond, and carbon nanotube (CNT) materials. The conductivity and flexibility of graphene make it a popular choice for bioelectronic applications, and graphene/graphene derivatives have been demonstrated to be effective in the stimulation[45,53] and recording[44,46] of neural cells. CNTs have been used as coatings for improving bioelectronic performance, demonstrating efficacy in the stimulation of neural cells.[54]

4.2.2 Insulating Materials

Bioelectronics require insulating material in addition to the electrode materials, and the tissue interface with these insulators is also a vital consideration in the device design. Table 4.2 lists common electrode materials, both for electrodes and insulators, in different types of neural interfaces.

The most commonly reported insulating materials used to interface nerve tissue are silicone elastomers and imide polymers. Silicone is a silicon-based polymer formed from siloxane bonds that has been widely used as both an electrode insulator and as a device encapsulant. Silicone elastomers have important properties that make them ideal as neural interface insulators.[75] Silicone polymers are not only highly resistant to electric flow and chemical abrasion, they can also be cured at low temperatures yielding a low modulus substrate that allows control over the shape and volume of the insulator. Also, they can withstand high temperatures after curing which allows for sterilisation of the device,[76] and they have been successfully tested in animals and humans.[77] Their elastic properties (up to 450% elongation) provide neural interfaces with a substrate that can be deformed continuously, a common mechanical environment presented to implanted neural interfaces.

Organic polyimide polymers have also been used as neural electrode insulators. Like silicone, polyimides are resistant to heat and chemicals and

Table 4.2 Common electrode and insulator materials used for neural interface devices.

	Interfacing tissue	Inorganic material	Insulating material	Type	Reference
CNS	Cortical slices (*in vitro*)	Pt	Silicone	Cortical	55
	Subthalamic nucleus	Pt, PtIr	Silicone	Deep brain penetrating electrode	56
	Subthalamic nucleus	Glassy carbon	Silicone	Deep brain penetrating electrode	57
	Cochlea, spiral ganglion neurons – VIII nerve	Pt, PtIr	Silicone	Cochlear implants	58
	Thalamic nucleus	Au/Cr	Polyimide	Intracortical	59,60
	Rat's cortex	Pt	Polyimide	Intracortical	61
	Motor cortex	Ti	SU-8/Parylene	Intracortical	62
	Retinal ganglion neurons – optic nerve	Pt	Silicone	Retinal implant	63
	Spinal cord	Pt	Silicone	Thin, flexible electrode	64
CNS/PNS	Intracortical/intraneural	Pt, Au, Ir	Polyimide or silicon nitride	Penetrating electrode	65
PNS	Sciatic nerve	Pt	Silicone	Flat interface nerve electrode	66,67
	Sciatic nerve Spinal cord	Au	Polyimide	Intracortical	55,68
	Vagus nerve Phrenic nerve Peroneal nerve	Pt	Silicone	Helical electrodes	69,70
	Sciatic nerve Epimysium	SS	Silicone	Cuff electrode	71
	Epimysium	SS	Silicone	Penetrating, intramuscular electrodes	72
	Vagus nerve	Au, Pt, Pt-Black	Polyimide	Cuff electrode	73
	Sacral nerve	PtIr	Silicone	'Book electrode'	74

allow for the formation of thin, flexible and conformable films.[55] An attractive feature of polyimides is their photosensitivity, which in contrast to silicone allows for the fabrication of complex micropatterned electrode arrays *via* photolithography. Silicones and polyimides are long-lasting materials with very limited biological reactivity and have been used clinically as neural electrode insulators in implanted neural prosthetic devices.

4.3 Biocompatibility

Biomaterials must be inherently 'biocompatible' and have no adverse immediate or long-term effects on the surrounding tissue. The interaction between a living cell and the electrode material is the bridge between the worlds of biology and electronics, and hence is important to understand and optimise. The mechanical trauma and long-term effects of implanted devices are also key considerations when discussing biocompatibility and will be discussed further in Section 4.6.2.1. Although the term 'biocompatible' has been widely used as a material property, the reader is reminded that biocompatibility testing is always context dependent and is specific to the final, intended application of the material and/or device.

4.3.1 Assessing Cellular Response

To assess the biocompatibility of a bioelectronic interface, the first step is to conduct *in vitro* experiments and observe cytocompatibility with relevant cell types. Typically, bioelectronic cytocompatibility is assessed using relevant immortalised cell lines, such as fibroblasts, glial cells, or rat neural cells derived from the pheochromocytoma (PC12).[14,19,78,79] Immortalised cells lines are easy to procure and their cell culture relatively simple, as they can be cultured for indefinite periods of time with genotypic stability.[80] The cells are seeded and cultured on the bioelectronic material, and then assessed for several important biological responses. The cytocompatibility of the cells can be simply observed using immunofluorescent stains, which identify live and dead cells to determine what percentage of the seeded cells survive. Additionally, the proliferation and differentiation of the cells can be assessed to determine how well the substrate supports normal cell function. Immortalised cell lines however do not mimic their *in vivo* counterparts due to the genetic mutations or modifications required to enable their indefinite proliferation, and hence may have reduced biological relevance.

Primary cell lines are cells which are obtained from living sources (animal or human) and provide a more reliable biocompatibility analysis due to the similarity to the cells found *in vivo*. Primary neuronal *in vitro* culture provides a window to understand how cells can interface with, and are influenced by, the bioelectronic material, both passively and with electrical stimulation. Dorsal root ganglion (DRG) neurons,[81,82] Schwann cells,[83] and neural stem cells[28,84,85] are primary cell lines that are commonly used to assess biocompatibility. Furthermore, since brain and nerve tissue function are highly dependent on the cross-talk between different cells, mixed or co-culture systems can provide a more relevant model to study biocompatibility, as this mimics the natural tissue environment more accurately.

Beyond two-dimensional cell (2D) culture, *ex vivo* models are a recent development in the field of medical device characterisation. Organotypic cultures are produced by taking slices of living tissue and maintaining the tissue *ex vivo*, providing biologically relevant tissue cultures without taking the step into *in vivo* studies with live animals. Organotypic cultures can be ideal models for understanding more long-term tissue responses to the presence of bioelectronics, and the performance of the electrode in turn can also be assessed.[80]

The development of organoids is a promising area of research for the future of bioelectronic biocompatibility. Organoids are grown from stem cells, and ideally create self-organised 3D tissue structures which have all the necessary phenotypes in the correct and functional structure.[86] The ability to create functional neural tissue from human derived stem cells moves the field towards more accurate models that do not require rare tissue donation, and can even be used for personalised medicine approaches for tailored testing platforms.[87]

Finally, *in vivo* models are the last stage of testing for neural implants before moving to human clinical trials, and are the more accurate models currently available to study the long-term biocompatibility, in addition to the efficacy of the implanted device.[88]

4.3.2 Improving Biocompatibility

4.3.2.1 Physical Modification

The biocompatibility of an inorganic biomaterial can be improved through physical treatments of the electrode surface to improve the tissue–electrode interface. A common approach is to modify the topography of the surface. Overly smooth surfaces inhibit cellular adhesion, and hence reduce the integration of the electrode into tissue. Roughening the electrode surface not only has electrical benefits, as discussed below, but also improves the ability of cells to adhere to the surface.[89] Metallic electrodes can be mechanically or electrochemically roughened to improve biocompatibility. TiO_2 synthesised *via* cluster-assembly growth has been shown to produce nano-structured surfaces which activate microglia to become pro-inflammatory, significantly more than planar TiO_2 surfaces.[90] Activated microglia assist with tissue repair and debris removal, beneficial processes at the electrode interface. Similarly, nanoporous gold electrodes, fabricated through selective dealloying, have been demonstrated to enhance neuron adhesion compared to planar gold electrodes.[20] Creating a more textured surface is beneficial for general cell adhesion, and taking this further to create a more ordered topographical surface can be highly beneficial in supporting the communication between neurons on an electrode. Micro-textured diamond electrode surfaces have been shown to promote axonal growth along the topographical features, leading to more ordered cellular growth.[48]

4.3.2.2 Chemical Modification

Biocompatibility can be chemically improved as well; controlling the hydrophobicity/hydrophilicity, introducing cell adhesion molecules or biomimetic peptides, and fully coating the electrode surface with extracellular matrix components have all been used to modify electrode surfaces.

The hydrophobicity/hydrophilicity of a surface can influence cellular adhesion through affecting protein adsorption and conformation. Chemical functionalisation of a surface, through introducing carboxyl, hydroxy, amide, or amino groups increases the hydrophilicity of the surface, which in turn supports better cellular adhesion. Furthermore, coating of electrode surfaces with a self-assembled monolayer (SAM) is a common technique used to control hydrophobicity/hydrophilicity through careful selection of monolayer composition, in addition to providing a controlled and well-defined surface functionalisation. Thiol-based SAMs are often used for gold electrodes, as they provide a relatively simple functionalisation process, and it has been demonstrated that controlling the gold electrode surface chemistry through amino-terminated SAMs can support neuronal attachment.[91]

Extracellular matrix components (ECM) are key components within normal cellular environments that assist cellular adhesion, migration, and development. Commonly used ECM molecules for surface coatings are fibronectin, laminin, and collagen, as they promote cellular adhesion. These molecules are quite large (200–900 kDA) and can be physisorbed or covalently bound to the substrate. Hence, while they promote cellular adhesion, they may also reduce the efficacy of the electrode performance.

As an alternative, biomimetic peptides can also be used to enhance cellular adhesion to electrode materials, typically using the RGD (Arg–Gly–Asp) sequence, as it is a common cell adhesion motif present in fibronectin and throughout the ECM. These biomimetic peptides are far smaller than whole proteins (*e.g.* fibronectin), which reduces the distance between the cell membrane and the electrode surface. Functionalising a gold surface with an RGD peptide has been shown to result in enhanced PC12 adhesion and improve the electrochemical signal measured from the cells, compared with bare gold electrode surfaces.[21] SAMs are often used as a surface coating to couple with peptides, for ease and control of functionalisation. Using SAMs to link a synthetic laminin-derived peptide sequence to a gold electrode surface has been shown to enhance the attachment and outgrowth of embryonic mouse hippocampal neural cells.[92]

4.4 The Tissue–Electrode Interface

It is important to consider that neurons send their electrical signals, or action potentials, using ions and ion fluxes to transfer charge. To efficiently deliver and record electrical signals in neural tissue, there must be good coupling between the organic 'ionic' signal of the neurons and the electrode at the tissue–electrode interface. This interface can be described as a simplified

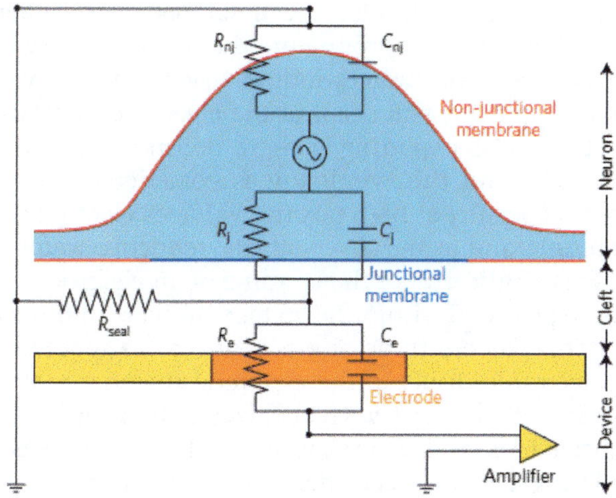

Figure 4.2 Schematic layout depicting the spatial relationships between a neuron and a substrate-integrated electrode and the analogue passive electrical circuit.
Reproduced from ref. 93 with permission from Springer Nature, Copyright 2013.

electrical circuit, as illustrated in Figure 4.2. The model in Figure 4.2 represents a neuron, the electrode, and the gap between the neuron and electrode surface. The junctional membrane interfaces with the electrode, and the non-junctional membrane interfaces with the surrounding solution; both generating resistance. The potentials generated by neurons produce current flows through both the junctional and non-junctional membranes. The gap between the neuron and electrode generates additional resistance.

With this simple circuit in mind, we now consider the role of the electrode, in stimulating cells with an electrical signal or recording the electrical signals from the cell. When assessing the electrical performance of an electrode interfacing with cells or tissue, there are several key parameters to discuss – the electrical impedance, the charge injection limit, and the charge storage capacity. The impedance of an electrical system is the effective resistance of an electrical circuit with an applied voltage. In a biological setting, the impedance is influenced by several factors; the surface area of the electrode, the culturing media, and the cell–electrode interface. The charge injection capacity (Q_c) is the maximum charge density that the electrode can deliver without causing irreversible chemical reactions at the electrode–tissue interface, and the charge storage capacity (CSC) is an estimate of how much charge an electrode can inject. These properties are dictated by the electrode material and design.

4.4.1 Stimulating Electrodes

Electrically stimulating cells and tissues requires careful consideration of the electrode material and stimulation parameters to be used, whilst still also delivering enough charge to induce a cellular response. When electrically stimulating in a biological environment, the voltage limits for electrical stimulation are dictated by the 'water window' – a voltage range which indicates at what point water splitting will occur at the electrode interface, typically at around −1 V to 1.2 V. Water splitting is highly detrimental to tissue and if the electrodes operate beyond this window it is considered unsafe. The water window is material specific; at high positive voltages electrodes facilitate oxidative water splitting, and at negative voltages reductive water splitting. The water window is also indicative of the Q_c value of an electrode; the maximum charge that can be provided within the voltage limits is in units of Coulombs per unit area. Hence, as electrode size is reduced (as required by more sophisticated bioelectronic devices), the electrode material may be unable to provide the required charge safely. Hence, when choosing the electrode material, it is important to consider the Q_c value for the required electrode design.

When stimulating *via* an electrode, a reaction at the electrode–tissue interface occurs to inject the charge into the tissue. There are two classes of charge injection mechanisms; Faradaic and capacitive. A capacitive injection mechanism occurs where charge is distributed through the presence of charged chemical species *via* a double-layer at the electrode interface. This double layer can be electrostatic with a simple ionic charge distribution, or electrolytic with charge stored across the electrode. Conversely, a Faradaic injection mechanism occurs when electrons are transferred between the electrode and tissue, resulting in a redox reaction at the electrolyte–tissue interface. Examples of the charge injection behaviour of different inorganic biomaterials are listed in Table 4.3.

A Faradaic injection mechanism can be problematic for bioelectronic interfaces, as new chemical species are generated at the electrode–tissue interface, and possibly irreversible chemical reactions, such as water electrolysis, changes in the local pH, or protein reduction/oxidation.[47] All of

Table 4.3 Characteristics of materials (including graphene) used as electrodes in neural interfaces. The values noted are as published in the respective reports. Adapted from ref. 2 with permission from John Wiley & Sons, Copyright 2017 WILEY-VCH Verlag GmbH & Co. KGaA, Weinheim.

Type	Material	Mechanism	Charge injection limit (mC cm^{-2})	Ref.
Metals	Gold	Faradaic	—	94
	Pt	Pseudocapacitive/ Faradaic	0.05–0.15	14,51
Metal Compounds	PtIr	Capacitive/Faradaic	0.13	95
	TiN	Capacitive	0.7–1	96
	IrOx	Faradaic	1–5	95
Inorganic Carbon Allotropes	Porous diamond	Capacitive	3	97
	CNTs	Capacitive	1.6	54
	Single layer graphene	Capacitive	—	98

these outcomes can contribute to tissue damage, increased impedance, and decreased long-term device sustainability. Conventional metal electrodes transfer charge in a faradaic manner, transferring electrons to the tissue. IrOx has a Faradaic injection mechanism, due to reversible Faradaic reactions between the Ir^{3+} and Ir^{4+} redox states, and the flow of counter ions from the physiological electrolyte.[29] Pt and PtOx can be described as pseudocapacitive, as electrons do pass between the electrode and tissue, but the redox reaction is confined to a monolayer on the electrode surface.[99] TiN has a capacitive charge injection mechanism; within the water window no electrochemical reactions occur at the electrode,[52] making the electrode stable when applying stimulation in a biological setting. Materials such as tungsten and tantalum have only been recommended for recording electrodes since issues such as electrode decomposition have been observed upon electrical stimulation.[49] Also, materials with nickel content have been reported to produce allergies in some subjects.[49]

For stimulating electrodes, a low impedance enables safe charge injection *via* the electrode.[51] Low impedance materials facilitate charge transfer in stimulating electrodes and demand less power from the device to deliver the required charge. A low impedance is vital to ensure when an electrical stimulation is applied, the electrical potential in the stimulated tissue will not damage the integrity of the cell membranes or introduce electroporation; safe stimulation levels for neural prostheses vary within the region of 0.5–4000 $\mu C\ cm^{-2}$, depending on the placement and application of the electrode. Low impedance is also important for efficient charge transfer between the electrode and living tissue. A high charge storage capacity is also desirable, as the electrode is then able to store a relatively large charge without undergoing irreversible and possibly cytotoxic Faradaic reactions.

4.4.2 Recording Electrodes

Depending on the application, at least two different types of voltage signals are required to be detected by recording electrodes, slow-varying local field potentials (LFPs), which reflect the collective behaviour from a small group of neurons near the electrodes, and action potentials, which last on the order of milliseconds ('spikes') from individual neurons or single units. Ideally, recording electrodes should be capable of measuring both types of signals from arrays of neurons, across large areas, and with high spatiotemporal resolution. While LFP recordings can be achieved using electrodes larger than 50 μm, single unit recordings require the use of smaller electrodes of cellular dimension (<20 μm). Impedance is a common metric that is used to measure the performance of an electrode. In general, a lower impedance value is associated with a higher recording quality and lower noise level, therefore leading to a higher signal-to-noise ratio. LFPs and single units can present in the raw recording traces from the same measurement, and then be decomposed by Fourier transform into different frequency bands. LFPs are typically below 100 Hz, while single unit action potentials are above

250 Hz. Similarly, action potentials generally have a magnitude in the range of 10 to several 100 μV, and LFPs can have amplitudes that are even larger. One electrode may be able to record action potentials from multiple neurons nearby (normally within 140 μm).[100] Spike-sorting is then necessary to decode the recording to obtain signals from single neurons.

4.5 Electrode Array Fabrication

The techniques for fabricating neural interfaces have advanced significantly over the past several decades. Sophisticated platforms are now available with utility not only in neuroscience research, but also in non-pharmacological approaches to treating diseases. The next generation of implantable devices requires large-scale, high density platforms that are capable of measuring and modulating neural activities at the single neuron level and across large 3D volumes. The development of functional electrode materials, together with the application of novel electrode designs and fabrication technologies, will enable long-term neural stimulation and recording with largely improved spatiotemporal resolution.[101]

Depending on the research topics and applications, electrodes with different designs have been fabricated and applied. Neurons can be recorded and stimulated either intracellularly or extracellularly.[102] Intracellular stimulation and recording can access more detailed information from neurons with the best selectivity, as only the targeted neuron will be monitored or modulated. However, intracellular techniques are normally more invasive than extracellular interfacing, which may damage the neurons and limit the time of application. Extracellular stimulation resolution depends on the both the electrode size and the electrode/neuron distances. With extracellular recording, the signals can be processed to extract information in different frequency bands, as described previously. Action potentials, also referred to as single-unit recordings, represent the structural and functional unit of the nervous system, therefore they hold the key to the understanding of the fundamental principles of neural networks. In this chapter, when discussing the design of recording electrodes, we focus on the efforts towards the application of single-unit recordings.

Electrodes with small diameters are normally preferred for neural interfaces. Small electrodes can improve the quality of recording signals from individual neurons, and at the same time, have the potential to selectively stimulate neurons with high spatial resolution.[101] However, there are several technical problems when fabricating small electrodes involving impedance and Q_c. First, the electrode impedance increases when the exposed area of the electrode decreases. For neural recording, an increase in electrode impedance leads to larger noise during recording, therefore reducing the signal-to-noise ratio. For neural stimulation, electrodes with larger impedance also require higher power consumption and voltage compliance, which may bring challenges to powering systems and integrated circuit designs. Second, for neural stimulation, smaller electrodes must inject a greater amount of charge for neural activation. Therefore, the charge required may exceed the safe charge injection limit of the stimulating materials. To overcome these difficulties, efforts have been made to improve the electrochemical properties of the electrode materials. One commonly used approach is to fabricate electrodes with large effective surface areas, while retaining a small geometric surface area.[103]

For long-term applications, neural recording and stimulation can both benefit from an intimate contact between the electrodes and the target neurons. Electrodes communicate with neurons *via* electrical fields, which decrease with distance by a factor of 2. Therefore, when the recording electrodes are far from the target neurons, the voltage signals drop rapidly, as do the recording signal-to-noise ratios.[104] Such scenarios also require the stimulation electrodes to deliver more charge for neural activation. In addition to the physical placement of electrodes, the formation of glial cells is another major factor separating the electrode from the neural tissues. Glial cells are non-neural cells found in the central nervous system (CNS) and can form a passive barrier around implants in neural tissue.[105] Therefore, the materials for electrodes are required to be biocompatible and to mimic neural tissue characteristics for minimal inflammation and to eliminate neuron loss. For *in vivo* brain studies, invasive penetrating electrodes are normally used to access the neurons deep in the tissue. In these applications, thin and small electrodes help to minimise brain injury during both implantation and their long-term presence. Conventional electrode fabrication normally uses rigid materials, such as silicon, which have Young's moduli much higher than soft neural tissues. Replacing the rigid materials with flexible components, discussed further in Section 4.5.4, can reduce the mechanical mismatch, therefore minimising the chronic response following electrode implantation and improve the ability for devices to interrogate and modulate neural activities.[106]

In addition to the use of novel materials, advanced fabrication techniques are also required to enable scalable fabrication of high-density electrode arrays. Fabrication is expected to be low cost and high efficiency, with minimal manual procedures and maximal yields. Different technologies have been developed based on the electrode materials and array configurations, which is introduced in detail in Section 4.5.4.

In this section, we introduce the design and fabrication of: 1) *in vitro* microelectrode arrays, used for experiments with cultured neurons and tissues such as explanted retina and brain slices; 2) *in vivo* microelectrode arrays, suitable for studies in both the central and peripheral nervous systems; and 3) multifunction electrode arrays, which can perform more than one function, not only electrically, but also optically and chemically interacting with the nervous system.

4.5.1 *In Vitro* Microelectrode Array

In vitro studies with cultured neurons and neural tissues, such as explanted retina and brain slices, have contributed significantly to our understanding of neural circuits and their function in the nervous system. Compared with *in vivo* studies, *in vitro* experiments require less in terms of animal handling, and when using cell lines they can also largely reduce the number of animals used. Here, we introduce different types of microelectrode arrays for extracellular or intracellular interfacing.

4.5.2 Extracellular Interfacing

Planar microelectrodes are the most commonly used extracellular platforms for *in vitro* applications. Conventional planar microelectrode arrays have a pattern of less than 300 electrodes, with each electrode typically around 30 μm (Figure 4.3a). The electrodes are made of metals such as gold, titanium nitride and platinum, insulated with organic or inorganic materials, such as epoxy resin or silicon oxide. More recent progress in planar microelectrode array fabrication uses carbon-based materials for electrodes, such as CNTs

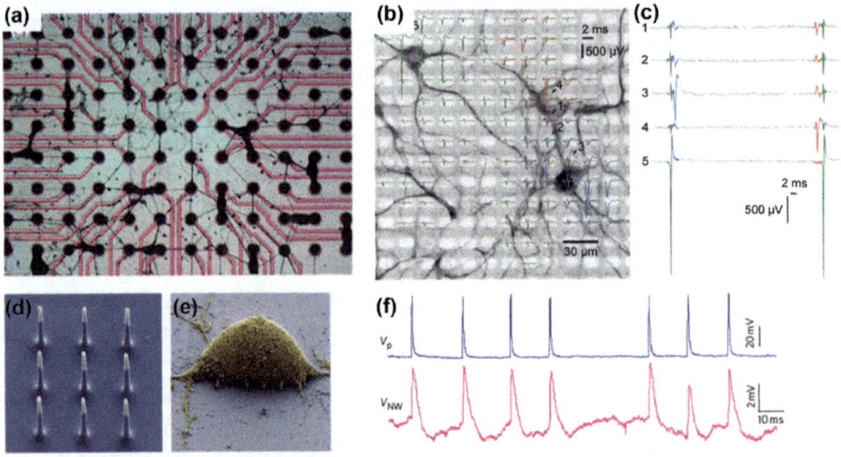

Figure 4.3 Microelectrode arrays for *in vitro* experiments. For extracellular interfacing, (a) shows a planar microelectrode array for interfacing cultured neurons. The array is composed of CNT electrodes (dark circles) and passivated interconnecting TiN lines (red lines). After several days of cell plating, the neurons on the array formed networks with cell aggregates on the electrodes and cell-free connections in between. Reproduced from ref. 40 with permission from Springer Nature, Copyright 2009. In (b), neurons were cultured on a high-resolution CMOS multielectrode array platform. The extracellular recording signals from three different neurons are drawn in green, blue and red. The raw data of 60 ms from five electrodes indicated in (b) are shown in (c). Published by the Royal Society of Chemistry.[107] For intracellular interfacing, (d) shows electrodes composed of silicon nanowires, which could interface the neuron membrane coupled on the surface, as shown in (e). (f) Signals recorded from nanowires (VNW) are consistent with those recorded using patch clamping (VP), a standard technique for intracellular neural recording. Reproduced from ref. 108 with permission from Springer Nature, Copyright 2012.

and graphene. For example, CNTs have been demonstrated to promote neuron growth and the neurons have been reported to show enhanced spontaneous synaptic activities.[40] The improved contact between CNT electrodes and the neurons results in a better signal-to-noise ratio for recording and higher efficiency in evoking a neuronal response than conventional metal electrodes.

For substrates, both glass and silicon wafers are commonly used. However, glass wafers are usually more often selected due to their transparency, which makes it easier to observe cells adhered on the array surfaces using transmitted light microscopy. Before plating cultured cells, the substrates are coated with poly-lysine or laminin to promote neuron adhesion. The mechanical mismatch between the array and neurons can be reduced by replacing the substrates with flexible polymer materials, such as polydimethylsiloxane (PDMS), polyimide, or silk.

Fabrication procedures for *in vitro* arrays normally involve multiple steps, including photolithography, materials deposition and chemical or physical etching. Depending on the materials used, the fabrication can be either top-down or bottom-up. Most of the fabrication techniques have been well established and are compatible with current silicon technologies. Furthermore, the fabrication can be easily scaled up to increase the density of electrodes on the arrays.

With improved data analysis algorithms and advances in the design of application-specific integrated circuits, it is now possible to simultaneously record from thousands of neurons *in vitro* with planar microelectrode arrays. Most of these platforms use amplifiers based on complementary metal–oxide–semiconductor (CMOS) technologies. For example, Ballini *et al.* reported a 1024-channel CMOS microelectrode array with 26 400 electrodes for recording and stimulation.[109] They demonstrated *in vitro* recording with both rabbit retina and cultured cortical neurons with recording resolutions down to the sub-cellular level (<20 μm) (Figure 4.3b, c).[107]

In addition to *in vitro* applications, flexible planar microelectrode arrays have also been used as surface electrodes *in vivo*. The techniques include electroencephalogram (EEG), electrocorticography (EcoG) and electroretinogram (ERG). However, due to the large distance between the surface electrodes and the target neurons, the recorded signals are normally LFPs. To conduct single-unit recordings *in vivo*, penetrating electrodes are more often used and will be discussed below.

4.5.3 Intracellular Interfacing

Extracellular interfacing is non-invasive and allows long-term and multiplexed measurements. However, extracellular recording can only provide the extracellular forms of action potentials, while the more sophisticated signals, such as subthreshold synaptic activities, are lost. Therefore, the development of intracellular recording techniques can provide additional insights into neuroscience research. To overcome the limitation of traditional microelectrode arrays and to enable intracellular integration, nanostructured electrodes have been fabricated to increase the contact area of the cells with the electrodes. The first demonstration of the intracellular recording electrode arrays was reported by Spira and colleagues.[110] They fabricated gold mushroom-shaped microelectrodes with a size and surface chemistry that activated endocytotic-like mechanisms. The cultured neurons were found to engulf the electrodes, which significantly increased the area of the junctional membrane and dramatically improved the seal with the electrode surface. This resulted in an increase in the neuron-microelectrode coupling coefficient by more than two orders of magnitude. Another example was the utilisation of an array of vertically aligned Si nanowires (Figure 4.3d–f).[108] In this work, electroporation was used to facilitate internalisation of the electrode through neural membranes, which enabled both intracellular recording and stimulation. Similar designs have also been demonstrated using iridium oxide nanotubes,[111] platinum nanopillars,[112] gold nanocylinders,[113] and platinum nanopores.[114] In addition to electroporation, access to the intracellular environment has also been demonstrated using surface functionalisation of the micro/nanostructures as well as plasmonic optoporation.[113]

4.5.4 *In Vivo* Microelectrode Arrays

In vivo recording and stimulation normally require that electrodes are implanted deep into the tissue, although surface electrodes are also used for LFP recording. To be used in the long term, it is important to minimise the chronic immune response, shear motion of implanted probes, chemical instability of electrodes, delamination and degradation of insulating layers, and potential infection through percutaneous exposure. Each of these events can potentially lead to unstable neural interfaces over time. In this section, we introduce two types of electrodes: penetrating electrodes and injectable flexible electrode arrays.

4.5.4.1 Penetrating Electrodes

Penetrating electrodes can be used for stimulation and recording action potentials and LFP in animal studies to study the central and peripheral systems. Clinically, they have been used in applications such as deep brain stimulation, cochlear implants and visual prostheses. This section will first introduce two traditional penetrating microelectrode arrays made of metal microwires and silicon-based technologies. Carbon-based materials usually show better mechanical compliance with neural tissues, therefore are now under development for building microelectrode arrays for *in vivo* experiments.

Metal-based Microwire Arrays. Metal-based microelectrodes are the most widely used probes for neural interfacing, and have enabled the start of many breakthroughs in neurosciences over the last sixty years. In 1957,

Hubel pioneered the fabrication of sharpened tungsten microelectrodes by electropolishing, which led to a tip of sub-micrometre dimensions (from 0.5 to 0.05 μm in diameter).[119] The sharpened electrodes made it possible to penetrate into animal cortexes and to record extracellular action potentials from much smaller neurons and axons in the mammalian brain. This later enabled the discovery of the receptive field of simple and complex cells in the primary visual cortex,[120] which was awarded the Nobel Prize in Physiology or Medicine in 1981. Today, these metal microelectrodes are still commonly used in labs. They are normally composed of electrochemically sharpened wires that are less than 100 μm in diameter and completely insulated with a small, exposed area at the tip which forms the recording or stimulating site (Figure 4.4a). The materials used for electrodes include tungsten, platinum, iridium, platinum–iridium, gold, stainless steel and molybdenum. To insulate the wires, quartz glass,

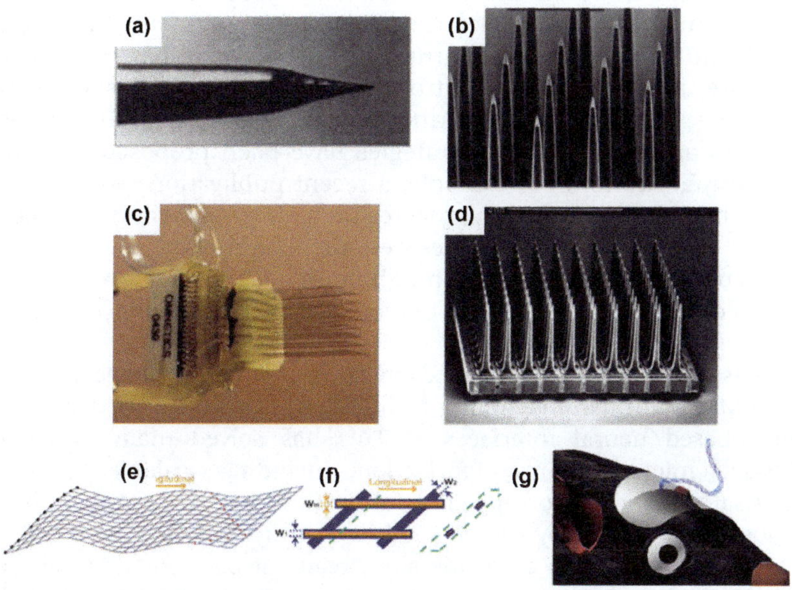

Figure 4.4 Microelectrode arrays for *in vivo* experiments. (a) A stainless steel wire electrode insulated in glass for extracellular recording. Reproduced from ref. 115 with permission from Elsevier, Copyright 1985. (b) Insulator-coated electrodes with exposed platinum tips in a Utah array. Reproduced from ref. 116 with permission from Elsevier, Copyright 1999. (c) A 64 channel, floating, discrete 8×8 microwire electrode array assembled into connector. Reproduced from ref. 117 with permission from Taylor & Francis, Copyright 2008. (d) SEM image of a full Utah array. Reproduced from ref. 116 with permission from Elsevier, Copyright 1999, Injectable mesh electrode (e) design, (f) 2D structure, and (g) schematic of mesh electronics injected into a mouse brain, with part of the mesh sagging between the brain and the needle. Reproduced from ref. 118 with permission from PNAS.

parylene-C, Teflon and polyimide are used, and the exposure of the electrode tips can be achieved with techniques such as laser cutting, plasma etching and photolithography.

To assemble metal-based microwire single electrodes into an electrode array is more challenging (Figure 4.4b, c). One method is to individually glue already insulated wires onto a connector and then expose the end of the tips. For example, Williams *et al.* showed the assembly of an electrode array with 33 microwires arranged in three parallel rows.[121] In each row, the insulated wires were first soldered onto a standard sub-miniature connector and then aligned using a specially designed fabrication jig. After assembly of all rows, the electrode tips were cut using fine surgical scissors. In another example, fabrication used laser machined ceramic substrates.[122] The wires were fit into holes drilled by lasers in the ceramic substrates and the tip was also cut open using a laser. Epoxy was then used to encapsulate the ceramic substrate. After carefully removing the ceramic, a flexible substrate surface was finally obtained. However, these conventional approaches for fabricating metal based microwire electrode arrays involve several manual procedures, making it difficult to build any array with a large number of electrodes. Furthermore, it is difficult to control electrode parameters, such as their sizes, pitches, and lengths, resulting in varying electrical characteristics from one to another. Potential strategies have been proposed to improve control over fabrication. For example, a recent publication reported an assembly technique for vertical microwires on a substrate using magnetic attraction.[123] In this work, the Ni wires were first magnetised, then assembled into a silicon substrate with feedthrough holes by magnetic force. The array was capable of recording action potentials *in vivo* from a mouse brain.

Silicon Based Electrode Arrays. Advances in micro-electromechanical systems (MEMS) and micromachining technology have led to the development of silicon-based neural interfaces.[124] This has solved many issues with metal-based microwire array fabrication, including scalable fabrication, miniaturisation, and uniformity. The fabrication normally starts with a silicon substrate, followed by the deposition of a metal layer. The metal layer is then patterned to form electrode sites, read-out pads for external circuit connection, and interconnecting traces between the electrodes and read-out pads. A second insulation layer is deposited over the whole device and patterned to expose both the electrodes and the read-out pads. The electrodes can be easily functionalised to promote their performance, and the fabrication procedures be modified to form different configurations. After implantation, silicon-based electrode arrays can be efficient for neural recording and stimulation for over several years. Among all the designs, Michigan probes and Utah arrays are the most widely used.

i) *Michigan Probes*

Michigan probes were the first developed silicon-based microelectrode arrays, reported in 1970. Nowadays, there are many different configurations available, ranging from single shaft, multi shaft to 3D-stacked multichannel structures.[125] The patterning of silicon defines the shaft shape of the probe and involves anisotropic etching with ethylene diamine pyrocatechol (EDP). This is done based on the fact that when silicon is doped with boron, the etching rate decreases dramatically compared with the undoped silicon. Various materials including gold, platinum and iridium can be used as the electrodes, which can be fabricated either by deposition or electroporation.[15,33] The insulation on top of silicon substrate is normally made with three layers of materials, including silicon dioxide, silicon nitride and another layer of silicon dioxide. Finally, flexible but thick polysilicon cables can be used for interconnection.

ii) *Utah Arrays*

Utah arrays (Figure 4.4d) are another class of extensively used penetrating arrays based on silicon technologies. There are mainly two types of Utah arrays, the Utah electrode array (UEA) for cortical applications, and the Utah slanted electrode array (USEA) for use in the peripheral nervous system.[116] UEAs consist of 100, 1.5 mm long silicon microneedles that project out from a 4 mm×4 mm×0.25 mm thick substrate, and USEAs have a similar architecture but different electrode length varying from 0.5 to 1.5 mm. The fabrication starts with a monolithic piece of boron-doped silicon followed by micromachining and acid etching to form the needle structure of the electrodes. Multiple layers of metals are then deposited onto the silicon needle tips to form the electrodes. The un-metalised silicon electrode shafts and the substrates are then insulated with a layer of polyimide. The connection to external circuits is finally achieved by depositing metal bonding pads from the backside of each electrode and interconnecting wires.

Utah arrays have electrodes all pointing opposite to the silicon substrate, while the electrodes on Michigan probes are on the same surface with the substrates. In Utah arrays, the probe lengths are restricted by the silicon substrate thickness, and only one recording site can be incorporated onto each probe shaft.

Carbon-based Electrode Arrays. Both metal and silicon are rigid materials, with Young's moduli much larger than those of soft neural tissues. Also, the thickness of metal wires and silicon probes are large, which may lead to the formation of glial scars after their long-term implantation. Therefore, carbon-based materials that are softer and with smaller diameters have also been used for fabricating penetrating electrode arrays.

The most commonly used carbon material for neural interfacing is carbon fibre, which has been demonstrated for both neural recording and chemical sensing.[126] After modification with other materials, such as conductive polymers, the charge injection capacity of the electrodes can be significantly improved, making it suitable for neural stimulation.[127] Ultrasmall dimensions (<10 μm) facilitate implantation and have been demonstrated to effectively reduce the chronic immune response after implantation.[127] In the same manner as metal-based microwire electrodes, carbon fibres can be insulated using Paralyene-C, Teflon, polyimide or silk and then the tips exposed by blade cutting, laser cutting or fire sharpening. The structure of the electrode arrays is normally similar to those of Utah arrays and metal-based microwire arrays. To assemble the arrays, Massey *et al.* reported fabrication using both silicon microfabrication processing and microassembly.[127] The resultant array has 32 channels, with a pitch of 38 μm. In this work, the authors first patterned silicon substrates and then assembled carbon fibres onto the substrate using silver epoxy. The tips were finally exposed using a blade and functionalised with poly(3,4-ethylenedioxythiophene) (PEDOT): polystyrene sulfonate (PSS) by electroporation.

Similar fibre structures have also been fabricated using many other carbon-based materials, which show capacities for both neural stimulation and high quality recording. These materials include carbon nanotube fibres,[37] laser-reduced graphene oxide fibres,[45] and graphene-based fibres.[44] These fibres all have small diameters (30–60 μm) and show good mechanical compliance with neural tissues, and therefore are expected to be biocompatible and induce less immunogenic response, beneficial for use in long-term applications. However, novel fabrication techniques are needed to assemble the fibres into higher density electrode arrays for future research.

4.5.4.2 Injectable Flexible Microelectrode Arrays

In addition to the use of softer electrode materials, there are also many designs that use flexible polymers as the array substrates. These array configurations are more-or-less similar in design to Michigan probes and Utah arrays, in which the micro-patterned silicon substrates are replaced with flexible polymers, such as PDMS, silk, and polyimide.

Among flexible polymer-based microelectrode arrays, injectable mesh electronic neural probes (Figure 4.4e–g) developed by Lieber and colleagues showed the greatest integration with neural tissues in 3D volumes.[118,128] The mesh electrodes led to minimal chronic immune response while providing good neural recording and stimulation performance. The fabrication of these arrays is based on standard photolithography on silicon wafers, as for Michigan probes, which are then released to an aqueous solution to adopt a suspended polymer-like 3D morphology. These mesh electronics have free-standing polymer–metal–polymer structures, with feature sizes similar to individual neuron soma and bending stiffness values similar to those of brain tissue. The 3D macroporosity of these arrays enables neurite interpenetration and prevents accumulation of pro-inflammatory signalling molecules. These electrodes have been implanted in cortex for up to 1 year with no chronic immune response and are capable of recording and electrical stimulation.[128]

4.5.4.3 Multifunctional Microelectrode Arrays

Multifunctionality is a general trend in neural interface fabrication; devices which can perform more than one function are expected to improve the efficiency of the tissue–electrode interface. For example, closed-loop feedback that relies on both neural stimulation and recording from the same device has been demonstrated to significantly improve the performance of cochlear prostheses and devices for deep brain stimulation. Multifunctional devices require the use of optimised materials as well as novel electrode technologies for device design and fabrication. In this section, some examples of devices fabricated to perform multiple functions are introduced.

Optical Integration. In addition to monitoring and modulating neural activities through electrical recording and stimulation, there have also been many other approaches using light. Optically, neural activities can be imaged after introducing ion sensitive or voltage sensitive indicators, or through functional magnetic resonance imaging (fMRI).[129] Compared with electrical recording, optical imaging is less invasive with better spatial resolution, allowing individual neurons in a large area to be monitored simultaneously. For neural stimulation, optogenetics enables precise deconstruction of neural circuitry with neuron-subtype specificity, where genetically engineered ion channels or pumps are used for optical stimulation.[130] Optoelectric devices that convert light into electrical fields have also been developed using metal or silicon-based materials.[38,131] There are many other materials, such as gold nanotubes, which have been demonstrated to be efficient for neural stimulation through photothermal effects.[132]

To optically stimulate and electrically record neural activities from the same device, Lee *et al.* fabricated flexible and transparent microprobe arrays using conducting semiconductor ZnO crystals.[133] In their arrays, ZnO served as an optoelectronic electrode for neural stimulation, and was used to read out the change in extracellular potentials. Neural recording and stimulation was demonstrated *in vivo* from mice cortex. Transparent electrode arrays have also been fabricated to enable simultaneous optical imaging and electrical interfacing using materials including CNTs[134] and graphene.[46]

Another type of array design for optical and electrical integration aims at facilitating light delivery for optogenetics. In these designs, micro light-emitting diodes (μLEDs) are incorporated into electrode probes, as demonstrated by Wu *et al.*[135] Using these novel probes, the authors demonstrated high spatiotemporal precision of parallel stimulation and recording in freely moving animals. They could independently control different cells of ~50 μm apart while recording high quality action potentials at the same time.

Chemical Integration. Neural activities can also be stimulated and recorded through chemical modification of electrode surfaces. Detecting neurotransmitters such as dopamine can provide additional information other than action potentials or LFP. The detection of neurotransmitters has been performed *via* electrochemistry. In addition to electrical or optical stimulation, neural activity can be modulated by delivery of channel-targeted or receptor-targeted biologics into a discrete brain region. Compared with optogenetics, these approaches can precisely manipulate biochemical pathways in the brain without genetic modification of the target neurons.

Multifunctional fibres have been fabricated for simultaneous optical, electrical and chemical interrogation of neural circuits.[136] The fibres were fabricated from polymers, exhibiting high flexibility and functionality under bending deformation. The micro-channels within the fibres can be used for delivering light as well as chemicals to modulate neural activities. At the same time, the electrodes incorporated in the fibres can be used for extracellular recording.

Other Integration. Other neural recording and stimulation approaches include magnetic or mechanical approaches. For example, both magnetic fields and ultrasound have been shown to evoke action potentials in neurons.[137,138] Magnetic recording of neural activity can be realised using nitrogen-vacancy centres in diamond or magnetic sensors based on spin electronics.[139] However, novel techniques are necessary for incorporating these functions into future devices.

4.6 Electrode Implantation

Electrode implantation is a necessary procedure to interface neural electrode technologies with target tissues. All current, clinically approved neural interfacing technologies require surgery for implantation, with the exception of surface electrodes for diagnostic purposes (for example electrocardiogram, electromyography and electroencephalogram) and a small number of experimental transcutaneous technologies. The technique used for the implantation of neural electrodes varies according to the tissue of interest (central nervous system, CNS, or Peripheral Nervous system, PNS).

4.6.1 Central Nervous System

Electrode arrays that aim to interface with the CNS can be categorised according to the degree of penetration; cortical, intracortical, depth electrodes and deep brain electrodes. All these systems require craniotomies to grant access to the brain tissue. This surgical intervention requires removal of the skin, the skull, and the dura matter over the area of interest. Using stereotactic frames supported by anatomical brain maps allows for localisation of the region of interest. Implantation can be further assisted by radiographic or magnetic resonance imaging techniques to coordinate electrode insertion and to confirm appropriate electrode positioning within the target tissue.

Figure 4.5 presents instances of clinically approved technologies that interface nerve tissue for electrical stimulation and/or recording. These electrode arrays are supported by either rigid or flexible substrates that penetrate nerve tissue presenting multiple electrode sites along the shanks. Rigid arrays can be inserted directly into the brain cortex, whereas flexible arrays require the support of a rigid shank to guide the insertion (Figure 4.5a and b). Electrode technologies such as neural probes (Figure 4.5b) are fabricated with silicon substrates stiff enough to allow for tissue penetration. However, some of these technologies have been modified into more flexible electrodes to allow for a closer proximity with neural tissue to be achieved, while maintaining the stiffness required for electrode insertion.[59–61]

Once insertion is complete, the craniotomy can be sealed using conventional techniques. Most technologies make use of wires to connect the electrode array with the stimulator source or the recording unit (Figure 4.5c). This is a critical limitation for these technologies since the wire bridges the interface between the brain tissue and the external environment, creating a risk of infection. Also, the wire maintains a physical support between the electrode array and the skull. Free movement of brain tissue within the skull combined with a tethered electrode creates a continuous mechanical stress at the neural interface. To address this limitation, electrode array technologies have incorporated the electronic components within the electrode array, with wireless adaptations to send and receive signals.[22,144] However,

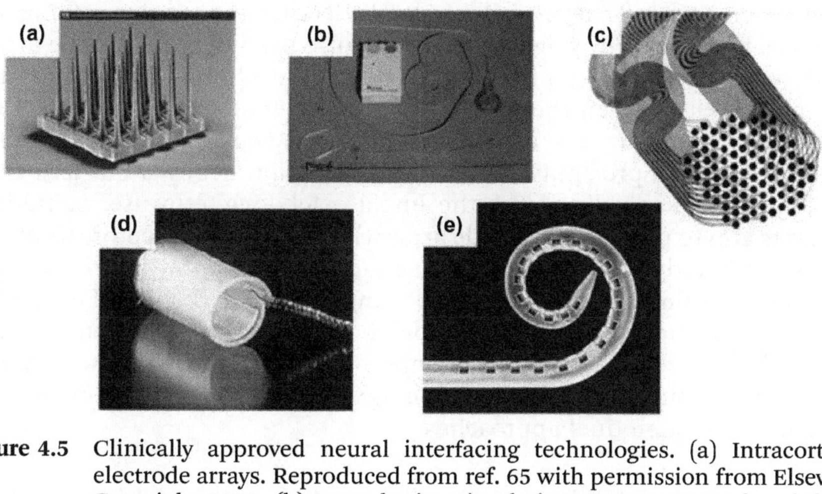

Figure 4.5 Clinically approved neural interfacing technologies. (a) Intracortical electrode arrays. Reproduced from ref. 65 with permission from Elsevier, Copyright 2000. (b) Deep brain stimulation system. Reproduced from ref. 140 with permission from John Wiley & Sons, Copyright 2008 International Center for Artificial Organs and Transplantation and Wiley Periodicals, Inc. (c) Retinal implant, Reproduced from ref. 143 with permission from IOP Publishing, Copyright 2004. (d) Cuff electrode. Reproduced from ref. 141 with permission from John Wiley & Sons, Copyright 2005 Peripheral Nerve Society. (e) Cochlear implant. Reproduced from ref. 142 with permission from John Wiley & Sons, Copyright 2005 The Triological Society.

these technologies are still in the early stages of testing for the biocompatibility, hermeticity and safe electrical performance of wireless platforms.

Neural prosthetic devices such as cochlear implants have managed to return auditory perception to profoundly deaf people (Figure 4.5e). These devices interface with spiral ganglion neurons that convey auditory sensory input to the brain. The spiral ganglion neurons are the soma of auditory nerve fibres, located in a structure known as Rosenthal's canal within the cochlea. Cochlear implants interface these neurons by inserting an electrode array in the scala tympani, a duct adjacent to the cochlear, which can be accessed through the round window or through a membrane that separates the scala tympani and the middle ear.[145] Interfacing this structure requires cutting skin and soft tissues behind the ear and drilling through the bone to access the middle ear in the cochlea, a process known as mastoidectomy. Once the round window is exposed, the electrode array is gently threaded into the cochlea. The array is connected to an implanted receiver–stimulator that is placed in between the skull and the skin and electromagnetically interfaced with an external sound processor.

The electrode technology of retinal implants is based on the working principle of cochlear implants. This consists of an electrode array that is placed near the output of the retina, which can produce a visual sensation upon electrical stimulation (Figure 4.5f). Retinal implants interface with retinal ganglion cells located within the membrane positioned in the internal back part of the eyeball. Several electrode–retina interfacing approaches have been investigated. These involve placing the electrode arrays either onto the retina (epiretinal), between the choroid and the retina (subretinal), or between the choroid and the sclera (suprachoroidal). Each interfacing approach has advantages and disadvantages. The epiretinal interface provides proximity to the retinal ganglion neurons but it is challenging to mechanically secure the implant for long-term use. Subretinal implants aim to place the electrode array where the photoreceptors would be in a healthy retina. Although ideal, surgical implantation is difficult and requires very thin electrode arrays ($<400\ \mu m$), adding to the risk of detaching or damaging the retina. Suprachoroidal electrode implantation is more mechanically stable and does not compromise the functional tissue layer of the eye but requires higher electrical charge to evoke a response compared to epiretinal and subretinal approaches.

4.6.2 Peripheral Nervous System

The surgical procedure for interfacing PNS tissue structures involves dissecting and displacing surrounding tissue (*e.g.*, skin, muscle, blood vessels and connective tissue) to reach the target nerve fibres that innervate the tissue of interest.[141] Depending on the degree of selectivity for stimulation and/or recording, the nerve will require further dissection. To effectively interface target nerves, electrode technologies are designed to conform to the tissue of interest while ensuring a stable mechanical and electrical interface. Different electrode technologies have been developed to surround, penetrate, or regenerate nerve tissue. These electrode technologies can be commonly categorised according to the mechanical interface conformed with the tissue. For instance, 'Cuff' electrodes, are made of split cylindric structures surrounding nerve fascicles (Figure 4.5(d)). Electrode sites placed in the inner face of the cuff electrically stimulate and/or record from nerve tissue. This PNS interfacing electrode is the least invasive but also the least selective format since nerve fascicles contain several nerve fibres that innervate different target organs. To increase selectivity, a variation of this technology consists of electrodes moulded into a 'T' shape extension that expands radially from the lumen (also known as book electrodes since they resemble a bookmark).[146] The T shaped endings are slowly pushed through the epineurium with the aim of segregating nerve fascicles.

Intrafascicular electrodes utilise penetrating probes to interface nerve bundles into and through the fascicles.[147] These electrode arrays can penetrate the nerve longitudinally or transversally. Longitudinal electrodes can interface with single fascicles whereas transversal electrodes have direct contact with the fascicle providing a higher level of stimulation and recording selectivity. The most invasive electrode interface aims to regenerate severed stumps of peripheral nerves. This system requires transecting the nerve and realigning it using a sieve. A mesh-like electrode array is placed transversally to the nerve in the centre of the sieve and, assisted by growth factors, axons grow through the holes of the array. However, successful regeneration depends on the size of the defect and the number of nerve fibres to match. Provided the sieve allows for nerve regeneration, the electrode array provides a direct interface to a high number of nerve fibres.

4.6.2.1 Challenges for Long-term Implantation

Neural interfaces operate by delivering electric charge to a target tissue using metallic electrodes. Despite the remarkable results of some of these technologies, such as cochlear implants that to an extent have returned hearing perception to profoundly deaf people, there is a continuous effort to improve these devices as they face several limitations (Figure 4.6). Perhaps the most challenging limitation to overcome is the inflammatory response against the implant. Kozai *et al.* have reported that microglia, the main active cell agent of the CNS immune system, are immediately activated following insertion of devices on brain tissue.[148] Despite designing and using materials that are regarded as biocompatible, the trauma caused during insertion and the persistence of the foreign object elicited a series of molecular and cellular events that targeted the electrode as a foreign object, ultimately isolating it from the target tissue *via* deposition of a collagenous rich scar tissue (fibrosis).

This limitation is further exacerbated by the mechanical mismatch that exists between stiff materials such as metallic electrodes (mostly Pt or PtIr)

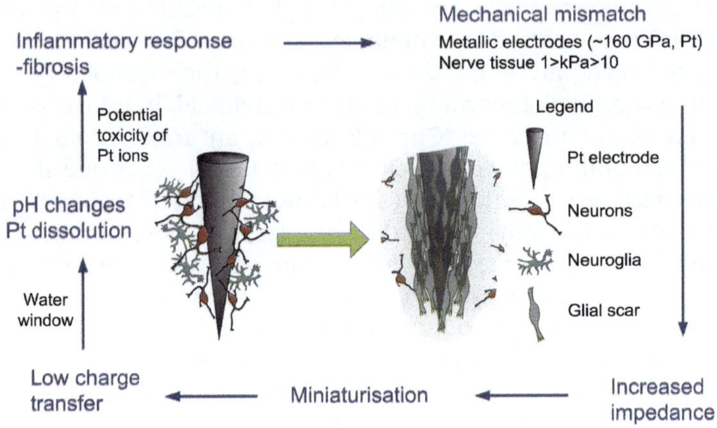

Figure 4.6 Schematic of feedback looping issues that affect the long-term performance of neural electrodes.

and soft substrates like nerve tissue. For electrode arrays that are mechanically tethered to the skull *via* lead wires, this represents a major issue given the continuous shear stress that maintains the neural interface in constant inflammatory conditions. Even wireless systems are not exempt from this limitation, as the continuous micromotion of the brain may create a shear stress between the device and the tissue and the inner skull.[149]

It is a well reported phenomena that, following surgery, electrodes undergo an increase in impedance.[150] The increase in impedance demands more power from the device to deliver the charge to effectively stimulate target nerve tissue, as previously discussed. The impedance increase is associated with the formation of extra layers of non-excitable scar tissue, which further affects the electrode's charge transfer. For instance, clinical and animal studies of cochlear implants have shown that electrodes undergo two distinct impedance fluctuation phenomena after implantation. At first, electrode impedance increases from 2–4 kΩ (baseline levels) up to 7–15 kΩ after 2–4 weeks post-implantation.[16,31,32,150,151] Second, upon device stimulation a drop in impedance of 1–2 kΩ has been observed, but electrode impedance does not return to baseline levels.[31,32,150,152] This change in impedance significantly affects the physicochemical stability of these electrodes and the power required to stimulate nerve tissue. As electrode impedance increases, more charge is required to stimulate nerve cells. For instance, for cochlear implants the minimum charge density levels required for auditory stimulation range is around 2–20 $\mu C\,cm^{-2}$.[16,153] An increase in impedance, such as the one observed in clinical studies (~3-fold baseline levels), could readily overlap with the charge injection limit of platinum electrodes (50–150 $\mu C\,cm^{-2}$),[154] resulting in potential tissue damage. This supports the proposed hypothesis that electrically active electrodes undergo a higher degree of fibrosis than unstimulated controls.[155,156]

For typical metal electrodes, such as platinum, corrosion resulting from electrochemical processes that occurs during stimulation has also been observed during long-term implantation. Pt particles, found as ions or protein complexes, are relatively common in long term users of cochlear implants. *In vitro* studies suggest that Pt ions and complexes are cytotoxic.[157] Tissue evaluation of cochlea from patients that have used cochlear implants for a showed that Pt particles can be found within the fibrotic capsule surrounding the electrode array and it remains unclear if these particles are able to reach other tissues.[158] In addition, previous studies support that Pt ions and complexes produced by corrosion can exacerbate the inflammatory response against the implant.[159] Overall, Pt dissolution creates a feedback loop whereby the initial inflammatory response results in further mechanical mismatch, and subsequent increased impedance (Figure 4.6). The increased impedance and tissue response against the electrode reduces the charge transfer and hence requires more power for stimulation, as described above, further exacerbating the dissolution of the Pt electrode and perpetuating this negative feedback loop. This results in issues that enable fibrosis, jeopardise neuronal survival, and present the need for using high charge electrical pulses to stimulate auditory neurons.

Additionally, there is a pressure to develop smaller and more densely packed electrode arrays for stimulation of specific populations of neurons, with the aim of inducing a more physiological response. Smaller electrodes are expected to minimise chronic responses after implantation as they result is less trauma to the tissue during implantation. However, decreasing the electrode size also results in an inherent decrease in the amount of charge that can be safely delivered. Pushing the charge injection limits beyond the water window results in electrolysis, pH changes, and the formation of toxic by-products that can detrimentally affect surrounding nerve tissue. Overall, these limitations present a great challenge in the design and performance of neural interfaces.

4.7 Hybrid Materials and Coatings

Great efforts have been devoted towards the optimisation of metal and semiconductor-based materials for neural interfacing to improve their electrochemical properties and biocompatibility. However, metal and semiconductors are generally stiff, with Young's moduli in the range of TPa, compared to nerve tissue with mechanical properties of almost ten orders of magnitude less. The significant mechanical mismatch between these conventional inorganic materials and the neural tissue can lead to chronic inflammation. The same challenges are also faced by most of the inorganic carbon allotropes. As discussed in previous sections, ideal materials for neural interfacing electrodes are expected to be biocompatible, soft, and mechanically conformable. They are also expected to transfer electric charge through capacitive mechanism *via* a low impedance, high charge storage capacity and high charge injection electrode–tissue interface. Unfortunately,

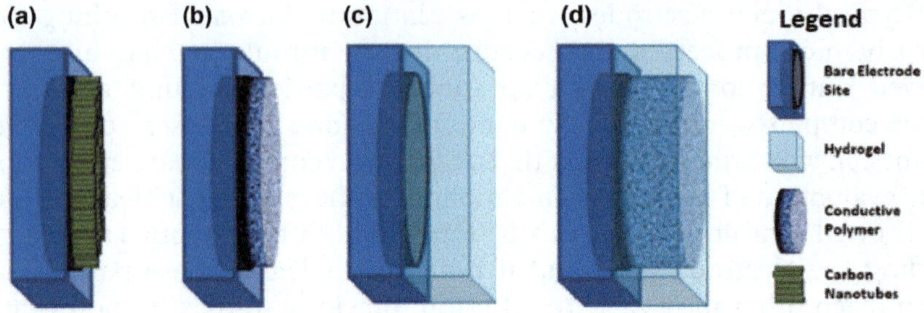

Figure 4.7 Schematic of coating approaches used for addressing the electrochemical and mechanical limitations of neural interfacing electrodes; (a) carbon nanotubes, (b) conductive polymers, (c) hydrogels, and (d) conductive hydrogels.
Adapted from ref. 50, https://doi.org/10.3389/fneng.2014.00015, under the terms of the CC BY 3.0 license, https://creativecommons.org/licenses/by/3.0/.

to date there is no single material that fulfils such design criteria and, to meet such specifications, significant efforts have been dedicated to identifying electrode modification technologies to address the gaps. Hence, the development and application of coatings to improve inorganic biomaterial electrodes can assist in improving the electrode–material interface. Furthermore, hybrid materials that incorporate the benefits of inorganic biomaterials and then modulate the materials properties to make them more compliant and beneficial for neural prostheses have also been investigated.

Neural interfaces can be fabricated at low cost using silicon-based techniques, as described in Section 4.5. The electrode materials used in these devices are commonly noble metals such as Au and Pt. Electrode coating technologies have been shown to significantly improve the electrochemical performance of traditional metallic and inorganic biomaterial electrodes.[50,160] These coatings can improve the electrode–tissue interface by providing more advantageous properties to facilitate good coupling.

Improving the electrical, mechanical, and biological properties of electrodes has been achieved in part through the development of electrode coating technologies. Coating technologies have made use of inorganic materials such as platinum-black (Pt-black), IrOx (iridium oxide), TiN, and some carbon-based materials including CNTs, as well as organic materials such as conductive polymers (CPs) and hydrogels (Figure 4.7).

4.7.1 Inorganic Coatings

4.7.1.1 Metal-based Materials

Pt-black, IrOx and TiN are three commonly used metal-based coating materials in neural interfacing devices. These materials can be deposited on metal surfaces *via* electroporation or sputtering. These coatings have been shown to significantly improve the charge transfer properties of traditional metallic electrodes, usually by creating a roughened surface that increases the geometric surface area, but each technology presents inherent limitations.

Pt-black has superior electrochemical properties than pure Pt. Pt can be blackened by sand blasting the surface and chemical treatment with hydrochloric acid, platinum chloride and lead, followed by applying an electrical potential. The resulting impedance of the electrode is ~30 times lower than that of Pt.[49] Although blackening Pt significantly enhances its charge transfer properties, this electrode modification has been found to produce a substantial adverse tissue response compared to Pt or PtIr alloys.[161]

IrOx films can be formed by electrochemical activation[18,162] or by reactive sputtering.[163] The former creates a hydrated oxide film by subjecting iridium to repetitive oxidation and reduction. The latter uses oxidising plasma on Ir substrates. IrOx films can also be thermally formed on Ti and PtIr substrates.[164] IrOx films formed on Ti and PtIr can increase their charge injection limit up to 1–5 mC cm^{-2}. Compared to Pt or PtIr alloy, this represents a 6–20 times increase in charge injection limit. However, the limitations of IrOx coatings include charge transfer through Faradaic mechanisms and the requirement for a positive bias for high charge injection. The effect of positive bias potential on tissue reactions and the long-term stability of electrodes is unknown and requires further exploration.

One electrode modification material that shows capacitive charge transfer without the need for potential bias is TiN. TiN can be sputter-deposited to create a highly porous geometry. This creates a high geometric surface with a charge injection capacity of 0.9 mC cm^{-2},[30] which is 6–18 higher than Pt electrodes (0.05–0.15 mC cm^{-2}). TiN is mechanically stable and can be patterned through photolithography and can also be adapted to current neural interfaces such as intracortical silicone probes, retinal prosthesis and electrode arrays.[36,51]

4.7.1.2 Carbon Nanotubes (CNTs)

CNTs are another commonly used coating material for improving the performance of electrodes. CNTs are formed from ultra-thin sheets of graphene (thickness ~1 atom) that can be assembled into single-walled or multi-walled tubes. CNTs can be applied onto metallic electrodes by electrochemical deposition,[39,165] immersion drying[166] or chemical vapor deposition.[167] Electrochemical deposition is the simplest way of forming CNT-coated metallic electrodes as it can be directly applied to an electrode site and avoids the requirement for toxic chemicals and high temperatures, such as those required for vapor deposition. CNTs can greatly enhance the electrical properties of metallic electrodes, resulting in higher signal-to-noise ratios in recording electrodes[39,41,168] and higher storage capacity in stimulating electrodes (~70 mC cm^{-2}).[169] However, in the latter study the enhanced charge storage capacity was most likely due to a combined effect caused by the incorporation of conductive polymer.

The enhanced electrochemical effects of CNTs can be attributed to their characteristic low volume and high surface area. However, CNTs have two major limitations. These materials are very stiff (~1.25 TPa),[170] further worsening the mechanical mismatch at the electrode–tissue interface. Also, their nanoscale dimensions and biologically inert features allow them to infiltrate tissues without being identified as foreign objects by the immune system.[171] Since high concentrations of CNTs can induce cytotoxicity,[172] the biocompatibility of these coatings is still controversial whenever the mechanical anchoring of CNTs cannot be guaranteed.

4.7.1.3 Organic Coatings

Organic coatings encompass conductive polymers (CPs), polymer composites and conductive polymer blends with hydrogel scaffolds (conductive hydrogels, CHs). These coatings have been used to significantly enhance the electrochemical profile of traditional neural interfacing electrodes. Specifically, they provide metallic electrodes with low impedance and high charge injection capacity compared to their uncoated counterparts. In addition to electrochemical properties, these organic coating materials have better mechanical compliance with nerve tissues than inorganic materials, therefore they are expected to reduce chronic inflammation following implantation.

Conductive Polymers. CPs are formed from monomers that when combined into a polymer chain generate a conjugated system that consists of alternating double and single bonds within their chemical structure.[173] The most common CP employed to coat neural electrode sites is PEDOT, due to its electrochemical stability within hydrated and oxygenated environments.[173–175] Other CPs that have been investigated include polypyrrole (PPy)[174,176,177] and to a lesser extent polyaniline[178,179] and polyterthiophene.[180]

CPs can be applied to neural prosthetic electrodes by chemical or electrochemical deposition, and the fabrication method and choice of CP material determine the final mechanical and electrochemical performance.[181] As for CNTs, electrochemical deposition is the method of choice since it allows controlled and conformal deposition on electrode sites. Table 4.4 lists common coating technologies electrodeposited on inorganic metallic substrates. The electrochemical polymerisation of the monomer must be undertaken in the presence of a dopant molecule to convey the polymerised substrate with electrical conductivity. The dopant is an electrically charged molecule, usually highly electronegative molecules (anions) for electrodeposition, however cations can also be used to dope CPs. Common dopants that have been widely investigated include poly(styrene sulfonate) (PSS), para-toluene sulfate (pTS), dexamethasone phosphate (Dex-P) and

Table 4.4 Coating technologies adapted to metallic substrates *via* electrodeposition.

Metallic substrate	Coating material	Reference
Gold	PPy	185
Gold	Polyterthiophene	180
Indium–tin oxide	CNT	39
Indium–tin oxide	PPy	185
Titanium nitride	CNT	168
Platinum	PEDOT	175,181
Platinum	PPy	186
Iridium	PEDOT	187
Iridium	PPy/PEDOT	188
Platinum–iridium	PEDOT	189

perchlorate (ClO_4^-).[50] Biomolecular dopants have also been incorporated into CPs, such as hyaluronic acid, chondroitin sulfate, heparin and taurine.[175,182–184] The properties of the final CP material can be greatly influenced by the chemical nature of the dopant, changing its roughness, stiffness, and conductivity.[183] CPs have been shown to significantly enhance CSC, the charge injection limit (CIL) and electrochemical impedance. Electrochemical deposition of PEDOT using small sulfonate dopants, such as pTS, has been shown to increase CSC by one order of magnitude and the CIL by up to two orders of magnitude, as well as showing low impedances across the frequency spectra.[175]

The improvement in electrical performance has been primarily attributed to the increase in surface area resulting from the inherent rough morphology of PEDOT. The thickness and roughness of PEDOT and other CPs can be controlled by varying the time over which electrodeposition occurs.[185,190] While CPs address electrochemical limitations that allow the size of Pt electrodes to be reduced, issues remain regarding mechanical stability and biocompatibility. These CP-coated inorganic electrode systems are still too stiff compared to nerve tissue (1–10 kPa). CP elastic moduli remain comparatively high, generally ranging from 1–3 GPa. In some cases, hydrated versions of PEDOT allow for a drop in mechanical modulus to 40 MPa, but this is still over two orders of magnitude higher than native nerve tissue. Also, *in vivo* reports have shown that CP coated neural electrodes can delaminate[191,192] and rough-surface variations of PEDOT become brittle and friable.[182,193] Several approaches have been explored to enhance the mechanical stability of CPs. These include the formation of hybrid composites *via* chemical tethering, or mechanical interlocking with carbon-based coatings such as graphene or with polymeric networks like hydrogels.[194,195]

Conductive polymer coatings, with the inclusion of CNTs and graphene, have also been demonstrated to be effective materials for electrode coatings. Zhou *et al.* reported the charge injection capacity of synthesised films to be as high as 6.2 mC cm^{-2}.[169] Here, CNTs are a physical and electrochemical constituent of the electrode coating but also serve as the dopant for increasing the conductivity of PEDOT. PEDOT/CNT films coated onto Pt electrodes support the growth of neuron-like cells (PC12), and demonstrate excellent long-term stability during prolonged electrical stimulation protocols.

Conductive Hydrogels. A small number of conductive hydrogel (CH) technologies have been developed for neural interfacing purposes. Examples include modified cochlear implants[196,197] and intracortical arrays (Figure 4.8).[198] Hassarati *et al.*[197] and Dalrymple *et al.*[196] showed that coating cochlear implants (Pt electrodes) with a CH based on PEDOT and poly(vinyl alcohol) (PVA) resulted in significantly enhanced CSC and CIL, as well as a significantly lower impedance. Similarly, Kim *et al.*[198] showed that electrodepositing PPy through an alginate coating on microprobes (Au electrodes) increased the CSC by approximately one order of magnitude (560 $mC\,cm^{-2}$, compared to Pt ~6 $mC\,cm^{-2}$) and decreased the impedance to a similar extent (~9 $\Omega\,cm^2$, compared to bare gold ~90 $\Omega\,cm^2$). Impedance was also demonstrated to be controlled by the electrodeposition time, indicating that further reduction of impedance in future research could be achieved through longer polymerisation times.

Despite the advantages provided by CH technologies, these coating constructs still require validation for safety and stability of the implanted coatings for their translation into the clinic. It is unknown whether the coatings can withstand insertion or resist delamination, as well as the potential for tissue reactions with the interfacing materials.

4.7.2 Hybrid Materials

Different from the coatings described above, where a layer of an alternative material covers the electrode, hybrid materials are formed by incorporating a

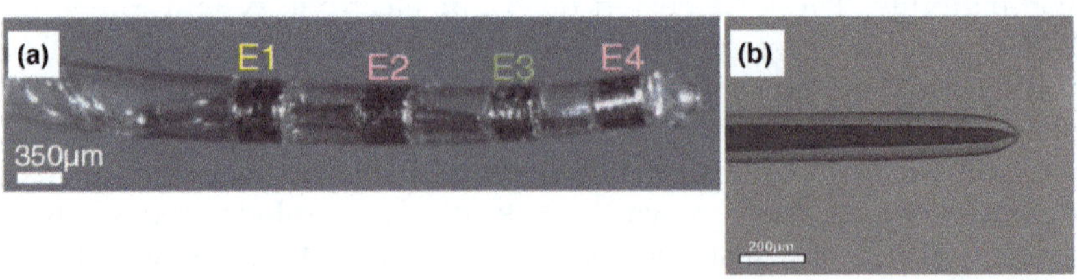

Figure 4.8 (a) Cochlear electrode array with four ring electrodes. A tripolar configuration was used to deliver electrical stimulation: electrode 3 (E3; green) was the active electrode at the centre of the tripole, delivering 2.0 mA pulses at a charge density of 200 $\mu C\,cm^{-2}$/phase; E2 and E4 (pink) flanking electrodes were shorted together and served as returns with a charge density of ~100 $\mu C\,cm^{-2}$/phase; E1 (yellow) served as an unstimulated control. Reproduced from ref. 196 with permission from IOP Publishing, Copyright 2019. (b) Intracortical neural probe modified with alginate/PPy conductive hydrogel. Reproduced from ref. 198 with permission from John Wiley & Sons, Copyright 2004 Wiley Periodicals, Inc.

second material into the electrode material matrix itself. The incorporation of the new material is used to improve the conductivity, electrochemical properties, or biocompatibility of the matrix.

In hybrid materials, the matrix is often organic, such as a CP, CH, or hydrogel. The incorporation of inorganic materials such metals, dielectrics, CNTs and graphene-based materials into an organic matrix can produce new hybrid materials suitable for use in bioelectronic applications. There are several reasons why the combination of these materials can produce materials suitable for bioelectronic electrodes. First, the synthesis of organic materials for neural interfacing, such as PEDOT, PPy or CHs normally starts with a liquid solution that then undergoes polymerisation. Inorganic suspensions are stable in various aqueous solutions, and therefore organic/inorganic hybrid structures can be easily synthesised in an aqueous environment. Furthermore, additional dopants can be required to adjust the conductivity of the organic materials, hence highly conductive inorganic materials can serve as the dopant in these systems. Lastly, inorganic biomaterials are typically non-toxic and may further improve the biocompatibility of the synthesised hybrid materials, or at minimum improve the conductivity without having any detrimental effects on biocompatibility.

4.7.2.1 *Inorganic-based Hybrid Materials*

Pt-black has been combined into a hybrid material as Pt-black/poly(ethylene glycol) (PEG) and Pt-black/Nafion. Compared to Pt-black, peripheral cuff electrodes fabricated using these hybrid materials demonstrate improved electronic stability and biocompatibility, with only minimal negative impact on their electrical properties.[83] The Pt-black/Nafion electrode exhibits high CSC and low impedance, and also improved rat Schwann cell proliferation and adhesion *in vitro*, demonstrating excellent properties for implantable electrodes.

Thin inorganic dielectric films, such as SiO_x, SiN_x, or SiC, have been incorporated into polymers, such as parylene-C or polyamide, to create flexible bioelectronics. The properties of these hybrid bioelectronics appear to be very beneficial for long-term bioelectronic implants, with good materials, electronic, and chemical stability in a physiological environment.[199–203] There have been some recent developments in this area in using these types of devices for bioelectronics, however there has not yet been progress in applying these materials *in vivo*, and challenges remain in using these materials due to their high modulus and low strain tolerance.

4.7.2.2 *Carbon-based Hybrid Materials*

To combine the complementary properties of organic and inorganic materials, researchers have been developing composite scaffolds with hybrid structures. The inorganic materials most commonly used in this research are CNTs and graphene-based materials. Blending hydrogels with CNTs has been shown to significantly improve both mechanical and electrochemical properties.[202,203] This has been achieved by mixing synthetic polymers such as PEG with single-walled CNTs. Following polymerisation of the hydrogel, the CNTs are physically entrapped within the PEG matrix. The blended hydrogels can then be used as coatings on metal electrodes. As shown by Xiao *et al.*[203] this modification can increase the conductivity of steel by over two orders of magnitude.

Improved stability of PEDOT/CNT films has been observed when compared with PEDOT/PSS.[169,204] *In vitro*, after 96 h electrical stimulation in PBS, the impedance of the PEDOT/CNT electrodes was found to be stable;[169] *in vivo*, after 7–8 weeks of implantation in rat cortex, the PEDOT/CNT electrodes performed with a higher signal-to-noise ratio when compared with electrodes made of PEDOT/PSS.[204] The PEDOT/CNT electrodes have also been used for retinal ganglion cell stimulation.[205] In this study, a lower voltage and charge injection was reported when using PEDOT/CNT electrodes compared with TiN. To further improve the performance of PEDOT/CNT electrodes *in vivo*, the anti-inflammatory drug dexamethasone has been added into the mixture prior to polymerisation.[169] The controlled release of the drug was found to reduce the immune response to the electrode after implantation, and the electrodes were used for *in vivo* rat dorsal ganglion cell stimulation.

Another CP matrix used is PPy, which can also be electrochemically synthesised, as discussed above. Electrosynthesis of PPy in water produces films with poorer electrical and mechanical properties than those produced in other solvents, such as nitromethane or acetonitrile, but the toxicity of these solvents makes then unsuitable for use in biomedical applications.[206] The addition of CNTs has been demonstrated to improve the electrical and mechanical properties of PPy, as well as their electrochemical stability. The fabrication of PPy/CNT electrodes is performed in a similar manner to that of PEDOT/CNT films, in that the electrode is normally electropolymerised from a solution containing both pyrrole and CNTs. Lu *et al.* showed that the CSC of PPy/CNT electrodes can reach 7.5 mC cm^{-2} and that the films support the growth of neurons.[207] In this study, they also tested the chronic response to the electrodes in rat cortex. After 6 weeks of implantation, a lower inflammatory response for the PPy/CNT electrodes was observed compared to the controls including PPy/Cl and PPy/PPS, with significantly lower levels of glial fibrillary acidic proteins and higher neuronal nuclei immunostaining found on the PPy/CNT electrodes. In another study, the controlled growth of PPy/CNT films with different porous morphologies was reported.[208] Here, the synthesis started with electrophoresis of CNTs onto stainless steel electrodes to create a porous CNT film coating. PPy was then polymerised onto the CNT film using pulsed electrochemistry. The PPy monomer could diffuse into the CNT network and preferentially polymerise around the nanotubes to form a coaxial composite. The films were resistant to surface delamination during surgical implantation and suitable for long-term use. The biocompatibility of the fabricated electrode was tested using *in vitro* culture of PC12 cells, and demonstrated strong viability and cell attachment.

Most of the graphene-based hybrid materials start with graphene oxide (GO) flakes, prepared from graphite powder following the modified version of Hummers' method.[209] The GO flakes are then used either directly or further reduced and mixed with polymers in an aqueous solution. The hybrid structures are finally formed through electropolymerisation onto other conductive substrates. Tian *et al.* reported that GO/PEDOT hybrid electrodes exhibit significantly improved electrochemical properties over bare gold substrates.[210] They found the charge injection capacity was up to 4.71 mC cm^{-2} and impedance 20 times lower than that on the gold electrodes. The biocompatible of the fabricated films was demonstrated through *in vitro* neuron cultures of both PC12 and NIH/3T3 cells. Luo *et al.* modified GO/PEDOT structures by introducing laminin peptides that covalently bind onto GO to enhance and increase neurite outgrowth on the material.[211] In this study, the impedance increased slightly after laminin peptide incorporation but remained significantly lower than that of the bare metal electrode. The introduction of laminin peptide improved the neuronal growth on the surface. As demonstrated using *in vitro* primary rat neuronal cultures, the neurite length was also found to be much longer on GO/PEDOT with laminin peptide-modified surfaces, compared to the surface without peptide or only bare metal. Similar fabrications were also demonstrated by replacing PEDOT with PPy, and/or replacing laminin peptides with dexamethasone as an anti-inflammatory drug.[212] Here, the charge storage capacity was found to be highest and the impedance the lowest on the GO/PPy–dexamethasone surface. The biocompatibility of the synthesised electrodes was shown using *in vitro* neuronal cultures.

In addition to GO, reduced GO has also been tested to form hybrid electrodes for neural interfacing. Huang *et al.* showed that the conductivity of the film is improved after introducing reduced GO into PEDOT nanotubes, compared to plain reduced GO, resulting in much lower impedance.[213] In another study, sulfonated graphene (SG)/PPy films were deposited from an aqueous solution containing a mixture of pyrrole monomer, SG flakes and dodecylbenzene sulfonic acid.[214] The introduction of the SG improved the conductivity and electrochemical properties of the PPy films. In addition, the stability of the films following charge/discharge cycles was also enhanced.

4.7.3 Biohybrid Materials

A step further in creating coatings or hybrid materials involves the incorporation of living cells in implantable devices, with the aim to improve the integration of the device into the tissue. The presence of the cells ideally will reduce the immune response to the implanted electrode, and thus reduce the fibrosis and encapsulation problems which impede the efficacy of the electrodes. This approach was taken to improve the function of cochlear implants, by coating the platinum iridium alloy electrode with fibroblast cells, which excreted brain derived neurotrophic factor (BNDF) prior to implantation. This in turn promoted the preservation of spiral ganglion neurons *in vivo* once implanted, leading to a new approach in improving the integration of the electrode with the patient's own cells.[215] Similarly, combining an alginate hydrogel coating with neural stem cells (NSCs) was demonstrated to improve the early tissue inflammatory response in rats over a 3 month time period. Parylene neural probes with a hollow well to incorporate the NSC hydrogel were implanted into rat brains, and a reduced tissue response was observed, demonstrating the viability of neural biohybrid probes for future neurophysiological research.[216]

Presenting neural cells at the electrode–tissue interface aims to minimise the immunogenic response against the implant and to repair the target nerve tissue damage following insertion. Different techniques have been explored to deliver cells at the neural interface. Richter *et al.* showed that neuronal cells can be seeded and cultured onto a polyimide-based DBS electrode. However, the cell layer can be readily removed from the electrode substrate upon undergoing mechanical stress, similar to that experienced by implanted electrodes in brain tissue.[217] To address this limitation, the cell-coated electrode can be supplemented with an outer layer of hydrogel such as fibrin. *In vitro* studies by Richter *et al.* showed that the fibrin coat can protect cells from the shear stresses induced during electrode insertion in an *in vitro* model, ultimately ensuring delivery of cellular constituents at the neural interface.[217] Moreover, the fibrin coating did not interfere with the cellular growth and differentiation potential of the electrode. However, these studies require further validation *in vivo* to assess the cellular survival, growth and differentiation of a cell-coated electrode array.

Alternative experimental approaches have explored cellular encapsulation within cyto-compatible hydrogel electrode coatings. Goding *et al.* developed a PVA-based bioactive hydrogel construct referred to as the 'living electrode'.[184] As shown in Figure 4.9a, the hydrogel construct consisted of two functional layers. The first layer, based on a non-degradable CH coating, was used to enhance electrochemical performance of traditional Pt electrodes. The second, overlaying layer consisted of a degradable chemically modified (methacrylate and sulfonic acid) PVA hydrogel that was tailored with biochemical and mechanical cues to support survival, growth, and differentiation of neural stem cells. The construct design aimed to stimulate target neurons *via* synapses instead of electric fields, by providing neuronal progenitors with appropriate cues to expand and form a neuronal network as the hydrogel degraded *in vivo*. While this technology is still in its early developmental stages, it was shown to support the survival of primary neurospheres (Figure 4.9b) and has been optimised to support the growth and differentiation of complex neural cultures consisting of neuron-like cells physically and biochemically supported by glia (Figure 4.9c).

The incorporation of cellular progenitors within electrode coatings is a promising approach by which to minimise the tissue foreign body response while providing a softer, more biologically suitable interface. However, this

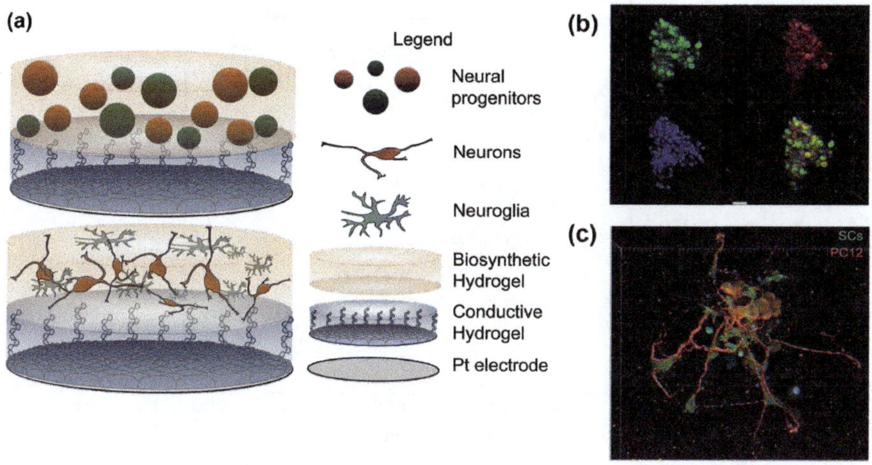

Figure 4.9 (a) Schematic of a 'living electrode' construct. The middle layer consists of a non-degradable conductive hydrogel (PVA/PEDOT). The upper layer consists of a degradable biosynthetic PVA hydrogel modified with gelatin and sericin (PVA-SG) to support the growth and differentiation of neural progenitor cells (top) and promote the formation of neuronal networks (bottom). Reproduced from ref. 50, https://doi.org/10.3389/fneng.2014.00015, under the terms of the CC BY 3.0 license, https://creativecommons.org/licenses/by/3.0/. (b) Neurospheres grown within the living electrode construct. Neurospheres were cultured *in vitro* for 14 days, expressing laminin (green) and collagen (red). Reproduced from ref. 218 with permission from Cambridge University Press, Copyright 2017 (c) Schwann cells (SC, green) and PC12 cells (red) grown in co-culture for 8 days within PVA-SG hydrogel. Reproduced from ref. 219 with permission form Elsevier, Copyright 2018.

technology is still in its early developmental stages and requires comprehensive *in vitro* characterisation to identify the optimal cell seeding densities and suitable cell culture conditions that will allow cells to grow and functionally integrate with native tissue *in vivo*.

4.8 Next Generation Materials

4.8.1 Liquid Metals

A key design criteria for neural interfaces requires flexibility, high electrical conductivity and soft mechanical properties. As described in previous sections, to date there is no single standalone material that meets such specifications, but it can be approached by incorporating flexible and conductive constituents to create a composite device. Recent attention has been paid to liquid metals for the development of soft neural bioelectronics, which are highly conductive and mechanically compliant (Figure 4.10).

These metals are liquid at or near room temperature, maintain the highly conductive properties of metals and due to their fluid state they are

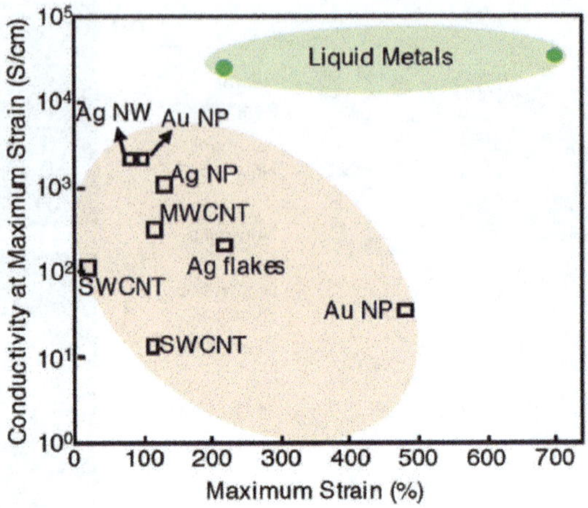

Figure 4.10 Comparison of electrical conductivity and maximum strain of various nanomaterials compared with liquid metals.
Adapted from ref. 220 with permission from John Wiley & Sons, Copyright 2017 WILEY-VCH Verlag GmbH & Co. KGaA, Weinheim.

extremely deformable,[220] including metallic elements such as caesium (Cs), gallium (Ga), mercury (Hg) and rubidium (Rb). Among these, only Ga has been identified as a potential candidate for electrode interfaces, since Rb and Cs are highly reactive when exposed to air, and Hg is very toxic.

The device design of electrodes that utilise liquid metal are typically an insulating polymer or elastomer matrix encapsulating liquid metal channels as a coaxial fibre or microchannel device due to the liquid, low viscosity nature of these materials. An elastomeric carrier for physical containment is required to facilitate the liquid metal, and this has been achieved by developing patterned elastomeric structures such as silicone[221] or polymeric hollow fibres (Figure 4.11).[222]

The injection of Ga into hollow elastomeric carriers is facilitated by the low viscosity of this material, and its high surface tension minimises the formation of bubbles and facilitates seamless channel infiltration. Overall, Ga-based liquid metals could be highly functional candidates for neuronal interfaces, provided that the metal can be physically and electrochemically contained within a carrier. However, the electrochemical stability and safe performance of these materials as interfacing neural electrodes has not been explored and still requires comprehensive characterisation in physiologically relevant environments.[220]

The physicochemical properties of Ga include high electrical conductivity (one order of magnitude higher than copper), low viscosity (~twice that of water), being a liquid at temperatures ranging from 15.7 °C (Ga alloy with In, EGaIn) to 30 °C (Ga), and forming alloys with several metals and organic materials.[221] However, one key limitation is that in the presence of oxygen it readily forms an oxide surface layer, which can introduce an unwanted

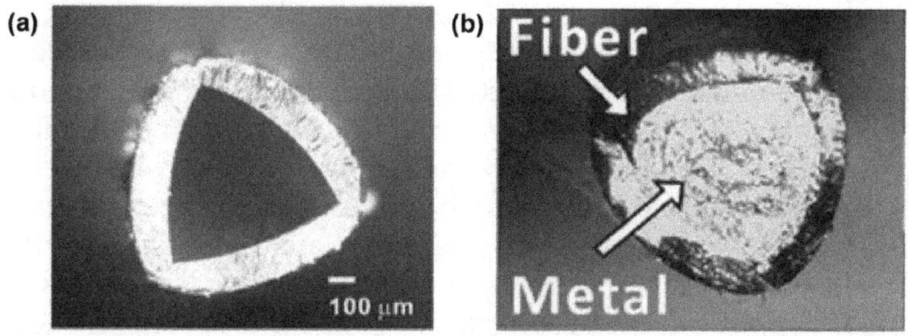

Figure 4.11 (a) Polystyrene-based fibre and the same type of fibre (b) filled with EGaIn.
Reproduced from ref. 222 with permission from John Wiley & Sons, Copyright 2013 Wiley-VCH Verlag GmbH & Co. KGaA, Weinheim.

electrical barrier at the electrode interface. The formation of oxide layer can be beneficial as it allows for chemical functionalisation and stabilises the mechanical properties, allowing for moulding into non-equilibrated shapes. However, the interface of a liquid metal electrode and neural tissue has not been explored in depth at this stage, with most devices working as insulated sensors rather than directly stimulating or recording living cells.

The toxicity of Ga and Ga alloys such as EGaIn or Ga–In–Tin (Galistan) has been observed to be much less than that of Hg.[223,224] However, these studies have been limited to *in vitro* toxicity tests that assess cellular viability of human cells (HeLa) exposed to Ga ions in solution.[225] It is important to highlight that, as observed by Kim *et al.*, naturally released ions of Ga under unperturbed conditions do not reduce cellular viability, but if the liquid metal is subjected to mechanical agitation the amount of metallic ions released into the cell media substantially increases and with it the degree of toxicity.[225] This toxic effect has been supported by studies were EGaIn nanodroplets were found to be toxic when exposed directly to human keratinocytes, whereas alginate-coated nanodroplets produced minimum toxicity.[226] Most of these materials have to date been explored for skin-electrode interfaces and in this regard are considered non-toxic.[227] However, this dermal toxicity cannot be translated into a neuronal environment lacking the protection from skin, therefore this technology still requires comprehensive assessment of biocompatibility with relevant interfacing neuronal tissue.[221,222]

4.9 Conclusions

Bioelectronics and neural prostheses are important technologies for innovative healthcare and have demonstrated efficacy at treating some of the most debilitating and chronic neurological conditions. To continue advancing this field, the design and fabrication of bioelectronics requires smaller electrodes, more sensitive recording, minimal implantation tissue trauma, and long-term durability. Furthermore, the interface between the 'bio' and the 'electronic' must be optimised to allow for the most efficient and effective transport of an electric charge. Here, we have discussed the use of inorganic biomaterials which continue to be a major component of bioelectronic devices. Inorganic materials, particularly metals and metal oxides, have long been the 'gold' standard for electrodes in bioelectronic devices. However, as electrode and device design demands increasingly smaller electrode arrays, and more sensitive and flexible recording electrodes, new approaches must be taken to overcome the limitations of these traditional materials. Creating hybrid materials that take advantage of the conductive properties of inorganic biomaterials and create a more biocompatible or compliant implanted device through the incorporation of organic materials, such as hydrogels or conductive polymers, is an exciting future for this area.

References

1. M. DiLuca and J. Olesen, *Neuron*, 2014, **82**, 1205–1208.
2. K. Kostarelos, M. Vincent, C. Hebert and J. A. Garrido, *Adv. Mater.*, 2017, **29**, 1700909.
3. British Cochlear Implant Group, https://www.bcig.org.uk/, (accessed 19/10/2020).
4. V. Kremen, B. H. Brinkmann, I. Kim, H. Guragain, M. Nasseri, A. L. Magee, T. P. Attia, P. Nejedly, V. Sladky and N. Nelson, *IEEE J. Transl. Eng. Health Med.*, 2018, **6**, 1–12.
5. W. H. Theodore and R. S. Fisher, *Lancet Neurol.*, 2004, **3**, 111–118.
6. M. J. Morrell, *Neurology*, 2011, **77**, 1295.
7. C. N. Heck, D. King-Stephens, A. D. Massey, D. R. Nair, B. C. Jobst, G. L. Barkley, V. Salanova, A. J. Cole, M. C. Smith, R. P. Gwinn, C. Skidmore, P. C. Van Ness, G. K. Bergey, Y. D. Park, I. Miller, E. Geller, P. A. Rutecki, R. Zimmerman, D. C. Spencer, A. Goldman, J. C. Edwards, J. W. Leiphart, R. E. Wharen, J. Fessler, N. B. Fountain, G. A. Worrell, R. E. Gross, S. Eisenschenk, R. B. Duckrow, L. J. Hirsch, C. Bazil, C. A. O'Donovan, F. T. Sun, T. A. Courtney, C. G. Seale and M. J. Morrell, *Epilepsia*, 2014, **55**, 432–441.
8. M. Alireza, T. Shervin, H. B. Jetan, D. W. Christopher, N. Farshad, J. J. O. Vincent, M. A. D. B. Rob, K. K. Suneil, H. Mojgan, P. M. Renato, F. Alfonso and M. L. Andres, *J. Neurosurg.*, 2018, **128**, 1199–1213.
9. G. Deuschl, C. Schade-Brittinger, P. Krack, J. Volkmann, H. Schäfer, K. Bötzel, C. Daniels, A. Deutschländer, U. Dillmann, W. Eisner, D. Gruber, W. Hamel, J. Herzog, R. Hilker, S. Klebe, M. Kloß, J. Koy, M. Krause, A. Kupsch, D. Lorenz, S. Lorenzl, H. M. Mehdorn, J. R. Moringlane, W. Oertel, M. O. Pinsker, H. Reichmann, A. Reuß, G.-H. Schneider, A. Schnitzler, U. Steude, V. Sturm, L. Timmermann, V. Tronnier, T. Trottenberg, L. Wojtecki, E. Wolf, W. Poewe and J. Voges, *N. Engl. J. Med.*, 2006, **355**, 896–908.

10. C. Buhmann, T. Huckhagel, K. Engel, A. Gulberti, U. Hidding, M. Poetter-Nerger, I. Goerendt, P. Ludewig, H. Braass, C.-U. Choe, K. Krajewski, C. Oehlwein, K. Mittmann, A. K. Engel, C. Gerloff, M. Westphal, J. A. Köppen, C. K. E. Moll and W. Hamel, *PLoS One*, 2017, **12**, e0178984.
11. H. Cagnan, D. Pedrosa, S. Little, A. Pogosyan, B. Cheeran, T. Aziz, A. Green, J. Fitzgerald, T. Foltynie, P. Limousin, L. Zrinzo, M. Hariz, K. J. Friston, T. Denison and P. Brown, *Brain*, 2016, **140**, 132–145.
12. N. C. Sinclair, H. J. McDermott, K. J. Bulluss, J. B. Fallon, T. Perera, S. S. Xu, P. Brown and W. Thevathasan, *Ann. Neurol.*, 2018, **83**, 1027–1031.
13. H. Cagnan, T. Denison, C. McIntyre and P. Brown, *Nat. Biotechnol.*, 2019, **37**, 1024–1033.
14. C. Boehler, T. Stieglitz and M. Asplund, *Biomaterials*, 2015, **67**, 346–353.
15. G. Márton, I. Bakos, Z. Fekete, I. Ulbert and A. Pongrácz, *J. Mater. Sci.: Mater. Med.*, 2014, **25**, 931–940.
16. M. Tykocinski, Y. Duan, B. Tabor and R. S. Cowan, *Hear. Res.*, 2001, **159**, 53–68.
17. E. K. Brunton, B. Winther-Jensen, C. Wang, E. B. Yan, S. Hagh Gooie, A. J. Lowery and R. Rajan, *Front. Neuroeng.*, 2015, **8**, 5.
18. D. B. McCreery, L. A. Bullara and W. F. Agnew, *Exp. Neurol.*, 1986, **92**, 147–161.
19. I.-S. Lee, C.-N. Whang, J.-C. Park, D.-H. Lee and W.-S. Seo, *Biomaterials*, 2003, **24**, 2225–2231.
20. C. A. R. Chapman, H. Chen, M. Stamou, P. J. Lein and E. Seker, *Cell. Mol. Bioeng.*, 2016, **9**, 433–442.
21. M. A. Kafi, T.-H. Kim, C.-H. Yea, H. Kim and J.-W. Choi, *Biosens. Bioelectron.*, 2010, **26**, 1359–1365.
22. S. Kim, R. Bhandari, M. Klein, S. Negi, L. Rieth, P. Tathireddy, M. Toepper, H. Oppermann and F. Solzbacher, *Biomed. Microdevices*, 2009, **11**, 453–466.
23. J. McHardy, D. Geller and S. Brummer, *Ann. Biomed. Eng.*, 1977, **5**, 144–149.
24. I. Gotman, *J. Endourol.*, 1997, **11**, 383–389.
25. P. T. McCarthy, R. Madangopal, K. J. Otto and M. P. Rao, 2009 Annual International Conference of the IEEE Engineering in Medicine and Biology Society, 2009, DOI: 10.1109/IEMBS.2009.5334429, 2062–2065.
26. E. M. Schmidt, F. T. Hambrecht and J. S. McIntosh, *J. Neurosci. Methods*, 1982, **5**, 33–39.
27. L. S. Robblee, E. M. Kelliher, M. E. Langmuir, H. Vartanian and J. McHardy, *J. Biomed. Mater. Res.*, 1983, **17**, 327–343.
28. C. Chen, S. Ruan, X. Bai, C. Lin, C. Xie and I.-S. Lee, *Mater. Sci. Eng., C*, 2019, **103**, 109865.
29. R. D. Meyer, S. F. Cogan, T. H. Nguyen and R. D. Rauh, *IEEE Trans. Neural Syst. Rehabil. Eng.*, 2001, **9**, 2–11.

30. J. D. Weiland, D. J. Anderson and M. S. Humayun, Biomedical Engineering, *IEEE Trans.*, 2002, **49**, 1574–1579.
31. G. Paasche, C. Tasche, T. Stöver, A. Lesinski-Schiedat and T. Lenarz, *Otol. Neurotol.*, 2009, **30**, 592–598.
32. G. Paasche, F. Bockel, C. Tasche, A. Lesinski-Schiedat and T. Lenarz, *Otol. Neurotol.*, 2006, 27.
33. R. K. Franklin, M. D. Johnson, K. A. Scottt, S. Jun Ho, N. Hakhyun, D. R. Kipket and R. B. Brown, *Sensors, 2005 IEEE*, 2005, https://doi.org/10.1109/ICSENS.2005.1597971, p. 4.
34. T. Rose, E. Kelliher and L. Robblee, *J. Neurosci. Methods*, 1985, **12**, 181–193.
35. N. P. Aryan, M. I. H. B. Asad, C. Brendler, S. Kibbel, G. Heusel and A. Rothermel, *2011 Annual International Conference of the IEEE Engineering in Medicine and Biology Society*, 2011, 2866–2869.
36. F. Rodrigues, J. Ribeiro, P. Anacleto, A. Fouchard, O. David, P. Sarro and P. Mendes, *J. Neural Eng.*, 2019, **17**, 016010.
37. F. Vitale, S. R. Summerson, B. Aazhang, C. Kemere and M. Pasquali, *ACS Nano*, 2015, **9**, 4465–4474.
38. L. Bareket, N. Waiskopf, D. Rand, G. Lubin, M. David-Pur, J. Ben-Dov, S. Roy, C. Eleftheriou, E. Sernagor and O. Cheshnovsky, *Nano Lett.*, 2014, **14**, 6685–6692.
39. I. Suzuki, M. Fukuda, K. Shirakawa, H. Jiko and M. Gotoh, *Biosens. Bioelectron.*, 2013, **49**, 1–6.
40. M. Shein, A. Greenbaum, T. Gabay, R. Sorkin, M. David-Pur, E. Ben-Jacob and Y. Hanein, *Biomed. Microdevices*, 2009, **11**, 495–501.
41. E. W. Keefer, B. R. Botterman, M. I. Romero, A. F. Rossi and G. W. Gross, *Nat. Nanotechnol.*, 2008, **3**, 434–439.
42. M. Vomero, E. Castagnola, F. Ciarpella, E. Maggiolini, N. Goshi, E. Zucchini, S. Carli, L. Fadiga, S. Kassegne and D. Ricci, *Sci. Rep.*, 2017, **7**, 1–14.
43. S. Nimbalkar, E. Castagnola, A. Balasubramani, A. Scarpellini, S. Samejima, A. Khorasani, A. Boissenin, S. Thongpang, C. Moritz and S. Kassegne, *Sci. Rep.*, 2018, **8**, 1–14.
44. K. Wang, C. L. Frewin, D. Esrafilzadeh, C. Yu, C. Wang, J. J. Pancrazio, M. Romero-Ortega, R. Jalili and G. Wallace, *Adv. Mater.*, 2019, **31**, 1805867.
45. N. V. Apollo, M. I. Maturana, W. Tong, D. A. X. Nayagam, M. N. Shivdasani, J. Foroughi, G. G. Wallace, S. Prawer, M. R. Ibbotson and D. J. Garrett, *Adv. Funct. Mater.*, 2015, **25**, 3551–3559.
46. D. Kuzum, H. Takano, E. Shim, J. C. Reed, H. Juul, A. G. Richardson, J. De Vries, H. Bink, M. A. Dichter and T. H. Lucas, *Nat. Commun.*, 2014, **5**, 5259.
47. K.-H. Yang and R. J. Narayan, *Curr. Opin. Biomed. Eng.*, 2019, **10**, 60–68.
48. Y. Cai, F. Edin, Z. Jin, A. Alexsson, O. Gudjonsson, W. Liu, H. Rask-Andersen, M. Karlsson and H. Li, *Acta Biomater.*, 2016, **31**, 211–220.
49. L. A. Geddes and R. Roeder, *Ann. Biomed. Eng.*, 2003, **31**, 879–890.

50. U. A. Aregueta-Robles, A. J. Woolley, L. A. Poole-Warren, N. H. Lovell and R. A. Green, *Front. Neuroeng.*, 2014, 7.
51. S. F. Cogan, *Annu. Rev. Biomed. Eng.*, 2008, **10**, 275–309.
52. D. M. Zhou and R. J. Greenberg, Proceedings of the 25th Annual International Conference of the IEEE Engineering in Medicine and Biology Society (IEEE Cat. No.03CH37439), 2003, vol. **1962**, 2, pp. 1964–1967.
53. P. C. Sherrell, B. C. Thompson, J. K. Wassei, A. A. Gelmi, M. J. Higgins, R. B. Kaner and G. G. Wallace, *Adv. Funct. Mater.*, 2014, **24**, 769–776.
54. K. Wang, H. A. Fishman, H. Dai and J. S. Harris, *Nano Lett.*, 2006, **6**, 2043–2048.
55. S. P. Lacour, S. Benmerah, E. Tarte, J. FitzGerald, J. Serra, S. McMahon, J. Fawcett, O. Graudejus, Z. Yu and B. Morrison Iii, *Med. Biol. Eng. Comput.*, 2010, **48**, 945–954.
56. Medtronic, https://www.medtronic.com/us-en/healthcare-professionals/products/neurological/deep-brain-stimulation-systems.html (accessed 19/10/2020).
57. D. A. Vajari, M. Vomero, J. B. Erhardt, A. Sadr, J. S. Ordonez, V. A. Coenen and T. Stieglitz, *Micromachines*, 2018, **9**, 1–15.
58. F. A. Spelman, *Audiol. Neurotol.*, 2005, **11**, 77–85.
59. Y. Y. Chen, H. Y. Lai, S. H. Lin, C. W. Cho, W. H. Chao, C. H. Liao, S. Tsang, Y. F. Chen and S. Y. Lin, *J. Neurosci. Methods*, 2009, **182**, 6–16.
60. P. J. Rousche, D. S. Pellinen, D. P. Pivin, J. C. Williams, R. J. Vetter and D. R. Kipke, *IEEE Trans. Biomed. Eng.*, 2001, **48**, 361–370.
61. C. Hassler, J. Guy, M. Nietzschmann, J. F. Staiger and T. Stieglitz, Proceedings of the Annual International Conference of the IEEE Engineering in Medicine and Biology Society, *EMBS*, 2011, 644–647.
62. J. P. Seymour and D. R. Kipke, *Biomaterials*, 2007, **28**, 3594–3607.
63. G. J. Suaning, N. H. Lovell and T. Lehmann, *IEEE 2014 Biomedical Circuits and Systems Conference,* BioCAS 2014 - Proceedings, 2014, DOI: 10.1109/BioCAS.2014.6981711, pp. 256–259.
64. R. Avery and J. Wepsic, *U. S. Pat.*, US3724467A, 1973.
65. A. Branner and R. A. Normann, *Brain Res. Bull.*, 2000, **51**, 293–306.
66. D. J. Tyler and D. M. Durand, *Neural Syst. Rehabil. Eng., IEEE Trans.*, 2002, **10**, 294–303.
67. D. J. Tyler and D. M. Durand, *Ann. Biomed. Eng.*, 2003, **31**, 633–642.
68. D. Borton, M. Bonizzato, J. Beauparlant, J. DiGiovanna, E. M. Moraud, N. Wenger, P. Musienko, I. R. Minev, S. P. Lacour, J. D. R. Millán, S. Micera and G. Courtine, *Neurosci. Res. Lett.*, 2014, **78**, 21–29.
69. G. Creasey, J. Elefteriades, A. DiMarco, P. Talonen, M. Bijak, W. Girsch, C. Kantor and P. H. Peckham, *J. Rehabil. Res. Dev.*, 1996, **33**, 123–132.
70. D. W. Strege, W. P. Cooney, M. B. Wood, S. J. Johnson and B. J. Metcalf, *J. Hand Surg.*, 1994, **19**, 931–939.
71. G. E. Loeb and R. A. Peck, *J. Neurosci. Methods*, 1996, **64**, 95–103.

72. Y. Shimada, K. Sato, H. Kagaya, N. Konishi, S. Miyamoto and T. Matsunaga, *Arch. Phys. Med. Rehabil.*, 1996, **77**, 1014–1018.
73. N. Xue, I. D. Martinez, J. Sun, Y. Cheng and C. Liu, *Biosens. Bioelectron.*, 2018, **112**, 114–119.
74. G. S. Brindley, C. E. Polkey and D. N. Rushton, *Paraplegia*, 1982, **20**, 365–381.
75. S. Braley, *J. Macromol. Sci., Part A: Pure Appl. Chem.*, 1970, **4**, 529–544.
76. S. Lerouge and A. Simmons, *Sterilisation of Biomaterials and Medical Devices*, Elsevier, 2012.
77. A. Rahimi and A. Mashak, *Plast., Rubber Compos.*, 2013, **42**, 223–230.
78. S. Persheyev, Y. Fan, A. Irving and M. J. Rose, *J. Biomed. Mater. Res., Part A*, 2011, **99A**, 135–140.
79. K. A. Moxon, S. Hallman, A. Aslani, N. M. Kalkhoran and P. I. Lelkes, *J. Biomater. Sci., Polym. Ed.*, 2007, **18**, 1263–1281.
80. M. Gulino, D. Kim, S. Pané, S. D. Santos and A. P. Pêgo, *Front. Neurosci.*, 2019, **13**, 689.
81. A. M. Belle, H. A. Enright, E. V. Mukerjee, D. A. Soscia, J. J. Osburn, E. A. Kuhn, K. S. Kulp, E. K. Wheeler and V. M. Tolosa, *Proceedings of the 2017 IEEE International Symposium on Medical Measurements and Applications*, 2017, 453–457.
82. J. Fei, X. Wen, X. Lin, Saijilafu, W. Wang, O. Ren, X. Chen, L. Tan, K. Yang, H. Yang and L. Yang, *Mater. Sci. Eng., C*, 2017, **78**, 1155–1163.
83. Y. J. Lee, H.-J. Kim, J. Y. Kang, S. H. Do and S. H. Lee, *Adv. Healthcare Mater.*, 2017, **6**, 1601022.
84. E. Stewart, N. R. Kobayashi, M. J. Higgins, A. F. Quigley, S. Jamali, S. E. Moulton, R. M. I. Kapsa, G. G. Wallace and J. M. Crook, *Tissue Eng., Part C*, 2014, **21**, 385–393.
85. C. L. Kolarcik, K. Catt, E. Rost, I. N. Albrecht, D. Bourbeau, Z. Du, T. D. Y. Kozai, X. Luo, D. J. Weber and X. Tracy Cui, *J. Neural Eng.*, 2014, **12**, 016008.
86. I. Kelava and M. A. Lancaster, *Dev. Biol.*, 2016, **420**, 199–209.
87. G. Rossi, A. Manfrin and M. P. Lutolf, *Nat. Rev. Genet.*, 2018, **19**, 671–687.
88. W. M. Reichert, *Indwelling Neural Implants: Strategies for Contending with the In Vivo Environment*, CRC Press, 2007.
89. A. S. G. Curtis, B. Casey, J. O. Gallagher, D. Pasqui, M. A. Wood and C. D. W. Wilkinson, *Biophys. Chem.*, 2001, **94**, 275–283.
90. S. De Astis, I. Corradini, R. Morini, S. Rodighiero, R. Tomasoni, C. Lenardi, C. Verderio, P. Milani and M. Matteoli, *Nanoscale*, 2013, **5**, 10963–10974.
91. O. Palyvoda, A. N. Bordenyuk, A. K. Yatawara, E. McCullen, C.-C. Chen, A. V. Benderskii and G. W. Auner, *Langmuir*, 2008, **24**, 4097–4106.
92. K. Jans, B. Van Meerbergen, G. Reekmans, K. Bonroy, W. Annaert, G. Maes, Y. Engelborghs, G. Borghs and C. Bartic, *Langmuir*, 2009, **25**, 4564–4570.
93. M. E. Spira and A. Hai, *Nat. Nanotechnol.*, 2013, **8**, 83.

94. X. Du, L. Wu, J. Cheng, S. Huang, Q. Cai, Q. Jin and J. Zhao, *J. Biol. Phys.*, 2015, **41**, 339–347.
95. S. Venkatraman, J. Hendricks, Z. King, A. Sereno, S. Richardson-Burns, D. Martin and J. Carmena, *IEEE Trans. Neural Syst. Rehabil. Eng.*, 2011, **19**, 307.
96. R. Gerwig, K. Fuchsberger, B. Schroeppel, G. S. Link, G. Heusel, U. Kraushaar, W. Schuhmann, A. Stett and M. Stelzle, *Front. Neuroeng.*, 2012, **5**, 8.
97. G. Piret, C. Hébert, J.-P. Mazellier, L. Rousseau, E. Scorsone, M. Cottance, G. Lissorgues, M. O. Heuschkel, S. Picaud and P. Bergonzo, *Biomaterials*, 2015, **53**, 173–183.
98. D. Kuzum, H. Takano, E. Shim, J. C. Reed, H. Juul, A. G. Richardson, J. De Vries, H. Bink, M. A. Dichter and T. H. Lucas, *Nat. Commun.*, 2014, **5**, 1–10.
99. B. E. Conway, *J. Electrochem. Soc.*, 1991, **138**, 1539–1548.
100. G. Buzsáki, *Nat. Neurosci.*, 2004, **7**, 446–451.
101. G. Hong and C. M. Lieber, *Nat. Rev. Neurosci.*, 2019, **20**, 330–345.
102. G. Buzsáki, C. A. Anastassiou and C. Koch, *Nat. Rev. Neurosci.*, 2012, **13**, 407–420.
103. S. Negi, R. Bhandari and F. Solzbacher, 2012 Annual International Conference of the IEEE Engineering in Medicine and Biology Society, 2012, DOI: 10.1109/EMBC.2012.6347151, 5142–5145.
104. A. B. Majji, M. S. Humayun, J. D. Weiland, S. Suzuki, S. A. D'Anna and E. de Juan, *Invest Ophthalmol. Visual Sci.*, 1999, **40**, 2073–2081.
105. T. Roitbak and E. Syková, *Glia*, 1999, **28**, 40–48.
106. M. Jorfi, J. L. Skousen, C. Weder and J. R. Capadona, *J. Neural Eng.*, 2014, **12**, 011001.
107. J. Muller, M. Ballini, P. Livi, Y. H. Chen, M. Radivojevic, A. Shadmani, V. Viswam, I. L. Jones, M. Fiscella, R. Diggelmann, A. Stettler, U. Frey, D. J. Bakkum and A. Hierlemann, *Lab Chip*, 2015, **15**, 2767–2780.
108. J. T. Robinson, M. Jorgolli, A. K. Shalek, M.-H. Yoon, R. S. Gertner and H. Park, *Nat. Nanotechnol.*, 2012, **7**, 180.
109. M. Ballini, J. Muller, P. Livi, Y. H. Chen, U. Frey, A. Stettler, A. Shadmani, V. Viswam, I. L. Jones, D. Jackel, M. Radivojevic, M. K. Lewandowska, W. Gong, M. Fiscella, D. J. Bakkum, F. Heer and A. Hierlemann, *IEEE J. Solid-State Circuits*, 2014, **49**, 2705–2719.
110. A. Hai, J. Shappir and M. E. Spira, *Nat. Methods*, 2010, **7**, 200.
111. Z. C. Lin, C. Xie, Y. Osakada, Y. Cui and B. Cui, *Nat. Commun.*, 2014, **5**, 3206.
112. C. Xie, Z. Lin, L. Hanson, Y. Cui and B. Cui, *Nat. Nanotechnol.*, 2012, **7**, 185.
113. M. Dipalo, H. Amin, L. Lovato, F. Moia, V. Caprettini, G. C. Messina, F. Tantussi, L. Berdondini and F. De Angelis, *Nano Lett.*, 2017, **17**, 3932–3939.
114. M. Dipalo, G. Melle, L. Lovato, A. Jacassi, F. Santoro, V. Caprettini, A. Schirato, A. Alabastri, D. Garoli and G. Bruno, *Nat. Nanotechnol.*, 2018, **13**, 965.

115. J. Wesson Ashford, K. L. Coburn and J. M. Fuster, *J. Neurosci. Methods*, 1985, **14**, 247–252.
116. R. A. Normann, E. M. Maynard, P. J. Rousche and D. J. Warren, *Vision Res.*, 1999, **39**, 2577–2587.
117. G. Lehew and M. A. Nicolelis, *Methods Neural Ensemble Recordings*, 2008, **2**, 361–371.
118. T. Zhou, G. Hong, T.-M. Fu, X. Yang, T. G. Schuhmann, R. D. Viveros and C. M. Lieber, *Proc. Natl. Acad. Sci.*, 2017, **114**, 5894–5899.
119. D. H. Hubel, *Science*, 1957, **125**, 549–550.
120. D. H. Hubel and T. N. Wiesel, *J. Physiol.*, 1962, **160**, 106–154.
121. J. C. Williams, R. L. Rennaker and D. R. Kipke, *Brain Res. Protoc.*, 1999, **4**, 303–313.
122. S. Musallam, M. J. Bak, P. R. Troyk and R. A. Andersen, *J. Neurosci. Methods*, 2007, **160**, 122–127.
123. T. Yasui, S. Yamagiwa, H. Sawahata, S. Idogawa, Y. Kubota, Y. Kita, K. Yamashita, R. Numano, K. Koida and T. Kawano, *Adv. Healthcare Mater.*, 2019, **8**, 1801081.
124. M. HajjHassan, V. Chodavarapu and S. Musallam, *Sensors*, 2008, **8**, 6704–6726.
125. K. D. Wise, D. J. Anderson, J. F. Hetke, D. R. Kipke and K. Najafi, *Proc. IEEE*, 2004, **92**, 76–97.
126. T. J. Webster, M. C. Waid, J. L. McKenzie, R. L. Price and J. U. Ejiofor, *Nanotechnology*, 2003, **15**, 48–54.
127. T. L. Massey, S. R. Santacruz, J. F. Hou, K. S. J. Pister, J. M. Carmena and M. M. Maharbiz, *J. Neural Eng.*, 2019, **16**, 016024.
128. G. Hong, X. Yang, T. Zhou and C. M. Lieber, *Curr. Opin. Neurobiol.*, 2018, **50**, 33–41.
129. J. Rivnay, H. L. Wang, L. Fenno, K. Deisseroth and G. G. Malliaras, *Sci. Adv.*, 2017, **3**, 6.
130. F. Zhang, L.-P. Wang, M. Brauner, J. F. Liewald, K. Kay, N. Watzke, P. G. Wood, E. Bamberg, G. Nagel and A. Gottschalk, *Nature*, 2007, **446**, 633–639.
131. T. C. Pappas, W. S. Wickramanyake, E. Jan, M. Motamedi, M. Brodwick and N. A. Kotov, *Nano Lett.*, 2007, **7**, 513–519.
132. E. C. Dreaden, A. M. Alkilany, X. Huang, C. J. Murphy and M. A. El-Sayed, *Chem. Soc. Rev.*, 2012, **41**, 2740–2779.
133. J. Lee, I. Ozden, Y.-K. Song and A. V. Nurmikko, *Nat. Methods*, 2015, **12**, 1157.
134. J. Zhang, X. Liu, W. Xu, W. Luo, M. Li, F. Chu, L. Xu, A. Cao, J. Guan and S. Tang, *Nano Lett.*, 2018, **18**, 2903–2911.
135. F. Wu, E. Stark, P.-C. Ku, K. D. Wise, G. Buzsáki and E. Yoon, *Neuron*, 2015, **88**, 1136–1148.
136. A. Canales, X. Jia, U. P. Froriep, R. A. Koppes, C. M. Tringides, J. Selvidge, C. Lu, C. Hou, L. Wei and Y. Fink, *Nat. Biotechnol.*, 2015, **33**, 277.

137. Y. Tufail, A. Matyushov, N. Baldwin, M. L. Tauchmann, J. Georges, A. Yoshihiro, S. I. H. Tillery and W. J. Tyler, *Neuron*, 2010, **66**, 681–694.
138. S. Hughes, S. McBain, J. Dobson and A. J. El Haj, *J. R. Soc., Interface*, 2008, **5**, 855–863.
139. D. J. Garrett, W. Tong, D. A. Simpson and H. Meffin, *Carbon*, 2016, **102**, 437–454.
140. R. J. Coffey, *Artif. Organs*, 2009, **33**, 208–220.
141. X. Navarro, T. B. Krueger, N. Lago, S. Micera, T. Stieglitz and P. Dario, *J. Peripher, Nerv. Syst.*, 2005, **10**, 229–258.
142. J. T. Roland Jr, *Laryngoscope*, 2005, **115**, 1325–1339.
143. L. Bareket, A. Barriga-Rivera, M. P. Zapf, N. H. Lovell and G. J. Suaning, *J. Neural Eng.*, 2017, 14.
144. A. Nurmikko, D. Borton, M. Yin and R. K. Shepherd, *Neurobionics: The Biomedical Engineering of Neural Prostheses*, 2016, pp. 123–161.
145. R. K. Shepherd, P. M. Seligman and M. N. Shivdasani, *Auditory and Visual Neural Prostheses, in Neurobionics: The Biomedical Engineering of Neural Prostheses*, 2016 pp. 187–212.
146. G. Brindley, *Spinal Cord*, 1994, **32**, 795.
147. D. J. Tyler and D. M. Durand, *IEEE Trans. Rehabil. Eng.*, 1997, **5**, 51–61.
148. T. D. Y. Kozai, A. L. Vazquez, C. L. Weaver, S. G. Kim and X. T. Cui, *J. Neural Eng.*, 2012, 9.
149. A. Sridharan, S. D. Rajan and J. Muthuswamy, *J. Neural Eng.*, 2013, **10**, 066001.
150. P. A. Busby, K. L. Plant and L. A. Whitford, *Cochlear Implants Int.*, 2002, **3**, 87–103.
151. M. L. Hughes, K. R. Vander Werff, C. J. Brown, P. J. Abbas, D. M. R. Kelsay, H. F. B. Teagle and M. W. Lowder, *Ear Hear.*, 2001, **22**, 471–486.
152. G. De Ceulaer, S. Johnson, M. Yperman, K. Daemers, F. E. Offeciers, G. M. O'Donoghue and P. J. Govaerts, *Otol. Neurotol.*, 2003, 24.
153. J. Xu, R. K. Shepherd, R. E. Millard and G. M. Clark, *Hear. Res.*, 1997, **105**, 1–29.
154. T. L. Rose and L. S. Robblee, Biomedical Engineering, *IEEE Trans.*, 1990, **37**, 1118–1120.
155. R. Ishai, B. S. Herrmann, J. B. Nadol and A. M. Quesnel, *Hear. Res.*, 2017, **348**, 44–53.
156. R. K. Shepherd, J. Matsushima, R. L. Martin and G. M. Clark, *Hear. Res.*, 1994, **81**, 150–166.
157. K. Wissel, G. Brandes, N. Pütz, G. L. Angrisani, J. Thieleke, T. Lenarz and M. Durisin, *PLoS One*, 2018, **13**, e0196649.
158. K. Spiers, T. Cardamone, J. B. Furness, J. C. M. Clark, J. F. Patrick and G. M. Clark, *Cochlear Implants Int.*, 2016, **17**, 129–131.
159. J. B. Nadol, J. T. O'Malley, B. J. Burgess and D. Galler, *Hear. Res.*, 2014, **318**, 11–17.
160. P. Fattahi, G. Yang, G. Kim and M. R. Abidian, *Adv. Mater.*, 2014, **26**, 1846–1885.

161. A. M. Dymond, L. E. Kaechele, J. M. Jurist and P. H. Crandall, *J. Neurosurg.*, 1970, **33**, 574–580.
162. W. M. Grill, N. Bhadra and B. Wang, *Brain Res.*, 1999, **836**, 19–30.
163. J. D. Klein, S. L. Clauson and S. F. Cogan, *J. Vac. Sci. Technol., A*, 1989, **7**, 3043–3047.
164. L. S. Robblee, M. J. Mangaudis, E. D. Lasinsky, A. G. Kimball and S. B. Brummer, *MRS Proc.*, 1985, **55**, 303.
165. S. Yang, B. Xia, X. Zeng, S. Luo, W. Wei and X. Liu, *Anal. Chim. Acta*, 2010, **667**, 57–62.
166. B. Haghighi and S. Bozorgzadeh, *Talanta*, 2011, **85**, 2189–2193.
167. M. Heim, B. Yvert and A. Kuhn, *J. Physiol.*, 2012, **106**, 137–145.
168. T. Gabay, M. Ben-David, I. Kalifa, R. Sorkin, Z. R. Abrams, E. Ben-Jacob and Y. Hanein, *Nanotechnology*, 2007, **18**, 35201.
169. X. Luo, C. L. Weaver, D. D. Zhou, R. Greenberg and X. T. Cui, *Biomaterials*, 2011, **32**, 5551–5557.
170. A. Krishnan, E. Dujardin, T. W. Ebbesen, P. N. Yianilos and M. M. J. Treacy, *Phys. Rev. B: Condens. Matter Mater. Phys.*, 1998, **58**, 14013–14019.
171. S. K. Seidlits, J. Y. Lee and C. E. Schmidt, *Nanomedicine*, 2008, **3**, 183–199.
172. C. P. Firme and P. R. Bandaru, *Nanomed.: Nanotechnol. Biol. Med.*, 2010, **6**, 245–256.
173. N. K. Guimard, N. Gomez and C. E. Schmidt, *Prog. Polym. Sci.*, 2007, **32**, 876–921.
174. X. Cui and D. C. C. Martin, *Sens. Actuators, B*, 2003, **89**, 92–102.
175. R. A. Green, P. B. Matteucci, R. T. Hassarati, B. Giraud, C. W. Dodds, S. Chen, P. J. Byrnes-Preston, G. J. Suaning, L. A. Poole-Warren and N. H. Lovell, *J. Neural Eng.*, 2013, **10**, 16009.
176. P. M. George, A. W. Lyckman, D. A. LaVan, A. Hegde, Y. Leung, R. Avasare, C. Testa, P. M. Alexander, R. Langer and M. Sur, *Biomaterials*, 2005, **26**, 3511–3519.
177. R. A. Green, N. H. Lovell, G. G. Wallace and L. A. Poole-Warren, *Biomaterials*, 2008, **29**, 3393–3399.
178. P. R. Bidez, S. Li, A. G. MacDiarmid, E. C. Venancio, Y. Wei and P. I. Lelkes, *J. Biomater. Sci., Polym. Ed.*, 2006, **17**, 199–212.
179. J. Huang, S. Virji, B. H. Weiller and R. B. Kaner, *Chem. – Eur. J.*, 2004, **10**, 1314–1319.
180. G. Stevenson, S. E. Moulton, P. C. Innis and G. G. Wallace, *Synth. Met.*, 2010, **160**, 1107–1114.
181. S. Baek, R. Green, A. Granville, P. Martens and L. Poole-Warren, *J. Mater. Chem. B*, 2013, **1**, 6670.
182. J. H. Collier, J. P. Camp, T. W. Hudson and C. E. Schmidt, *J. Biomed. Mater. Res.*, 2000, **50**, 574–584.
183. A. Gelmi, M. J. Higgins and G. G. Wallace, *Biomaterials*, 2010, **31**, 1974–1983.
184. J. Goding, A. Gilmour, P. Martens, L. Poole-Warren and R. Green, *Adv. Healthcare Mater.*, 2017, **6**, 1601177.

185. J. M. Fonner, L. Forciniti, H. Nguyen, J. D. Byrne, Y. F. Kou, J. Syeda-Nawaz and C. E. Schmidt, *Biomed. Mater.*, 2008, 3.
186. D.-H. Kim, M. Abidian and D. C. Martin, *J. Biomed. Mater. Res., Part A*, 2004, **71A**, 577–585.
187. S. Wilks, S. Richardson-Burn, J. Hendricks, D. Martin and K. Otto, *Front. Neuroeng.*, 2009, **2**, 7.
188. M. R. Abidian, J. M. Corey, D. R. Kipke and D. C. Martin, *Small*, 2010, **6**, 421–429.
189. S. Venkatraman, J. Hendricks, S. Richardson-Burns, E. Jan, D. Martin and J. M. Carmena, *2009 4th International IEEE/EMBS Conference on Neural Engineering,* NER '09, 2009, **4**, 383–386.
190. S. Baek, R. A. Green and L. A. Poole-Warren, *J. Biomed. Mater. Res., Part A*, 2013, 1–12.
191. T. Boretius, M. Schuettler and T. Stieglitz, *Artif. Organs*, 2011, **35**, 245–248.
192. N. Peixoto, K. Jackson, R. Samiyi and S. Minnikanti, Engineering in Medicine and Biology Society, 2009. EMBC 2009. Annual International Conference of the IEEE, 2009, 658–661.
193. R. A. Green, N. H. Lovell and L. A. Poole-Warren, *Acta Biomater.*, 2010, **6**, 63–71.
194. D. Mawad, A. Lauto and G. G. Wallace, in *Polymeric Hydrogels as Smart Biomaterials*, ed. S. Kalia, Springer International Publishing, Cham, 2016, pp. 19–44.
195. W. Guo, X. Zhang, X. Yu, S. Wang, J. Qiu, W. Tang, L. Li, H. Liu and Z. L. Wang, *ACS Nano*, 2016, **10**, 5086–5095.
196. A. N. Dalrymple, M. Huynh, U. A. A. Robles, J. B. Marroquin, C. Lee, A. Petrossians, J. J. Whalen, D. Li, H. C. Parkington and J. S. Forsythe, *J. Neural Eng.*, 2019, **17**, 016015.
197. R. T. Hassarati, W. F. Dueck, C. Tasche, P. M. Carter, L. A. Poole-Warren and R. A. Green, *IEEE Trans. Neural Syst. Rehabil. Eng.*, 2014, **22**, 411–418.
198. D. H. Kim, M. Abidian and D. C. Martin, *J. Biomed. Mater. Res. A*, 2004, **71**, 577–585.
199. A. Hogg, S. Uhl, F. Feuvrier, Y. Girardet, B. Graf, T. Aellen, H. Keppner, Y. Tardy and J. Burger, *Surf. Coat. Technol.*, 2014, **255**, 124–129.
200. J. Li, R. Li, H. Du, Y. Zhong, Y. Chen, K. Nan, S. M. Won, J. Zhang, Y. Huang and J. A. Rogers, *ACS Nano*, 2019, **13**, 660–670.
201. C. Diaz-Botia, L. Luna, R. Neely, M. Chamanzar, C. Carraro, J. Carmena, P. Sabes, R. Maboudian and M. Maharbiz, *J. Neural Eng.*, 2017, **14**, 056006.
202. R. A. Green, R. T. Hassarati, J. A. Goding, S. Baek, N. H. Lovell, P. J. Martens and L. A. Poole-Warren, *Macromol. Biosci.*, 2012, **12**, 494–501.
203. Y. Xiao, L. He and J. Che, *J. Mater. Chem.*, 2012, **22**, 8076–8082.
204. T. D. Kozai, K. Catt, Z. Du, K. Na, O. Srivannavit, R. U. Haque, J. Seymour, K. D. Wise, E. Yoon and X. T. Cui, *IEEE Trans. Biomed. Eng.*, 2016, **63**, 111–119.
205. R. Samba, T. Herrmann and G. Zeck, *J. Neural Eng.*, 2015, **12**, 016014.

206. G. Z. Chen, M. S. P. Shaffer, D. Coleby, G. Dixon, W. Zhou, D. J. Fray and A. H. Windle, *Adv. Mater.*, 2000, **12**, 522–526.
207. Y. Lu, T. Li, X. Zhao, M. Li, Y. Cao, H. Yang and Y. Y. Duan, *Biomaterials*, 2010, **31**, 5169–5181.
208. H. Xiao, M. Zhang, Y. Xiao and J. Che, *Colloids Surf., B*, 2015, **126**, 138–145.
209. W. S. Hummers Jr and R. E. Offeman, *J. Am. Chem. Soc.*, 1958, **80**, 1339, 1339.
210. H.-C. Tian, J.-Q. Liu, D.-X. Wei, X.-Y. Kang, C. Zhang, J.-C. Du, B. Yang, X. Chen, H.-Y. Zhu and Y.-N. NuLi, *Biomaterials*, 2014, **35**, 2120–2129.
211. X. Luo, C. L. Weaver, S. Tan and X. T. Cui, *J. Mater. Chem. B*, 2013, **1**, 1340–1348.
212. C. L. Weaver, J. M. LaRosa, X. Luo and X. T. Cui, *ACS Nano*, 2014, **8**, 1834–1843.
213. T.-Y. Huang, C.-W. Kung, J.-Y. Wang, M.-H. Lee, L.-C. Chen, C.-W. Chu and K.-C. Ho, *Electrochim. Acta*, 2015, **172**, 61–70.
214. A. Liu, C. Li, H. Bai and G. Shi, *J. Phys. Chem. C*, 2010, **114**, 22783–22789.
215. D. Rejali, V. A. Lee, K. A. Abrashkin, N. Humayun, D. L. Swiderski and Y. Raphael, *Hear. Res.*, 2007, **228**, 180–187.
216. E. Purcell, J. Seymour, S. Yandamuri and D. Kipke, *J. Neural Eng.*, 2009, **6**, 026005.
217. S. M. Rajguru, A. I. Matic, A. M. Robinson, A. J. Fishman, L. E. Moreno, A. Bradley, I. Vujanovic, J. Breen, J. D. Wells and M. Bendett, *Hear. Res.*, 2010, **269**, 102–111.
218. J. Goding, A. Gilmour, U. A. Robles, L. Poole-Warren, N. Lovell, P. Martens and R. Green, *MRS Commun.*, 2017, **7**, 487–495.
219. U. A. Aregueta-Robles, P. J. Martens, L. A. Poole-Warren and R. A. Green, *Acta Biomater.*, 2019, **95**, 269–284.
220. M. D. Dickey, *Adv. Mater.*, 2017, **29**, 1–19.
221. M. D. Dickey, *ACS Appl. Mater. Interfaces*, 2014, **6**, 18369–18379.
222. S. Zhu, J. H. So, R. Mays, S. Desai, W. R. Barnes, B. Pourdeyhimi and M. D. Dickey, *Adv. Funct. Mater.*, 2013, **23**, 2308–2314.
223. J. E. Chandler, H. H. Messer and G. Ellender, *J. Dent. Res.*, 1994, **73**, 1554–1559.
224. Y. Lu, Q. Hu, Y. Lin, D. B. Pacardo, C. Wang, W. Sun, F. S. Ligler, M. D. Dickey and Z. Gu, *Nat. Commun.*, 2015, **6**, 10066.
225. J. H. Kim, S. Kim, J. H. So, K. Kim and H. J. Koo, *ACS Appl. Mater. Interfaces*, 2018, **10**, 17448–17454.
226. X. Li, M. Li, L. Zong, X. Wu, J. You, P. Du and C. Li, *Adv. Funct. Mater.*, 2018, **28**, 1–8.
227. R. Guo, X. Sun, S. Yao, M. Duan, H. Wang, J. Liu and Z. Deng, *Adv. Mater. Technol.*, 2019, **4**, 1–11.

CHAPTER 5

The Effects of Surface Properties on the Antimicrobial Activity and Biotoxicity of Metal Biomaterials and Coatings

KATHRYN A. WHITEHEAD,*[a] MOHAMED EL MOHTADI,[a] ANTHONY J. SLATE,[a,b] MISHA VAIDYA[a] AND JOELS WILSON-NIEUWENHUIS[a]

[a] Microbiology at Interfaces, Faculty of Science and Engineering, Manchester Metropolitan University, Manchester M1 5GD, UK;
[b] Department of Biology and Biochemistry, University of Bath, Claverton Down, Bath BA2 7AY, UK
*Email: k.a.whitehead@mmu.ac.uk

5.1 Introduction

The antimicrobial properties of metals have been recognised throughout the history of medicine and healthcare.[1] Before the discovery of antibiotics, silver salts were used to control eye and wound infections and copper and mercury salts were used to prevent leprosy, gonorrhoea, tuberculosis and syphilis.[2] The wide application of metals in medicine was predominant until the discovery of antibiotics, after which the use of metals as antimicrobials began to diminish. Advances in surface engineering and surgical technique have resulted in a wide array of metals which can be utilised as biomaterials. Although metals and their alloys are selected primarily for their structural

Inorganic Materials Series No. 7
The Chemistry of Inorganic Biomaterials
Edited by Christopher Spicer

Published by the Royal Society of Chemistry, www.rsc.org

integrity, in some cases they are also designed to have antimicrobial properties. However, regardless of their use, once implanted these substrates are foreign to the body and their effect on bacterial interactions and host:human interactions (biotoxicity) is unclear. This chapter discusses the properties of metal surfaces and their alloys and their use in biomaterials, with a specific focus on their antimicrobial properties and their cytotoxicity effects.

Metals are widely used in part, or entirely, for the construction of biomaterial implants, with such examples including plates, staples, knee, ankle, elbow, shoulder and hip implants, Kirschner (K) wires and fixation pins. Due to an ever-aging population, large numbers of biomaterials are now utilised. It has been estimated by the National Joint Registry that in 2018 alone, an average of 248 hip replacements and 265 knee replacements were recorded per hospital orthopaedic unit. For ankle replacements the average number of replacements carried out per orthopaedic unit was 6, whilst the average number of shoulder replacements was 20, with an average of 5 for elbow replacements.

5.1.1 Definition of a Biomaterial

The use of biomaterials has increased at an exponential rate over the past 50 years, due to an aging population, injuries and congenital diseases.[3] Due to this there has been a huge increase in the diversity and range of biomaterial applications.[4] The definition of a biomaterial is any material type that is used to produce a pharmacologically inert device, which is then utilised to aid or replace a function of the body, in a reliable, safe and physiologically-acceptable manner.[5,6]

5.1.2 Issues Associated with Biomaterial Infection

Treatment of biomaterial infections is costly and places a huge burden upon healthcare services around the world. The prevalence of external fixation infections has been reported to be as low as 3%, to over 80% with pin tract surgeries.[7] In the US, it is estimated that the cost of such infections will rise to $1.6 billion annually in 2020.[8] Many implants that become infected need to be removed or require further surgery.[9] Alongside the psychological and physiological trauma to the patient, bacterial infection of implants is more difficult to detect than incidences of aseptic loosening, thus making the process of infection detection and treatment time consuming, and costly.[9,10]

5.1.3 Bacterial:Surface Interactions

Despite much research, an understanding of the relationship between the effects of the surface properties of metallic biomaterials (chemistry, topography and physicochemistry) and how these affect bacterial binding and biofilm formation is still relatively poorly understood. This is in part because although the surfaces used for biomaterial production may be well characterised, the bacterial properties are not always considered. The bacterial morphological and biochemical properties will vary according to the environmental conditions, such as temperature, antimicrobials present and intraspecies relationships.[11,12] This interplay between the specific bacterial species present, the materials used in the biomaterial production, the environmental considerations and the biochemistry of the surrounding milieu all need to be considered to understand such relationships.

5.1.4 Basic Requirements of Biomaterials

With the use of implants increasing, methods to prevent and reduce biomaterial infection are required.[13] The design of a successful biomaterial requires some key considerations and these include the surface characteristics of the biomaterial, and the effect of the surface properties on the interactions with the surrounding environment. There are many types of biomaterials, including metals, glasses, ceramics, polymers and natural products that are used to produce implants,[14] although this work considers an overview of only metal surfaces and their alloys. In order for a material to be of use as a biomaterial, it needs to be pharmacologically and systemically inert once in place within the body. Due to the incidence of infection, biomaterials with antimicrobial properties have become more prominent. Unfortunately, the factors that result in a surface being antimicrobial, *e.g.* the leaching of metal ions, can also be considered a process of corrosion, which can induce cellular toxicity in the host. It may be argued that even when the bactericidal effect of the metal surface is by a mode of action known as 'contact kill', that the metal ions still have to leach from the metal surface directly into the bacterial membrane. Thus, an understanding of the surface:bacterial:host interactions is necessary. This chapter will consider the effect of the surface properties on microbial binding, the antimicrobial effects of metallic surfaces and the potential cytotoxicity of such materials.

5.2 Biomaterial Surface Properties

The general biomaterial properties (topography, chemistry, physicochemistry) that control bacterial interactions will first be considered before discussing the specific use of metallic surfaces/coatings.

5.2.1 Attachment, Adhesion and Retention of Bacteria to Biomaterials

The attachment, adhesion and retention of bacteria to metal surfaces is the initial step that enables the bacteria to bind to the surface. Once retained, if bacterial growth and subsequent biofilm formation occurs, infection of the biomaterials and surgical implants may ensue.[15,16]

Bacterial attachment, adhesion and retention are different phenomena, although the descriptions are often used interchangeably. Although bacteria may be in planktonic form, they will preferentially attach to a surface. The first step in bacterial attachment to a surface will be influenced by the physicochemical properties of both the bacteria and the surface. In this instance, the term physicochemistry refers to the hydrophobicity or wettability of a surface. On both surfaces and bacterial cells, hydrophobicity and charge can affect the attachment of bacteria to surfaces so that bacteria can arrive at the biomaterial by overcoming energy barriers.[17] However, at this stage, bacterial binding to the surface is reversible. The attachment process is followed by bacterial adhesion, which occurs due to specific and non-specific interactions between a microorganism and a surface, and this is an irreversible process, although the bacteria may be removed from the surface under the conditions of force or flow. The retention of bacteria on a surface occurs once the bacteria are bound and cannot be removed, even with a force such as flow. The prerequisite of biofilm formation is successful bacterial attachment, adhesion and retention of the bacteria to a solid surface substratum.

5.2.2 Biofilm Formation on a Surface

Bacteria present in a mature biofilm, may demonstrate an increase in the antimicrobial resistance by as much as 100–1000-fold, when compared to planktonic bacteria.[18,19] Thus, the growth of bacteria within a biofilm adds physical protection that makes biofilm removal and breakdown more challenging. It is generally considered that biofilms confer protection to microorganisms from antibiotics and antimicrobials, the innate immune response and the phagocytotoxic action of the immune system.[20] Thus, if bacteria are in a biofilm, the treatment of an infection is more difficult.[21] Bacteria that form biofilms are often found on the surfaces of tissues and on biomaterials in areas of persistent infection.[22] Medical implants are particularly susceptible to biofilm development, since they are a foreign material in the body, and this is a major concern for implant failure, since it frequently restricts the lifespan of many indwelling medical components.[23,24] Thus, biofilm colonisation on the metal and metal-based coatings of biomedical implants may result in an increase in device-associated infections, and subsequently this may lead to an increase in patient morbidity and mortality.[25]

5.2.3 Microbial Contamination of Biomaterials

5.2.3.1 Mono-species Biomaterial Infections

The biomaterial systems to be discussed in this chapter include both closed (Figure 5.1) and open systems (Figure 5.2). Closed systems can be defined as an internally sealed environment, for example a hip, elbow, ankle, shoulder

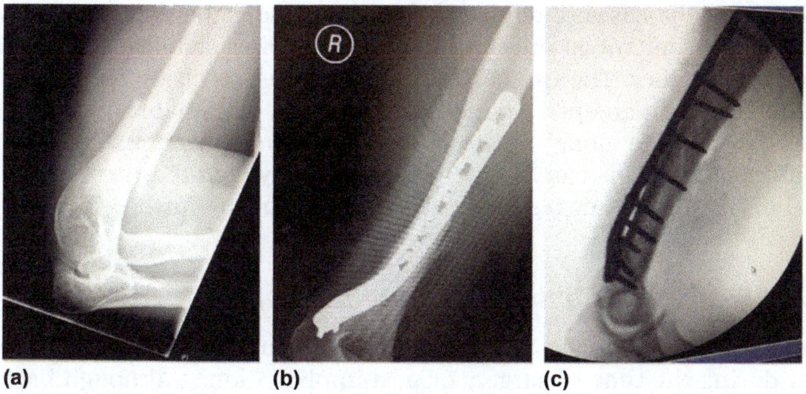

Figure 5.1 Insertion of biomaterial plates and screws used to support the humerus. (a) X-ray of a fractured humerus (b) X-ray demonstrating the positioning of the plate used to support the fractured humerus (c) X-ray demonstrating a view of pins and plate supporting the humerus.
Images courtesy of Miss J. Jefferies.

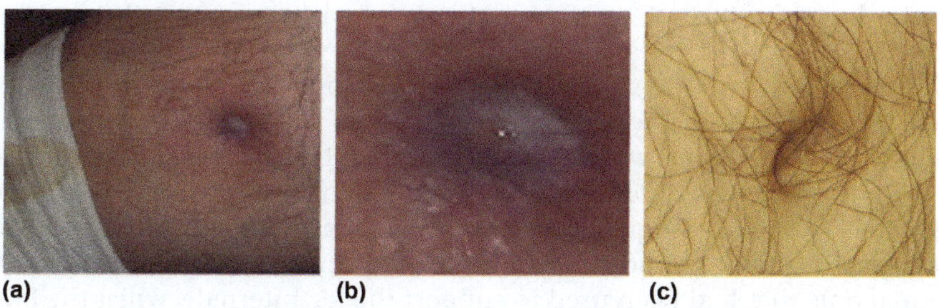

Figure 5.2 (a) Bacterial pus seeping from a wound in the leg, (b) magnified image of infection in the leg, and (c) following removal of the biomaterials (wound site after 2 years).
Images courtesy of Mr R. Brown.

or knee implant or plates, screws or nails. In closed systems, the majority of biofilm infections are usually mono-microbial, and approximately 80% are attributable to *Staphylococcus aureus* or coagulase negative Staphylococci (CNS), such as *Staphylococcus epidermidis*.[26–30]

5.2.3.2 Polymicrobial Biomaterial Infections

Polymicrobial infections are more usually found in association with open fracture fixation devices[28] (Figure 5.2). Open systems involve biomaterials that are open to the external environment, for example, pins, Kirschner (also known as K) wires or staples. In some cases, the use of a biomaterial results in an 'open wound'. *Staphylococcus aureus* is the most common pathogen isolated in osteomyelitis (an infection of the bone) since damaged or dead bone can act as a substratum for bacteria to attach to, and in greater than two-thirds of the cases of adult osteomyelitis studies have shown the incidence of a polymicrobial infection.[26] Kirschner wires stabilise broken bone fragments together. The use of K wires results in protrusion from the skin and this results in an open wound. Such a wound may enable opportunistic bacteria to migrate along the structure of the pin or wire, resulting in potential infection. This can lead to osteomyelitis, which may necessitate removal of the implant or result in the patient requiring further surgery.[31–33]

5.2.3.3 *Bacterial Contamination of the Implant*

Bacteria can contaminate prosthetic and implantable biomedical devices either during the time of surgery or post-implantation.[34] Although bacteria may be present in a wound, they may not immediately result in a notable infection. They may remain dormant on the surface until such time as the environment and conditions (*e.g.* illness, stress, fatigue in the host) enable their proliferation, or they may remain as a low grade infection. An example of such contamination is shown in Figure 5.2. An infection was observed as the result of a previous injury in the leg whereby following serious bone fracture of the femur, the leg had been repaired using biomaterials. Over 10 years later, yellow pus seeped from the leg. Upon further investigation, it was demonstrated that the underlying biomaterials had become infected and the resulting seepage was thought to be bacteria exuding from the previous wound site. The patient had not reported any issues with the leg but had stated that they were feeling fatigued prior to the leg beginning to seep. A surgical procedure ensued and the biomaterials inside the leg were removed and the wound healed. In this case, removal of the biomaterials that had been used to support the leg internally whist the bone healed was possible. Unfortunately, if the biomaterials must be replaced, such as might be the case for a joint replacement, the bacteria may still be present in the localised tissues and even with debridement and/or intensive antibiotic therapies, it may be that the replacement biomedical device will also become colonised.[35] Other physiological factors also influence the incidence of infection. For example, toe deformities require orthopaedic foot surgery to correct the positioning of the toe.[36] However, such surgical correction has a high incidence of infection, which has been primarily attributed to the contracture of scar tissue that may occur as a result of subclinical infection.[37]

5.2.3.4 *Antimicrobial Resistant Bacteria and Antibiotic Impregnated Biomaterials*

Vast amounts of research have been conducted to reduce implant-associated bacterial infections, such as impregnating antibiotics into the surface of biomedical devices.[38,39] However, with the ever-increasing prevalence of antimicrobial resistant (AMR) bacteria, there have been suggestions that a move away from using antibiotic-based approaches is now imperative to slow the development of AMR. Indeed, if continuous sub-lethal doses of antibiotics are released from such systems, the development of AMR bacteria in the tissue surrounding the biomaterial may be enhanced, thus making subsequent infections more difficult to treat.

5.2.4 Surface Properties of Biomaterials

5.2.4.1 Definitions of Wettability

Surface wettability is an interchangeably used term but may be used to describe if a surface is wettable or non-wettable when using water as the test solvent. There is some confusion over terminology since 'non-wettability' may be used interchangeably with hydrophobicity. In regards to thermodynamics, a hydrophobic effect is the free energy charge of water surrounding the solute; a positive free energy charge (of the surrounding solute) indicates hydrophobicity, whilst a negative free energy charge implies hydrophilicity.[42–44] Polar molecules are often hydrophilic due to the presence of an electric dipole/multipole moment.[45] It is thought that in biological systems the strongest non-covalent interactions are hydrophobic interactions.

5.2.4.2 Defining Surface Wettability

The threshold for determining a non-wettable surface, is often based on the water contact angle. It has been suggested that if a contact angle is $\leq 65^\circ$ then this will denote a surface being wettable, whereas a contact angle of $>65^\circ$ indicates a non-wettable surface.[46] This has become known as the 'Berg limit', with the name taken from the author who first defined this hypothesis.[46] However, in other studies, a contact angle of 90° is used as the threshold between the definitions.[47]

5.2.4.3 Considerations of Implant Surface Wettability

When designing a medical device surface, surface wettability or hydrophobicity plays an important role, as it has been suggested that it can be modified to reduce bacterial attachment.[48] As bacteria approach a surface, it is the physicochemical factors that influence the initial attraction or repulsion of the bacteria. Molecules and bacteria will initially approach a surface through a number of natural forces when in an aqueous environment, and these include gravity, diffusion and/or Brownian motion. Once the molecules and bacteria get closer to the surface in the aqueous environment, the physicochemical forces can affect the interactions at the bacterial:surface interface. In the initial attachment of the bacteria to a surface, van der Waals and Lewis acid–base interactions predominantly affect these interactions, which dictates the electrostatic and wettability/

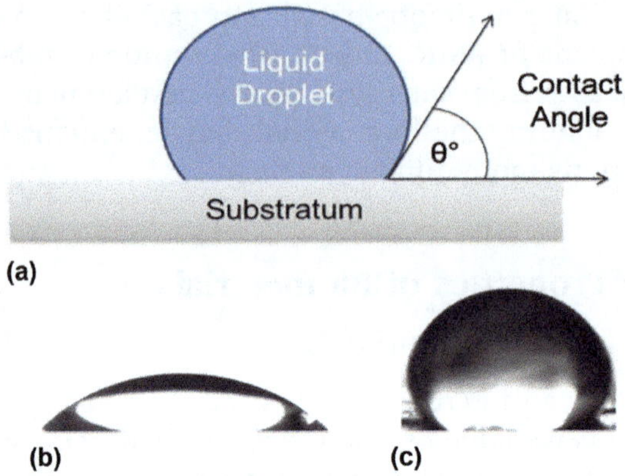

Figure 5.3 (a) Schematic of a water droplet on a solid surface, displaying the measured contact angle. Examples of (b) a hydrophilic surface (glass) and (c) a hydrophobic surface (polystyrene).

hydrophobic properties of a surface.[49] The hydrophobic or hydrophilic properties of a surface occur due to a combination of Lewis acid–base interactions.[43,50] The Lewis acid–base component encompasses two parameters; the electron accepting potential and the electron donating potential. It is the electron accepting potential and the electron donating potential that describe the polar charge of the surface tension component.[42] Low energy Lifshiftz van der Waals forces are also a component part of hydrophobic/hydrophilic interactions, and are apolar.[42] A standard method for the physicochemical characterisation of surfaces, inert or organic, is measuring the contact angles, usually of three liquids, two polar solvents, and one non-polar solvent,[42] the results of which can then be used to calculate the physicochemical parameters of the surfaces or bacteria (Figure 5.3).

5.2.4.4 *Wettability of the Bacteria*

The bacterial cell surface is a complex structure involving both biological and chemical interactions at the surface to aid its survival. At all stages, bacterial properties such as the presence of bacterial surface appendages/adhesins and the potential to produce extracellular polymeric substances (EPS) affect bacterial attachment, adhesion or retention to surfaces.[51,52] It is the interaction between the physicochemistry of both the cell and the surface that are thought to determine initial binding behaviour. The correlation between the hydrophobic and hydrophilic parameters of bacteria cells and surfaces has been described in some studies as the primary factor in driving the attachment of microbes to surfaces.[53–55] Thus, the measurement of the physicochemistry of the bacterial cell surface is of paramount importance when assessing the propensity of a surface to fouling and when attempting to reduce or prevent microbial colonisation.[49,52] The surface of a bacterial cell is thought to comprise islands of different physicochemistries due to the differing chemical moieties of the molecular components that make up the cell surface. Some researchers have argued that bacteria with an overall negative charge will preferentially adhere to surfaces that have an overall negative charge.[52,54] However, studies into the effect of surface physicochemistry on bacterial binding and a consensus on the influence of physicochemistry on bacterial:surface interactions are varied.[53,54,56,57] Further research is thus required to fully elucidate this phenomenon.

5.2.4.5 *Surface Wettability and Binding*

With regards to biomaterial surfaces, the effect of surface hydrophobicity on bacterial binding has been widely investigated using titanium surfaces. The effects of modified laser-etched titanium surfaces on reducing bacterial binding have been suggested to be attributable to the combined effects of reduced surface hydrophobicity, thicker and more stable oxide films and the presence of the laser-induced nanofeatures of the treated surfaces.[8] On laser-treated cobalt–chromium–molybdenum alloy (Co–Cr–Mo), surfaces, no similar antibacterial effects to those demonstrated on titanium surfaces were observed.[9] In 2009, Gallardo-Moreno *et al.* used ultraviolet (UV) irradiation at a wavelength of 258 nm to treat Ti-6Al-4V alloy surfaces.[58] The UV irradiation of surfaces is thought to degrade hydrocarbon contamination and increase the incidence of –OH groups, resulting in a more hydrophilic surface.[59] UV treatment of the surfaces indicated that the physicochemical changes on the UV treated surface caused a reduction in the adhesion rate of *S. aureus* and *S. epidermidis* bacteria. Similarly, in 2013, Kawano *et al.*[60] used a UV laser with a wavelength of 365 nm to treat commercially pure titanium. This study suggested that the exposure of commercially pure titanium to a UV laser, prior to exposure to bacteria, decreased the number of attached *Porphyromonas gingivalis* bacterial cells.[60] They ascribed the effect observed to the decrease in the water contact angle and increase in the anatase phase of the surface layer on the treated titanium surface.[60] Such results demonstrated that the manipulation of surfaces to enhance hydrophilicity results in a reduction in bacterial adhesion and attachment.

Porous titanium structures or coatings are of special interest in the field of osteo-implants as it is critical to establish anchorage of the implant in the host bone and this may occur by encouraging bone growth into the porous structure.[61] However, such surfaces increase surface roughness and provide larger surface areas which have the potential to increase bacterial attachment. In 2013, Braem *et al.*[61] used porous titanium coatings to determine their effect on bacterial colonisation relative to titanium implant surfaces that had been prepared in a number of ways. Using *S. aureus* and *S. epidermidis* they found that topological surface features influence biofilm formation, as discussed in the subsequent section, but also that surface hydrophobicity exerted a strong influence.[61]

5.2.5 Surface Topography of Biomaterials

5.2.5.1 Describing Surface Topography

Surface topography and roughness are important parameters to consider when investigating microbial retention on surfaces.[62] However, there is confusion in the terminology used. Although nanofeatures tend to be considered to be on a scale of <0.5 μm, other definitions of the range of roughness size are dependent on the scientific discipline. In microbiological terms, the surface topography may be defined in respect to microbial sizes and split into three categories: macrotopography (1–10 μm; microtopography 500 nm^{-1} μm; and nanotopography <500 nm).[63]

The combined structures of the surface topography are referred to as the surface texture.[64] Surface roughness is defined as 'the irregularities in the surface texture which are inherent in the production process but excluding waviness and errors of form'.[64,65] Surface roughness values may be further categorised into groups based on their functionality, which include hybrid, amplitude and spacing parameters. Amplitude parameters are commonly used in hygiene specification in industries when characterising surface topography (Table 5.1).[66–73]

The surface texture developed may be due to both the physical components of the material and/or the production process.[65] Current methods used to describe surface topography may be responsible in part for the differences reported in their effect on bacterial attachment. It has been

Table 5.1 Surface roughness amplitudes used to quantify the topography of a surface.[67–69,72,73] *R* values are measurements taken from a mean centre line, whereas *S* values are measurements taken across an area.

Parameter	Description
S_a R_a	Arithmetic average height: the average absolute deviation in topography. These values are insensitive to small changes in the surface topography.
R_q/S_q RMS	Root mean square: standard deviation of the distribution of the surface height. This has an increased sensitivity to large infrequent deviations when compared to R_a or S_a.
R_z/S_z	Ten-point height: average difference in height between the five highest peaks and the five lowest valleys taking into account multiple peaks.
R_p/S_p R_{pk}/S_{pk}	The maximum height of the surface peaks.
R_v/S_v	Lowest point of the surface profile.
R_t/S_t or R_{max}/S_{max}	Maximum height of the profile: overall height of the profile between the lowest valley and the highest peak. S_t/R_t is also equal to S_p/R_p and S_v/R_v. This parameter is extremely sensitive to surface anomalies such as deep scratches or high peaks.
Average height	The sum of all height values divided by the number of data points. This parameter gives a description of the average height of the deviations throughout the scan area.
Max range	The difference between the single largest peak and the single deepest valley similar to S_t/R_t.

suggested that although the amplitude parameters enable the comparison of surface roughness, they are deficient in describing the shape of the features of the surface topography.[71,74] Surface topography is composed of a series of valleys and peaks and these may be regularly or irregularly spaced in terms of width, height and spacing.[60] Although surface topographies are usually measured using vertical roughness parameters, this means that the shape of the surface features and the distances between them may not be considered.[75] Further, it may be speculated that to analyse the topography of a surface, alongside the evaluation of a range of amplitude parameters, the consideration of the shape, size and spatial distribution of the surface features should also be determined.[76]

5.2.5.2 The Effect of Surface Topography on Bacterial Binding

It has been proposed that an increase in surface roughness results in an increase in retained bacteria, due to the protection (*e.g.*, from shear forces) that topographical features provide to adherent microbes.[54,75,77–79] Further, a surface with a greater surface area may provide more contact points which can promote bacterial attachment. Thus, it may be that the selection of the correct surface topography would minimise the number of contact points available for bacterial binding, resulting in an anti-adhesive surface. In 2016, Cunha *et al.* used a femtosecond laser with a wavelength of 1030 nm to create nanofeatures on commercially pure titanium surfaces.[80] Using surfaces with nanotopographical features, they found that the adhesion of *S. aureus* was reduced, which suggested that it may have been due to the lowered contact area reducing the binding potential between the substrate and the bacteria (Figure 5.4).[80] However, it should be noted that, in terms of bacterial binding, surface nanofeatures such as pits may also encourage bacterial attachment or provide substrates for the attachment of microbial features such as flagella and pili.[81,82] In work conducted in 2005 by Whitehead *et al.* it was demonstrated that when using microorganisms of varying average sizes, the size of the microorganism with respect to the size of the surface defect was important in influencing the amount of bacterial retention to a surface, whereby surfaces with features similar to those of the bacterial size resulted in enhanced bacterial retention.[83] Thus, when characterising surface topography, the shape and size of features in regard to microbial size should be considered. In 2001, Edwards and Rutenberg carried out work to determine the effect that the cross-sectional shape of a groove had on the binding potential of microorganisms and they found that bacterial binding was reduced if the groove was too small for the bacteria to fit into, whereas grooves of bacterial size increased the points of bacterial:surface binding and hence the number of bacteria retained.[84] In 2002, using atomic force microscopy (AFM), Medilanski *et al.* demonstrated that when using linear featured surfaces, microbial adhesion was minimal on a surface with a particular roughness ($R_a = 0.16$ μm).[78] However, the use of surfaces with both increased and decreased roughness resulted in increased

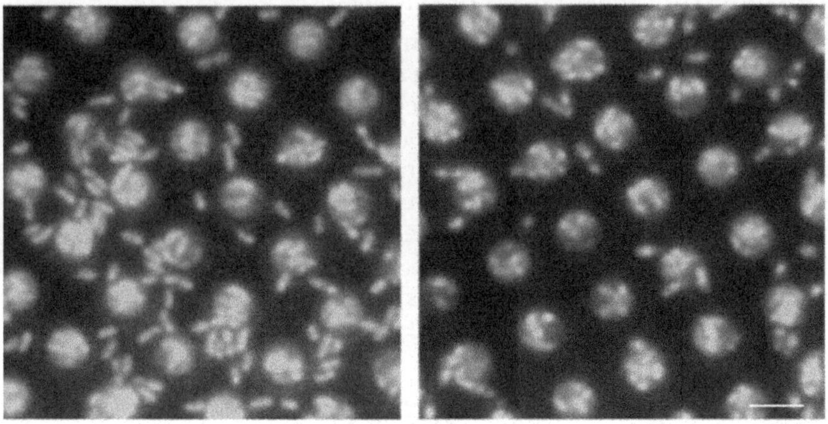

Figure 5.4 *Bacillus subtilis* spores retained on regular surface features of (a) 1 μm and (b) 2 μm surface topographies, demonstrating that more spores were captured within the 2 μm features, but more spores were retained overall on the surface with the 1 μm surface features. The scale bar denotes 2 μm.

microbial adhesion.[81] This leads to an interesting concept, that there may be optimal ranges of surface roughness that result in reduced bacterial attachment to a surface.[85]

5.2.5.3 *Surface Topography and Biomaterials*

It has been suggested that to reduce bacterial adhesion, modifying the surface topography of an implant may reduce the incidence of infection on orthopaedic implants.[86–88] In orthopaedic implants, the nanoroughness of a surface is extremely important, since such surface topography affects both the behaviour of the attachment and the growth of osteoblasts, particularly their osteointegration into the implant.[82] In addition, it has been suggested that particular topographies may simultaneously decrease competitive cell integration by cells such as fibroblasts, which are essential for bone integration.[88–91]

5.2.5.4 *Bacterial Binding to the Implant*

With regards to biomaterials, there have been controversial results regarding the ability of smoother or rougher implants to impede bacterial binding to a surface. The surfaces of orthopaedic implants are designed for different applications, and as such their tribological and chemical specifications require different design criteria.[9] The acetabular cup (back cup) or stem of a hip implant, which are usually made of a titanium alloy (Ti–6Al–4V), are designed with a rough texture to encourage osseointegration, while the femoral head, usually made of a cobalt–chromium alloy, is smooth, with the aim of reducing the friction between interacting components.[9] From the perspective of smoother surfaces being more beneficial for reducing bacterial binding, it has been suggested that when titanium and stainless steel plates are treated with thermal cycling (a temperature modulation process that improves the performance and longevity of materials) and tested with *S. aureus*, there is a reduction in the number of bacterial colonies on the smoother, more uniform surfaces.[92]

5.2.5.5 *Relationship Between the Surface Topography, Bacterial and Cellular Binding*

Changing the topography of a surface may result in decreased bacterial binding. However, there may also be a need to increase eukaryotic cell binding at the same time. In one study, electrophoretic deposition was used to treat titanium surfaces to produce two distinct nanotopography surface textures and it was demonstrated that surfaces with a texture consisting of smaller particle sizes demonstrated a statistically significant 1 log reduction in *S. aureus* and *Pseudomonas aeruginosa* attachment. However, the surface that had topographical features that were produced with larger particle sizes resulted in an increase in osteoblast proliferation for up to 5 days, when compared to the smoother control surface.[93] Thus, surfaces with larger particle sizes were optimal for osteoblasts, but also increased bacterial retention. In another study, *S. epidermidis* adherence was investigated to determine the effects of different levels of topography on a cobalt–chromium–molybdenum alloy (Co–Cr–Mo), oxidised zirconium–niobium alloy (Oxinium), 316L stainless steel, commercially pure titanium and a titanium alloy (Ti–6Al–4V). The number of bacteria that adhered to the biomaterials in the coarse group (R_a 7.2–30 nm) was higher than those in the fine group, and notable differences were observed for the Oxinium, stainless steel and titanium alloy ($p < 0.05$). Although the fine and course Co–Cr–Mo samples exhibited significantly lower numbers of adhered bacteria than the Ti–6Al–4V, commercially pure titanium and 316L, of the topographically fine group, the surface of Co–Cr–Mo exhibited the highest water contact angle,[94] suggesting that the surface hydrophobicity of the Co–Cr–Mo surface also plays a part in bacterial attachment. When copper, titanium, molybdenum, silver and iron surfaces were tested to determine their effect on microbial retention, it was demonstrated that their surface chemistry had an effect on microbial retention, although it was found that the surface nanotopography and shape of the surface features did not influence microbial binding for these materials.[95] However, *E. coli* retention on the coatings was affected by the surface physicochemistry, whilst *S. aureus* retention was affected by the surface micro-topography, with no relationships observed between the amount of bacterial retention and the bacterial or surface physicochemistry.[95] These studies demonstrate the importance of considering how the binding of different types of bacterial cells are affected by, once again, the critical interplay between the different surface parameters that effect bacterial binding.[94,95] Therefore, understanding bacterial binding to a surface is complicated by the properties of both the cells and the surface and the need to consider multiple interactions at the surface interface. Consideration also needs to be given to the application of the material and its surrounding environment. It may be that the conflicting results observed are due to the diversity of the materials and bacteria used. An alternative approach could be to consider using a more targeted approach to look for trends in an environment, within specific materials properties and/or the bacterial species used.

In terms of the efficacy of rougher surfaces reducing bacterial binding, Ti–35Nb–7Zr–6Ta surfaces were modified using fibre laser surface treatment and were assessed for their performance with *S. aureus*.[96] The results demonstrated that laser surface treatment created a rougher, spiky surface with homogenous elemental distribution and decreasing peak-to-peak distance between the ripples, which resulted in a surface with distinct micron and nanoscale features.[96] These surfaces showed improvements in the eukaryotic formation of bone-like nodules and cell spreading.[96] Furthermore, a reduction in bacterial attachment was demonstrated that was attributed to the surface features produced by the fibre laser treatment.[96] Using *S. epidermidis*, the adhesion and biofilm formation on smooth gold and well-defined nanostructured gold surfaces was tested, and it was found that more live biofilm was observed on the smooth gold surface, whilst there was a reduction in cell viability on the structure of the nanostructured gold surface.[97] Human monocyte activation and adhesion was also studied on the same smooth and nanostructured gold surfaces and it was found that the surface features did not have an effect on monocyte activity.[97]

5.2.5.6 *Translation of* In Vitro *Studies into* In Vivo *Applications*

The translation of *in vitro* work into functional products with increased technology readiness level (TRL) is confounded by a number of parameters, but perhaps the greatest issue is the understanding across the disciplines of the requirements of the final products and the experimental design. Many *in vitro* experiments are carried out under precisely controlled conditions, and this enables the production of well-defined and clear mechanisms of action. For example, Four hole Synthes® locking compression plates were investigated to determine their predisposition to infection.[98] The compression plates are composed of a titanium–aluminium–niobium alloy or clinically available commercially pure titanium and were tested in either a polished form or in their standard microrough form, against clinically available electropolished stainless steel.[98] The locking compression plates were fixed onto the tibia of mature, female New Zealand white rabbits. Following insertion, a clinical strain of *S. aureus* was added to the implantation site.[98] It was found that polishing the surface of both commercially pure titanium and the titanium–aluminium–niobium alloy did not influence the susceptibility to infection in the animal model.[98] Thus, care needs to be taken as *in vitro* findings may not translate directly into *in vivo* studies.

The caveat to *in vitro* research studies is that such conditions rarely occur in 'real' environments. However, research that is carried out with scale-up in mind, usually results in trends in the data, which can then be used to infer how a surface is more likely to react *in vivo*. One such example is the addition of conditioning films for surface:cell or surface:bacteria interaction studies. Once a conditioning film is added into a methodology, the spectrum of the surface and bacterial characteristics becomes altered, and the data produced does not demonstrate such clear conclusions. However, such methodologies are essential to demonstrate the efficacy of a product, before its TLR is increased. Thus, there needs to be a clear demarcation of work that enables the understanding of the final environmental conditions in which the surface is to be used, and an emphasis on projects that investigate industrially and medically relevant processes so that once a system is scaled up, there is a high probability that they will be successful for use in real applications.

5.2.6 Surface Chemistry of Biomaterials

There are a range of metals used in the production of biomaterials. Gold, cobalt–chromium alloys, stainless steel, titanium and its alloys, among other metallic biomaterials, have been in use as orthodontic materials since the twentieth century.[99] External fixation devices are mostly made of stainless steel or titanium.

5.2.6.1 Effect of Surface Chemistry on Bacterial Attachment

The results regarding the effect of surface chemistry (in the context of this chapter, reference is being made to the chemical structure of the metal at the surface) on bacterial attachment and biofilm formation are complex. One of the best known effects of surface chemistry on bacterial infection is demonstrated by different species of *Staphylococcus* (*Staphylococcus* spp.). Common sources of biomaterial infections are opportunistic pathogens that are part of the body's natural flora. However, *S. aureus* infections are most commonly found on metal biomaterials, although S. *epidermidis* is the second most isolated pathogen from metal implant infections. On polymeric biomaterials, *S. epidermidis* is more frequently associated with infections.[13,100,101] This could be due to a myriad of factors, including the bacterial surface physicochemistry and cell surface hydrophobicity, since for example *S. epidermidis* has been shown to be more hydrophobic than *S. aureus*.[102]

5.2.6.2 Heterogeneity in Surface Composition and the Effect on Bacterial Binding

In chromium steels, the areas of the surface containing chromium demonstrate a strong affinity for carbon which may lead to the formation of carbides, specifically at grain boundaries.[103] This leads to heterogeneities in the surface chemistries on the stainless steel. Stainless steel can be polished to produce a range of aesthetic finishes. However, polishing a chemically heterogeneous surface results in 'microniches' of different underlying chemistries, which may favour microbial retention.[104] Chemical elements present on the surfaces and in inclusions on the stainless steel may include silicon, sulfur, calcium, manganese and carbon (Figure 5.5). The presence of trace elements on surfaces may also be important, and previous studies have shown preferential microbial retention at grain boundaries.[105,106] However, on surfaces considered to be chemically homogeneous, such as titanium, the surface chemistry may be more complex. The surface of titanium has been shown to consist of a thin oxide of TiO_2, which is covered by a contaminating layer containing carbon and trace amounts of nitrogen, calcium, phosphorus, chlorine, silicon, sodium or sulfur.[107] When the surfaces of titanium microimplants were evaluated, the detection of some traces of metals such as lead, zinc and copper were also found on the oxide surfaces.[108] Thus, the 'microniches' of different chemical compositions may cause microorganisms to behave differently in comparison to the bulk surface properties. Although small, such areas of chemical heterogeneity may act as 'adhesion points' to which bacteria preferentially attach.[109] This means that if a surface is not completely chemically homogeneous consideration has to be given to the possibility that even the smallest of surface chemical heterogeneities may influence bacterial attachment to a surface.

Microimplants are becoming more widely used in clinical orthodontics to provide skeletal anchorage. The determination of the growth of biofilms on five commercially available micro-implants was carried out *in vitro*.[108] The R_a values were 182 and 69 nm for the titanium microimplants and the stainless steel micro-implants, respectively. Differences in the surface topographies were evidenced on the implants when sourced from different manufacturers, and these included pits, pores and grooves (due to the machining process).[108] Multiple linear regression was used to analyse the extent of biofilm growth, and it was demonstrated that biofilm growth was influenced both by the oxygen and carbon rich components and the topography of the surface of the implant.[108] It may therefore be suggested that residues of these elements could be utilised by bacteria to maintain growth.

5.2.6.3 *Interplay Between the Surface Chemistry and Topography on Bacterial Binding*

The effects of surface chemistry may dominate over the surface topographical effects, and their influence on bacterial retention. Stainless steel surfaces with comparable hydrophobicities and a range of R values (R_a 0.3–0.5 μm, R_z 2.5–3.8 μm) and surface topographies (304 coarse abraded, 304 2B finish, 304 fine polished and 316 2B) were coated with titanium. Retention assays (whereby the surfaces were submersed into a bacterial solution for 1 h) were carried out using *P. aeruginosa*, and it was demonstrated

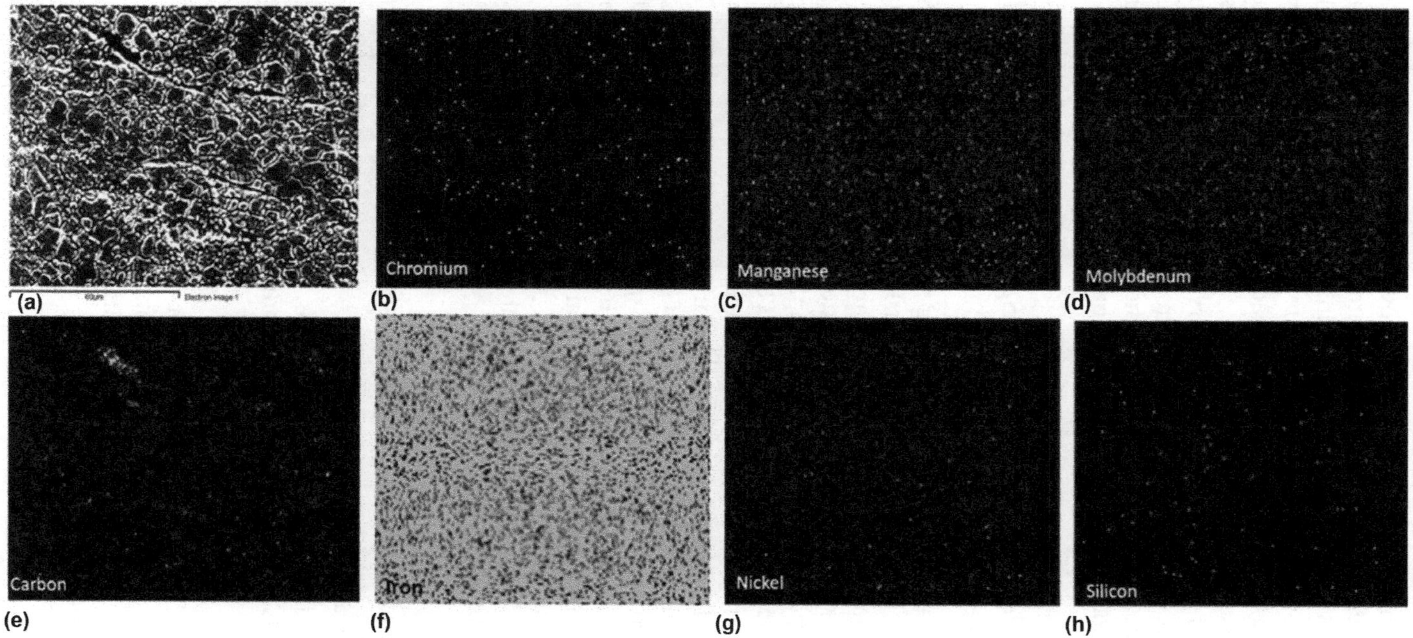

Figure 5.5 Distribution of different elements on a 316 2B finish stainless steel surface measured using energy dispersive X-ray spectroscopy of (a) a stainless steel surface, (b) the distribution of chromium, (c) manganese, (d) molybdenum, (e) carbon, (f) iron, (g) nickel and (h) silicon in the stainless steel surface.

that the pattern of microbial retention on both uncoated and coated surfaces was affected by the surface topography. However, following the use of adhesion assays, whereby cells were sprayed onto a surface, only on the stainless steel with the 2B topographies (with grains and grain boundaries) were the same distribution patterns of the bacteria noted, as was observed in the retention assays. On the surfaces that had been abraded or polished the bacteria were observed to form in clusters. When the same surfaces were coated with titanium, the bacterial numbers were found to be higher, and the cells were spread across the surface in a more homogenous fashion. This work illustrated that a combination of the topography, chemistry and the application method affected bacterial distribution across the surface.

5.2.6.4 *Complexity of the Chemical Species on the Metal/Alloy Surface*

A further complicating issue is the reporting of the chemical species that are on the surface of the material. When the surface of titanium microimplants was analysed, it was observed that the primary oxide may be TiO rather than TiO_2.[108,110–114] Previous studies have also demonstrated differences in the surface compositions of titanium microimplants and these have included the detection of carbide (TiCx) or nitride (TiNx) (possibly from the fabrication process) and standard stoichiometries of TiO_2 and non-standard stoichiometries of Ti_2O_3 and TiO.[113,114] Furthermore, a three-layer model has been proposed to describe the surface of titanium microimplants, which were suggested to consist of titanium substrate, an intermediate suboxide TiO_{2-x} layer, followed by a TiO layer with the outermost layer composed of TiO_2.[108,115] Such differences in the chemical species may affect the surface properties, such as the physicochemistry, and this will invariably effect bacterial binding. Thus, it is imperative that the composition of the surface chemistry is clearly identified.

5.2.6.5 *Summary of the Effect of Surface Chemistry on Bacterial Binding*

Since there are a large number of materials used in the fabrication of biomaterial implants, a summary of the findings of each material type is given below.

5.2.6.5.1 Aluminium. Aluminium is not used as a biomaterial in its native metallic form but is generally used in metal alloys and is one of the most common elements used to stabilise titanium alloys to be used as biomaterials.[99] Aluminium is used to stabilise the α-phase of titanium, which results in an increase in the α- to β- transformation temperature, since the α-phase results in enhanced surface properties, for example, oxidation resistance, good weldability and excellent strength characteristics.[116]

5.2.6.5.2 Chromium. Chromium is not used in its pure form, due to its cytotoxicity, but it is used in numerous alloys, such as cobalt chromium (CoCr) alloys, due to its high wear resistance and stiffness.[117,118] For use in orthopaedic implants, specifically total joint replacements, CoCr alloys are becoming increasingly used.[118,119]

5.2.6.5.3 Copper. Copper tends to be used as a coating to prevent the transmission of bacteria in hospitals rather than in biomaterials. Hospital surfaces can be a reservoir of potential pathogens and play a significant role in healthcare infections. Copper surfaces tend to be used for a range of surfaces within healthcare environments including door knobs, bed rails, touch plates and intravenous stands.[120–123] It has been demonstrated that bacteria present on surfaces can persist for months.[124,125] It has been suggested that the usage of copper surfaces can reduce the bacterial load by greater than 90% within 24 h and reduce the rate of the recolonisation of the surface.[126,127] However, since copper is not corrosion resistant like stainless steel, some copper surfaces may have an added layer of corrosion inhibitor added, but this can adversely affect the antimicrobial efficacy of the copper surface.[128,129] Not all uses of copper are as hard contact surfaces and other usages of copper are being explored, such as for wound dressings. However, such surfaces are beyond the scope of this chapter.

One suggestion as an alternative use of copper is within nanocomposite scaffolds for bone tissue engineering, which has been demonstrated to aid the morphological, physical and biocompatibility of the scaffolds.[130] Such work indicated that copper could be used as part of a biomaterial, especially as copper has been demonstrated to be involved in bone mineralisation, stimulation of collagen production, osteoblast cell adhesion and proliferation.[131–134]

5.2.6.5.4 Gallium. Gallium is a metal that is liquid at near-room temperature and thus it has rarely been used as a surface component. Gallium has instead been applied to treat burns/wounds and in drug delivery applications.[135,136] Often the antimicrobial efficacy of gallium has been assessed in the form of gallium nitrate.[137] Studies have assessed gallium as opposed to silver in the production of antimicrobial surfaces, and it has been found to be effective against *S. aureus*.[138] Gallium titanate surfaces have also been demonstrated to have antimicrobial activity against *Acinetobacter baumannii*.[139]

5.2.6.5.5 Gold. Gold is one of the best known biocompatible metals and it also demonstrates low solubility. In dentistry and medicine, gold has been used with varying results.[140–142] For many years, gold has been used in orthodontic applications, such as in the use of ligatures, bands and wires, and more recently gold has also been used in lingual brackets.[143–145] Interestingly, gold or gold–palladium coated grafts have been shown to prevent infection by *S. epidermidis*.[146,147] Gold in its monovalent form has anti-inflammatory properties and has been used to treat a number of inflammatory disorders.[148] In assays using astrocytes, the use of nanoporous gold surfaces demonstrated that the nanotopography, not the surface chemistry, maintained neuronal coverage whilst reducing astrocyte coverage.[149] Thus, it was suggested that such surfaces may result in nanostructure-mediated suppression of scar tissue formation.[149]

5.2.6.5.6 Molybdenum. The alloys that are currently most often used in the manufacture of orthopaedic implants contain molybdenum, since it provides corrosion resistance and bulk to the implant. Such alloys include Co–Cr–Mo and Co–Ni–Cr–Mo alloys. In hip prothesis, joint surfaces may be composed of Co–Cr–Mo alloys. In the femoral stems of hip replacements Co–Ni–Cr–Mo are often utilised.[150] New titanium alloys containing molybdenum have been developed and they are considered to be non-toxic and contain non-allergenic elements, and these include Ti–7.5Mo, Ti–15Mo, Ti–12Mo–6Zr–2Fe and Ti–12Mo–5Ta.[150–154]

5.2.6.5.7 Nickel. Nickel is used in alloys since it forms an oxide film that is protective to a surface, which increases the corrosion resistance of a material. Nickel also stabilises the formation of austenite in iron.[103]

5.2.6.5.8 Niobium. Niobium in its pure form demonstrates mechanical weakness, and it is an expensive raw material.[118] A niobium–tantalum alloy coating has demonstrated higher corrosion resistance than an uncoated CoCr alloy.[118] Titanium-based alloys containing niobium include Ti–13Nb–13Zr,[155] Ti–24Nb–4Zr–7.9Sn,[156] and Ti–29Nb–13Ta–4.6Zr.[157]

5.2.6.5.9 Palladium. For over 20 years, palladium-based alloys have been utilised in dental restorative materials (dental crown implants and bridge alloys).[140,142] Palladium is known to be biocompatible and have low solubility. Palladium alloys have also been considered as potential materials for use in temporary implants to prevent cardiovascular disease-linked infections.[158]

5.2.6.5.10 Platinum. Platinum is present at a low abundance in the earth's crust (0.003–5 ppm) making it an extremely valuable and expensive metal. After rhodium, it may sometimes be considered the second most expensive metal.[159] Despite its cost, platinum possesses the potential to be utilised in such devices as knee and hip implants and cardiovascular defibrillators.[147,160] Platinum is thought to be of use in biomedical implants, since it is inert, durable and biocompatible. Furthermore, platinum also demonstrates excellent radiopacity and electrical conductivity properties.[160]

5.2.6.5.11 Silver. It has been suggested that the rate of implant infections may be reduced using silver. Silver is used as it is antimicrobial and thought to be non-toxic, although due to its lubricious properties in surface form it is usually used in a composite coating for biomedical devices.[161–163] Silver coatings have been used in a plethora of biomedical devices, including external fixation pins, endotracheal tubes, vascular grafts, indwelling catheters, artificial joint replacements and dental implants.[101,164–168]

5.2.6.5.12 Stainless Steel. The main benefits of using stainless steel include economic profits, robust mechanical properties and ease of production. The family of stainless steels comprise a range of different iron-based alloys. Stainless steels contain between 11–30 wt% of chromium. Stainless steels also contain nickel in varying amounts. These alloys can be categorised into the chromium–nickel or chromium groups. Stainless steels are also grouped according to their microstructures, and these include the duplex (austenitic plus ferritic), ferritic, martensitic, and austenitic families.[103]

Different grades of stainless steel may be given different surface finishes. These include a bright annealed (2R) finish, or a pickling finish (2B).[169] Additional surface treatments such as electropolishing or mechanical polishing can be used to alter the characteristics of the stainless steel surface, for example its surface roughness.

Stainless steel is a material that has been used in a wide variety of applications, including orthopaedic prostheses and cardiovascular stents/valves. Stainless steel has been widely used because it demonstrates good resistance to metal fatigue and corrosion, whilst being extremely malleable. However, although stainless steel is known to have good mechanical properties, it, along with some other metal-based materials, lacks biofunctionality.[170] A key component of stainless steel is chromium, since it enhances the corrosion resistance of the material. The chemistry of stainless steel is variable due to its composition and various metal carbides are formed at the grain boundaries.[171] Another key component used in the production of stainless steel is nickel. Nickel enhances stainless steel as it strengthens the metal crystal structure and thus reduces the tendency of the metal to crack.[171] Carbon can also be used make stainless steel stronger.

In many applications, 317L stainless steel is preferred for use over 316L stainless steel, since the 317L stainless steel has a higher molybdenum content (3–4%) than the 316L steel (2–3% molybdenum). The incorporation of molybdenum into stainless steel enhances its corrosion resistance.[172] 317L stainless steel has also been modified to contain copper alloy (4.5%) (317L–Cu) and this material has been compared to titanium (Ti–6Al–4V) alloys and standard 317L stainless steel (317L) to determine its antimicrobial potential.[172] It was demonstrated that the 317L–Cu material does not affect the formation of osteoblasts on surfaces, but it does inhibit bacterial colonisation by bacteria both *in vitro* and *in vivo*.[172]

5.2.6.5.13 Tantalum. Tantalum in its pure form demonstrates mechanical weakness and is an expensive raw material.[173,174] Tantalum can be processed for use in fabrications from its pure or bulk form. However, tantalum has an extremely high melting temperature and thus it is difficult to use.[175,176] However, dense tantalum coatings possess high fatigue resistance between the coatings and substrate since they produce a sharp interface, which results in an absence of porosity.[177] The other properties of tantalum include its high density and high elastic modulus. Unfortunately, this makes tantalum unsuitable for use in load-bearing implants.[178,179] Medical grade tantalum clips are used in a range of biomedical applications, including as pliable sheets and plates for cranioplasty and reconstructive surgery, for use in staples for abdominal surgery and the ligation of vessels.[180–183] The potential use of tantalum is increasing for use in non-load bearing applications as it has been shown to biologically bond to bone and be bioactive.[175–177] Nanocrystalline surfaces with average grain sizes of ≤20 nm have been fabricated on coarse-grained tantalum and it was found that the nanocrystalline surface possessed higher surface hydrophilicity and enhanced corrosion resistance, adsorbed a higher percentage of protein and enhanced osteoblast adherence and spreading than was demonstrated on the course-grained surface.[184]

5.2.6.5.14 Titanium. Titanium is a material that has the potential for use in a wide range of applications in medicine due to properties including corrosion resistance, low specific weight, biocompatibility and low toxicity.[185] Due to its wear resistance properties, titanium is a favourable material for use as a biomaterial.[186] There are a variety of implants such as dental restorations, artificial joints and bone screws that use titanium and its alloys.[187]

Titanium exists in two forms, as a hexagonal closed pack structure (α-Ti) up to 882 °C and as a body-centered cubic structure (β-Ti) above this temperature, thus making it an allotropic material.[99] Titanium alloys can be further subdivided into α, near-α, α–β, and β alloys. Ti–6Al–4V is an α–β alloy and is the most widely used titanium alloy.[103] In the α-phase of titanium, aluminium is used to stabilise the α phase of the biomaterials.[99] There are many more elements solubilised in β-phase titanium than in α-phase titanium, with these including vanadium, molybdenum, niobium, chromium, iron, copper, nickel, palladium, cobalt, manganese, and certain other transition metals.[99]

It is suggested that titanium implants can contribute towards a reduction in bacterial infection over stainless steel, due to their chemical surface properties, rendering titanium a more suitable material for use in *in vivo* applications, although in some countries it is not always favoured over medical grade stainless steel.[188] Titanium has been shown to suppress microbial growth.[189] This is because titanium dioxide can generate reactive oxygen species (ROS) that can damage bacterial cell membranes and interfere with biochemical processes that can sometimes lead to the death of the bacteria.[190] However, titanium has been associated with high bacterial colonisation compared to other surfaces, limiting its sole application in biomedical devices.[191] Interestingly, under humid conditions increased growth of certain bacteria on titanium surfaces has been reported.[192] The susceptibility to infection using stainless steel or titanium dynamic compression plates in rabbit tibiae was tested by carrying out local bacterial challenge at various concentrations of *S. aureus*, following wound closure. It was demonstrated that the infection for stainless steel plates (75%) was significantly higher than that for titanium plates (35%).[188] Thus, the effect of the surface chemistry of titanium on the attachment and proliferation of bacteria on surfaces is conflicting, and this could be in part due to the methodologies used and the physiological and morphological properties of the bacterial species tested.

5.2.6.5.15 Tungsten. Tungsten is added to materials such as cobalt–chromium based alloys to improve solid–solution strengthening. Tungsten addition to such alloys is also used to control the size and distribution of carbides. Tungsten can also be used to impede corrosion fatigue strength and resistance.[103]

5.2.6.5.16 Zirconium. Zirconium has excellent anti-corrosion properties towards alkalis, acids and sea water and so is often used by the chemical industry. Due to the properties of zirconium, such as its extreme hardness, it is also used as an implant material.[193] Zirconium has been shown to exhibit properties very similar to those of titanium, such as low cytotoxicity and good osteointegration and it is used both as a solid material in its pure form or as a coating on polymers.[6,194,195]

5.2.6.5.17 Shape Memory NiTi Alloy. Martensitic NiTi has advantages over other alloys due to properties such as a high damping capacity and low elastic modulus, which result in its good response to stress.[196] NiTi alloys have been laser treated and it has been demonstrated that the surface changes imposed resulted in an effective reduction in bacterial attachment and biofilm formation.[196] Surface characteristics induced by laser treatment include increased hydrophilicity, crystallinity and surface homogeneity of the oxide surface.[196]

5.2.6.5.18 Surface Oxide Layers. The thickness of the oxide layer has also been demonstrated to make a difference to cell and bacterial binding to a surface. The composition of the oxide surface is dependent on the structure, thickness and the chemical properties of the bulk material, which in turn is influenced by the manufacturing procedures. For example, stainless steel contains chromium, and it is known that this element has an affinity for oxygen. This natural combination of elements enables a chromium-rich oxide film to form (about 2 nm thick) on the surface of stainless steel.[197] It has also been described that titanium samples with oxide thickness less than 100 nm were found to have a surface chemistry that demonstrated an amorphous chemical state, whereas when the oxide layer thickness exceeded 100 nm, a structured rutile layer was observed.[107] Thus, the depth of the oxide layer changed the chemical composition of the surface. Similarly, it has been suggested that due to chemical element segregation, the atomic layer (up to 10 nm) of the surfaces of alloys, for example Ti–6Al–4V, may be different in chemical composition to the bulk phase.[198] Differences in the chemical species on surfaces have been suggested to be more heterogeneous on implantable devices composed of TiAlV, or TiAlNb alloys or commercially pure titanium.[199,200] Surface oxide layers may also be incomplete and may contain pinhole defects or non-metallic elements in the grain boundaries,[27] and this may further result in chemical heterogeneity at the surface (Figure 5.6). As a result, materials that promote the passive formation of chromium oxide layers on their surface, such as cobalt–chromium based alloys, have been deemed superior to stainless steels in terms of their corrosion resistance.[197,201,202]

5.3 Conditioning Films

5.3.1 Adsorption of Proteins onto a Surface

Prior to the insertion of a medical device, a biomaterial surface is both clean and sterile. *In vivo,* the implant surface is coated immediately after insertion by a monolayer of adsorbed interstitial or serum proteins.[203,204] Similarly, during external fixation, a layer of organic material is adsorbed onto the surface of the implant. This layer of organic material is commonly known as a conditioning film or conditioning layer. The conditioning film will be composed of the surrounding tissue fluid and will contain a range of molecules, including albumin, fibrinogen, collagen, and fibronectin. The covering of the biomaterial with such molecules provides ligand sites that may promote adhesion for specific bacterial species.[26,205–208]

The surface chemistry (along with the surface topography and physicochemistry) and how it specifically interacts with biomolecules is of extreme importance, because the surface properties will influence the orientation of the conditioning film molecules, which in turn will allow ligands to be available for bacterial attachment (Figure 5.7).[209,210] However, much of the current research on the antimicrobial and anti-adhesive abilities of surfaces fails to consider the inclusion of a conditioning film.[211] The inclusion of conditioning films on surfaces that are to be used for specific applications is vital, in order to determine if following their binding to the surface, bacterial attachment is increased or reduced.[212]

5.3.2 Effects of Conditioning Films on Bacterial Binding

The substratum properties will influence the binding confirmation of the proteins to the surface, which then influences the increase or decrease in bacterial attachment (Figure 5.8).[210–215] The adsorption of proteins onto substrata involves electrochemical interactions between charged species on

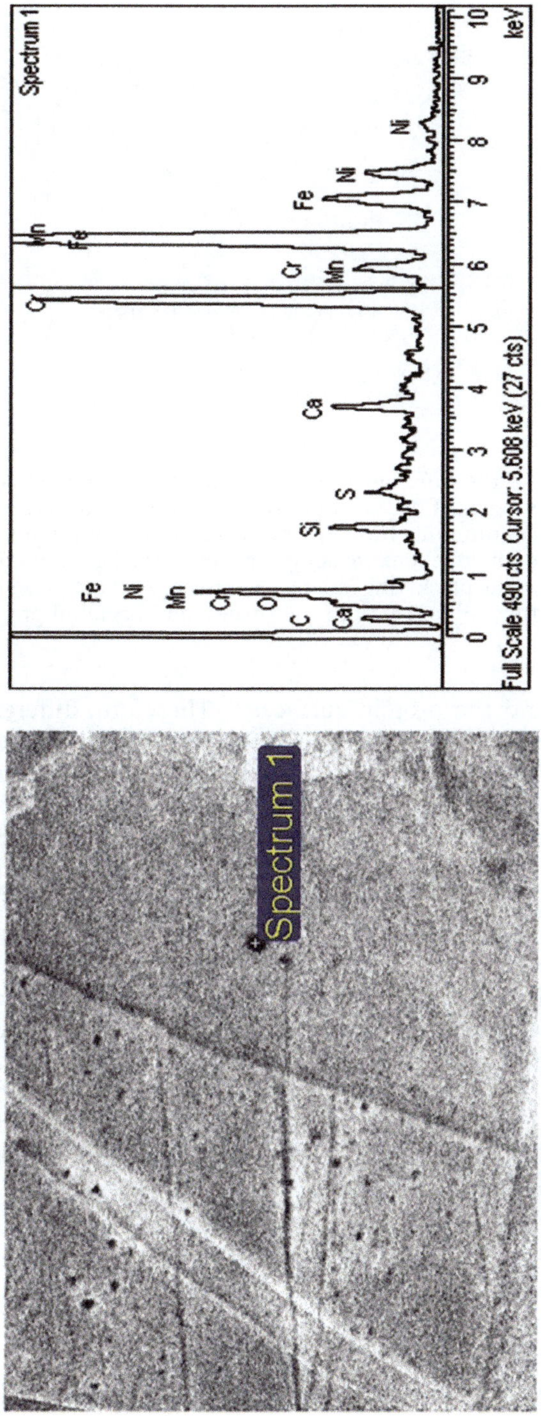

Figure 5.6 (a) SEM image of nitric acid stripped 316 stainless steel demonstrating the underlying surface structure and chemical heterogeneity of the surface. Spectrum 1 denotes an area of a non-metallic inclusion in the steel. (b) Energy dispersive X-ray spectrum of the inclusion, demonstrating the presence of calcium and other elements present in the stainless steel.

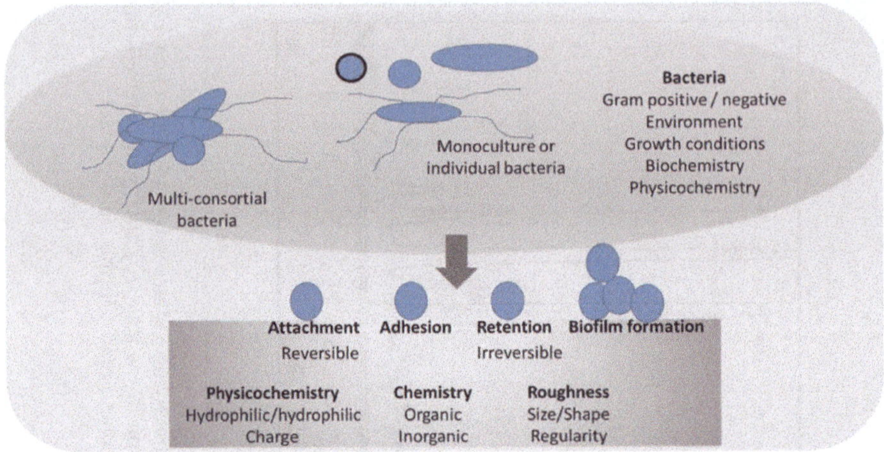

Figure 5.7 The interplay of the different morphologies and combinations of bacteria, with the stages of bacterial binding and biofilm formation, in the presence of a conditioning film. The presence of the conditioning film is denoted by the whole schematic being covered in the light pink film, and specifically by the black ring on the cocci-shaped bacteria, since the conditioning film will affect both the surface and bacterial properties.

the substrata surface and the protein surface.[216] Therefore, different conditioning films will adhere to different surfaces based on the physicochemical and chemical composition of the conditioning film and substrata components. The confirmation of the proteins will affect the specific receptor–ligand interactions that subsequently form between the bacteria and the conditioning film, and this is of importance since bacterial selection will be largely dependent on ligand–adhesin interactions.[34,217] The hydrophobicity of a surface can also affect the adsorption of proteins, such as albumin at the surface.[218] Surface specific characteristics provide certain properties, for example titanium surfaces display amphoteric characteristics, similar to ceramics, which make them more attractive to the binding of both acidic and basic compounds[216] and hence such surface properties will influence initial organic materials adsorption. In addition to conferring specific binding sites for specific bacterial species, the conditioning film also provides nutrients that may support bacterial proliferation. If the conditioning film is adequately dense and thick, then this may decrease the bacterial susceptibility to antimicrobial agents leached from the surface matrix, conferring protection to the bacteria present.

5.3.3 Conditioning Films and Bacterial Adhesins

The interaction of bacterial cells with the surface adsorbed molecules is regulated through the presence of species-specific adhesins located on the

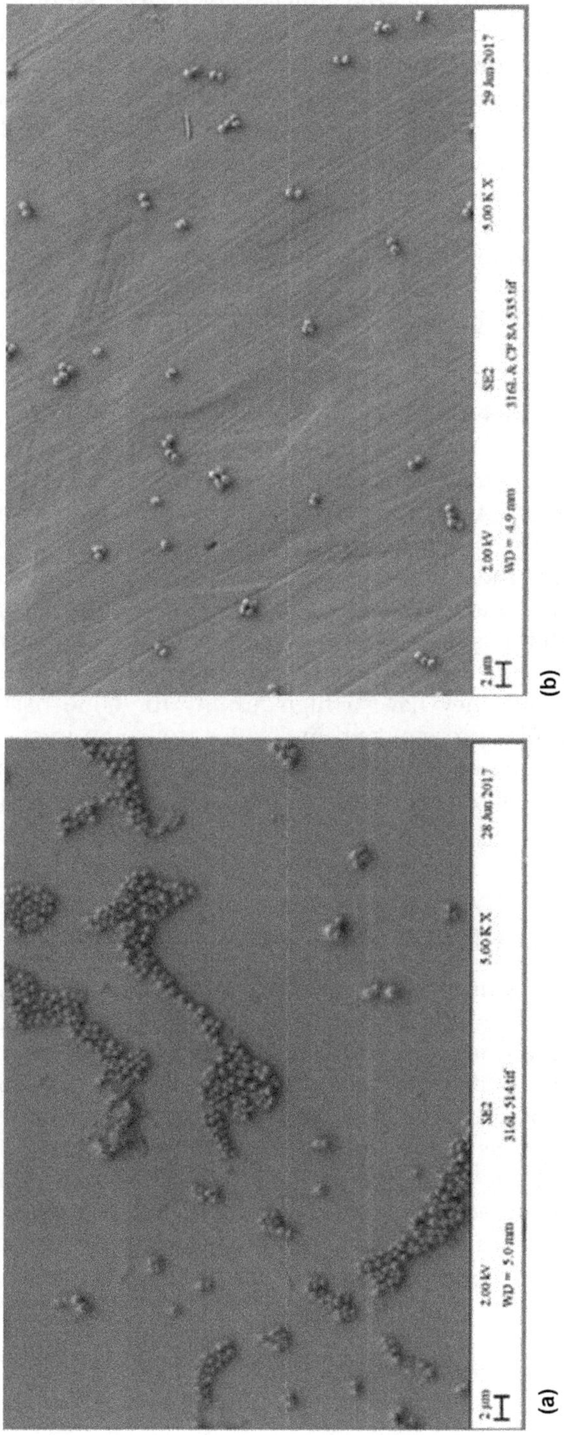

Figure 5.8 SEM images (5000× magnification) of *S. aureus* on (a) 316L stainless steel and (b) 316L stainless steel in the presence of 10% bovine plasma, demonstrating the differences in bacterial binding on the surface in the absence and presence of a 10% bovine plasma conditioning film (5000× magnification) (images courtesy of Dr Louise Tetlow).

surface of the bacterial cell. Bacterial adhesins may be either protein or non-protein based, and may be presented on the end of fimbriae or flagella on the cell surface. Such adhesins are of importance to the bacteria since they are important virulence factors, helping the bacteria to evade immune responses and to adhere, invade and damage host cells.[219,220] The highly specific and selective nature of the adhesin target molecules on the surface of the host cell, play a key role in bacterial pathogenesis, particularly in regards to the effect they have upon the host's immune system.[220] It is speculated that the removal of the bacterial ability to adhere and colonise specific tissues through removal of its adhesin-mediated receptor recognition would render the bacteria non-virulent.[219,221]

Bacterial strains and species can express a range of different types and quantities of adhesins, which might explain the differences seen in the virulence profiles observed between different bacteria.[222] In *S. aureus,* there are a group of adhesins known as microbial surface components that recognise adhesive matrix molecules (MSCRAMMs).[223] The pathogenicity and virulence of each MSCRAMM is strain specific, for example some bacteria have more fibronectin and fibrinogen binding proteins (Fnbp-A and -B). Bacteria that express these proteins are associated with the incidences of infective endocarditis and abscesses, since they are able to attach particularly well to epithelial tissue. However, bacteria that have adhesins that are specific for fibronectin, collagen or sialoprotein are associated with infectious arthritis since they have a high affinity for bone tissue.[223] This specificity for particular proteins and molecules is also an important factor when considering the components of conditioning films, as different strains of bacteria have preferential binding to specific components of the conditioning film.

5.3.4 The Effect of Proteins on Bacterial Binding

On mixed titanium, aluminium, vanadium and niobium surfaces, both fibronectin and albumin adsorption have been shown to be significantly greater on the non-aluminium surfaces.[224] Investigations into the application of conditioning films for the control of microbial attachment have demonstrated positive results. Research into single protein source conditioning films, such as albumin, has demonstrated reductions in bacterial adhesion.[225] Early work by Fletcher[226] showed that the addition of albumin (from bovine serum albumin) to form a conditioning layer on polystyrene Petri dishes was effective at impairing the attachment of *Pseudomonas* spp.[225,227] Serum proteins found in salivary pellicles such as albumin, amylase and immunoglobulins adsorbed onto titanium and zirconium surfaces have also been shown to reduce bacterial adherence to surfaces.[216] In contrast, work by others has investigated the role of fibronectin and found it to have a stimulatory effect towards bacterial adhesion.[225,227–229]

5.4 Antimicrobial Activity of Metals

The antimicrobial properties of metals have been recognised throughout the history of medicine and healthcare.[1] The wide application of metals in medicine was predominant until the discovery of antibiotics, after which the use of metals as antimicrobials began to diminish. However, at the commencement of the 21st century a rapid increase in AMR and a lack of new available antibiotics was apparent. In order to slow down the transmission of potentially infectious microorganisms, there has therefore been a revival in the utilisation of metals as antimicrobial/biocidal agents.[2]

Although the antimicrobial effects of metals have been established for hundreds of years, the mechanisms of why and how to best utilise metals is still not clear, and even today some toxic metals are still used for first line treatment against certain infections.[40] Furthermore, although many metals may be considered antimicrobial or toxic, this is largely dependent on the form and structure in which they are in (*e.g.* ions, oxides, nanoparticles) and their mode of action.

The antimicrobial action of metals begins with the affinity of metal ions towards cellular components and biomolecules, which might form stable complexes and damage vital bacterial cell processes.[40,230] Metals demonstrate toxic effects in several ways, such as displacement/damage of essential enzymes, blocking vital biomolecule functional groups or participating in cellular chemical reactions.[2] One or more of these processes may damage proteins, denature DNA, induce oxidative stress or affect biological walls/membranes.[2] Usually metals demonstrate more than one type of antimicrobial mode of action simultaneously. For a metallic implant to have ideal antimicrobial properties, it would have to have long term antimicrobial activity and continued structural integrity, whilst maintaining good biocompatibility.[231]

5.4.1 Antimicrobial Efficacy of Pure Metal Surfaces

5.4.1.1 Chromium

Chromium ions have been shown to be capable of killing pathogenic microorganisms and exhibit increased antimicrobial properties when utilised in a polymer.[232]

5.4.1.2 Copper

Copper, and its alloys such as brass or bronze, are natural biocidal metals that were exploited long before the concept of antimicrobials existed. Copper is well recognised as being an antimicrobial agent, with many studies confirming these properties.[227–233] It is effective against both Gram positive and Gram negative bacteria. However, differences in bacterial morphology, for example if the bacterial cell wall consists of a Gram negative (a thin layer of peptidoglycan) or Gram positive (a much thicker layer of peptidoglycan) composition, can affect the antimicrobial mechanisms involved in bacterial toxicity.[233] Copper remains effective at reducing bacteria, even after re-contamination, oxidisation or wet and dry abrasion.[234,235] The mode of action of copper surfaces is thought to be due to the cellular ultrastructure being damaged due to the accumulation of copper ions.[236] Unusually, dry copper surfaces have also been shown to be more effective than wet copper surfaces.[235] However, whilst the antimicrobial properties of copper are still being actively researched, they are suggested to be due to the ability of copper to change its oxidation state between Cu^+ and Cu^{2+}.[235] This enables copper to produce reactive oxygen species (ROS), which includes the production of superoxide anions and hydroxyl radicals that can damage DNA, proteins and lipids.[231,237,238] Copper, which is tightly regulated in native cell environments, displays higher binding affinity to biomolecules, based on the Irving–Williams series,[239] when compared to nickel and zinc.[240] When copper, titanium, molybdenum, silver and iron surfaces were tested to determine their levels of ion leaching and their antimicrobial effect on bacteria, it was demonstrated that copper and zinc had the greatest propensity to leach from coatings, whilst silver, titanium, iron and molybdenum leached at lower rates. Further, copper demonstrated the greatest antimicrobial potential, followed by silver and zinc.[192] Although copper has been shown to demonstrate a high amount of leaching in solution, more than is leached by zinc,[192] conflicting reports by Yasuyuki *et al.* have shown that zinc leaches four times the amount of metal ions than copper.[240] However, this could be due to the use of pure metal substrata (>99.5%) by Yasuyuki *et al.*, as opposed to Akhidime *et al.*, who used metal coated surfaces in their study.[192,240] Thus, even results from coated surfaces compared with bulk metals surfaces may produce conflicting results.

5.4.1.3 Gallium

Ga^{3+} ions have been purported to be ideal substitutes for silver in antimicrobial surfaces.[241,242] In 2017, Yamaguchi *et al.* incorporated gallium ions into the surface of titanium metal and demonstrated antimicrobial activity against *A. baumannii*.[240] In 2018, Wadge *et al.* produced novel gallium titanate surfaces, but in contrast to Yamaguchi's work found that biofilm assays showed no significant difference in biofilm growth.[139,240] Thus, the results of the antimicrobial efficacy of gallium are conflicting.

5.4.1.4 Gold

It has been indicated that metals such as gold and amalgam (mixtures of mercury with silver, tin, copper and other trace metals) exert an influence against adhering biofilms by damaging or killing bacteria.[243] Work by Dittmer *et al.* carried out in 2015 demonstrated that initial biofilm formation seemed to be increased on stainless steel surfaces compared with gold in short-term observations, suggesting that the gold surfaces have a short term effect on biofilm reduction.[244]

5.4.1.5 Molybdenum

In a study comparing copper, titanium, molybdenum, silver and iron surfaces, it was shown that molybdenum had a low leaching potential in the media and displayed limited antibacterial effects in all the methods utilised.[93,192] In 2008, Kawakami *et al.* reported similar results using molybdenum flakes as the substrata rather than metal coatings.[245] Studies on the concentrations of chromium, cobalt and molybdenum in patients with metal-on-metal total hip replacements and hip resurfacing arthroplasties have shown that the level of molybdenum in serum is generally low, compared with chromium and nickel.[246,247] Molybdenum disulfide has also demonstrated some antimicrobial efficacy when tested against *E. coli* and has been shown to significantly decrease their survival rate at high concentrations.[248] However, there have been some conflicting results, with studies also showing that at certain concentrations molybdenum disulfide may not efficiently kill bacterial cells.[249] However, molybdenum disulfide nanostructures have been shown to inhibit *P. aeruginosa* biofilm formation by up to 60% at 150 $\mu g\, mL^{-1}$,[250] which may be due to a surface topographical effect as opposed to a surface chemical effect.

One commonly tested compound involving molybdenum is the metal trioxide MoO_3. It has been shown that MoO_3 exhibits efficient antimicrobial activity towards opportunistic pathogens such as *S. aureus* and *P. aeruginosa*. Molybdenum trioxide surfaces have also been found to be non-cytotoxic.[251] Another study investigated the antibacterial efficiency of MoO_3 nanoplates against *E. coli*, *Salmonella typhimurium*, *Enterococcus faecalis*, and *Bacillus subtilis* (minimal inhibitory concentration between 8–16 $\mu g\, mL^{-1}$). This study demonstrated that MoO_3 has antimicrobial efficacy equivalent to the standard antibiotic kanamycin.[252] Tests of the antimicrobial activity against *E. coli* were performed using a molybdenum coating and it was demonstrated that there was complete killing of bacterial colonies.[253]

5.4.1.6 Nickel

When ninety three clinical isolates belonging to the species *E. coli*, *P. aeruginosa* and *S. aureus* were tested against metals, nickel was found to be the least effective antimicrobial metal. This is not surprising since the use of nickel as an antimicrobial agent is not acclaimed.[240,254]

5.4.1.7 Palladium

Palladium-based dental biomaterials are considered biocompatible for oral environments and are thought to control oral pathogenic infections.[249]

The antimicrobial efficacy of palladium ions has been determined and shown to control bacterial growth.[255]

5.4.1.8 Platinum

Platinum has metallurgic properties and is unreactive, so it can be potentially used as an antimicrobial in medical implants, such as cardiovascular defibrillators, hip and knee implants, catheters, pacemakers, stents or neuromodulation devices. Platinum ions have previously demonstrated antimicrobial properties against bacterial pathogens.[255,256]

5.4.1.9 Silver

Elemental silver requires ionisation for antimicrobial activity.[257] Silver is antimicrobial in the soluble form of silver ions, which exhibit antibacterial and antifungal properties.[258,259] Silver can be present in one of four oxidative states, Ag^0, Ag^+, Ag^{2+} and Ag^{3+}, with Ag^+ being the most common.[260,261] Ag^0 has no overall ionic charge and therefore demonstrates no antimicrobial activity. In comparison, Ag^+ cations are antimicrobial, inhibiting bacterial growth between 8 and 80 ppm.

The mechanism of antimicrobial activity of silver (and its derivatives) is yet to be fully elucidated.[262,263] It is thought that the activity is a multifactorial process, rather than the result of a single mechanism, which differs depending on the organism type, cell physiology, strain and methods used. Silver is known to target multiple components within the bacterial cell.[264] As silver needs to be ionised to demonstrate antimicrobial activity, as positively charged Ag^+ ions are attracted to the negative polarity of the bacterial cell membrane.[265] Gram negative bacteria are deemed to be more susceptible to silver ions than Gram positive bacteria, due to the reduced thickness of the outer peptidoglycan layer.[266] Silver has antimicrobial activity against both planktonic and biofilm bacterial phenotypes, with Ag^+ ions being able to inhibit the initial bacterial attachment process.[267]

Although elemental silver is antimicrobially inert, ionised silver exhibits high reactivity due to its interactions with electron donor groups (N, O or S atoms), which are present in bacteria in both proteins and in DNA.[161,268] Silver ions can displace essential metals in enzymes, leading to a deleterious effect on protein synthesis, bacterial DNA and cell membrane integrity.[41] Furthermore, the ability of ionised silver to bind to thiol groups can result in the release of iron from proteins, which can in turn generate ROS such as hydroxyl radicals, superoxide radicals and hydrogen peroxide.[238,268] Another antimicrobial mechanism of silver is the ability to inter-chelate with the phosphate backbone of DNA. This leads to strand breaks and mutations in DNA repair genes, which can result in a loss of replication and expression of ribosomal subunits and other cellular proteins.[269] Strains of bacteria that are sensitive to silver can adsorb silver from dilute solutions, in doses of as small as 1 ppm. Experimental studies demonstrated that a 60 ppm concentration of silver is sufficient to inhibit the majority of bacterial pathogens.

5.4.1.10 Tantalum

The design of porous structures using tantalum has resulted in surfaces with improved mechanical properties. In 2019, Rodríguez–Contreras *et al.* designed antimicrobial coatings for tantalum implants using polyhydroxyalkanoates (PHAs) as matrices and found that they were non-toxic, well-biointegrated homogeneous biopolymer coatings, which showed antibacterial properties.[179] A hydrophilic crystalline β-Ta_2O_5 coating was fabricated by rapid thermal annealing of a deposited tantalum oxide coating and was tested against *S. aureus* and *Actinobacillus actinomycetemcomitans* and it was found to exhibit good antibacterial performance.[270] Depending on the coating structures, the amorphous Ta_2O_5 exhibited good antimicrobial performance whilst the hydrophilic β-Ta_2O_5 coatings showed excellent cell viability of human skin fibroblast cells.[270] These results demonstrate that tantalum metal may be a good substrate for osteogenesis in animal implantation tests, and suitable for cell adhesion, proliferation and differentiation *in vitro* studies,[271,272] whilst also demonstrating antimicrobial activity.

5.4.1.11 NiTi

The antibacterial performance of martensitic NiTi is thought to be due to oxides formed by laser processing, which bring about changes to the surface chemistry. However, it has been suggested that the changes in surface topography, on a micrometre scale, play a minor role in influencing bacterial attachment.[196]

5.4.1.12 Metal Oxides

Recently, metal oxides have been studied as they have shown promise as broad spectrum antimicrobial agents.[273] It is thought that metal oxides damage bacteria by having positive charges that are attracted to the negative surface charge of most bacteria. Due to this attraction, they accumulate at the cell surface of the bacteria, which eventually allow them entrance to the cell. This leads to the metals causing toxicity inside of the cell as a result of the difference in charges, which leads to bacterial death. Metal oxides are also theorised to be more effective against Gram negative bacteria due to them having higher surface negative charges.[273]

5.4.2 Antimicrobial Efficacy of Metal Ions

A brief overview of the literature demonstrates that some metal ions demonstrate antimicrobial activity. For example, when the antimicrobial activity of metal ion solutions (silver, copper, platinum, gold, palladium, rhodium or ruthenium ions) were determined individually and in combination against *Enterococcus faecium*, *A. baumannii* and *Klebsiella pneumonia*, overall, platinum, palladium and gold metal ion solutions demonstrated the greatest antimicrobial efficacies against planktonic or biofilm bacteria.[255,274] However, again, the bacterial species tested influences the outcome of the results. When ninety three clinical isolates (composed of *E. coli, P. aeruginosa* and *S. aureus*) were tested against six metal salt solutions, (aluminium sulfate, copper sulfate, gallium nitrate, nickel sulfate, silver nitrate and zinc sulfate), the results demonstrated that there were wide antibacterial differences demonstrated between the isolates and that only several of the isolates were either resistant or sensitive to all of the six metals.[275] The *E. coli* strains demonstrated the greatest antimicrobial profile similarity between the isolates. In contrast, the susceptibilities of *S. aureus* and *P. aeruginosa* were more highly dispersed. A study conducted by Slate *et al.* in 2019 demonstrated that when the metal ions, molybdenum, rhenium, thallium and yttrium were tested against both *Klebsiella pneumoniae* and *A. baumannii* in planktonic and biofilm forms, overall, rhenium or yttrium ions were the most effective antimicrobial agents.[276] However, these results considered a range of antimicrobial assays (and conditions), and not just the results of one assay, for example minimal inhibitory concentrations (MIC). This work suggests that metals are not equal in their efficacies and hence, their use may be tailored against a particular microorganism and care should therefore be taken to ensure the use of the correct concentration[275] so as not to select for bacterial persisters or viable non culturable bacteria.

5.4.3 Effects of Conditioning Films on Ion Release

Work on the effect of conditioning films on antibacterial surfaces is conflicting. Silver is only effective as an antimicrobial agent when moisture is present, and so is not useful in all applications. In the presence of moisture (*e.g.*, serum proteins) the metallic silver becomes ionised, activating the antimicrobial properties.[264] In 1997, Wassall *et al.* found that the presence of silver in a stainless steel surface reduced the adherence of *S. aureus* cells to pin tract surfaces[277] and thus the presence of the conditioning film made the surface antiadhesive rather than antimicrobial.

An *in vitro* experiment analysing the metal release from silver-coated venous catheters showed that serum proteins have an impact on the dissolution of silver.[278] Around 5–10% of the silver detached into the surrounding water, while 20% detached into a 5% BSA solution. Over time, the release rates declined with a greater percentage recorded in the BSA solution (33%) than in the water (20%) after 14 days.[278] Therefore, the conditioning film influenced the dissolution rate of the silver. This affects the properties of the material and may compromise its functionality and integrity.

It has been demonstrated that the use of a mixed protein solution rather than a single protein solution in conditioning film assays is more representative of real-life conditions. On adsorption to a surface, the proteins organise themselves into an equilibrium state, with the smaller proteins being replaced by larger proteins.[279,280] Thus, it is important to investigate the effects that conditioning films have on bacterial retention and adhesion, but also to assess the role of proteins from complex protein sources, such as whole plasma or serum, rather than individual protein sources, which do not offer 'real world' environmental simulations.

5.5 Metal Toxicity

Biomaterials need to interact with a particular host tissue in a manner that is specific to their application, causing minimal inflammatory response, immunogenicity, or cytotoxicity.[281] However, the level of toxicity of metallic materials is very much determined by the form in which the metal is present. The environment inside the human body is physically and chemically different from ambient conditions and thus a metal that performs well (is inert or passive) in the air may suffer severe corrosion in the body.[103] The release of metal ions originates *via* dissolution, wear, or principally alloy corrosion, and although these ions may provide antimicrobial activity, they may also result in unacceptable levels of cytotoxicity.[150]

5.5.1 Nontoxic Metals

Zirconium and niobium are nontoxic metals that are widely used as biomaterials, since they have good biocompatibility. They may be used in artificial heart valves, replacement implants in shoulders, knees, hips, elbows and in orodental structures.[282,283] The transition metals rhenium and yttrium are also considered to be practically nontoxic (LD_{50} (single oral dose for rats) 5000–15 000 mg kg^{-1}).[276,284,285] Hornez and co-workers found that palladium showed no cytotoxicity in a study on dental implant alloys, when compared to a wide range of other metals.[286] Titanium-based alloy coatings containing niobium have demonstrated increased cytocompatibility when compared to CoCr alloy, with higher cell proliferation of MG-63 cells, and a noticeable improvement in haemocompatibility.[118]

The toxicity of silver is controversial. It has been stated that silver has no known biological function in the human body and the toxicity of silver is a subject yet to be fully understood. Some studies suggest that silver-coated biomedical implants show little evidence of potential toxic effects,[269,287] whilst other studies suggest that silver has toxic effects after prolonged exposure.[288,289] Bosetti *et al.* determined the biocompatibility of orthopaedic external fixation pins and compared stainless steel and silver coated pins.[290] The genotoxic effect of silver was determined using human peripheral blood lymphocytes, and fibroblasts and osteoblast-like cells were used for cytocompatibility and cytotoxicity studies. It was determined that silver was neither genotoxic nor cytotoxic when compared to stainless steel.[290]

5.5.2 Biotoxicity of Metals

Metal release processes can be governed both by corrosion and dissolution of the surface, and may result in potential biotoxicity issues.[291] Although a number of metals are used in the production of biomaterials, ion release may result in a number of physiological and inflammatory conditions, including fibrous encapsulation and metallosis (a blood poisoning condition).[281] Wear of the femoral head–neck area of a total hip replacement (trunnionosis), may result in arthroplasty failure, which may also result in local soft-tissue reactions as a result of the collection of corrosion debris.[292] Furthermore, for a large number of metal surfaces, the effect of bacterial biofilms resulting in microbial-induced corrosion is documented.[293,294] For example, studies on the effect of the corrosion resistance of CoCrMo surfaces to *S. aureus* and *P. aeruginosa* biofilms, demonstrated that *P. aeruginosa* influenced corrosion at the fretting interface due to changes in the electrochemical properties of the surface.[295] A short overview is given below on the biotoxicity findings on sheet metals and their corrosion products (Table 5.2).

5.5.2.1 *Aluminium*

Aluminium cytotoxicity has a profound effect on several diseases (and disease progression).[103] The release of aluminium has been demonstrated to cause neurotoxic effects, and soft tissue sarcomas have been demonstrated to occur due to the wear debris of aluminium oxide.[291,296] Aluminium has also been shown to affect osteoblasts, by stimulating the growth of the stem cells that subsequently differentiate into osteoblasts, and the growth of osteoblasts themselves. However, aluminium has also been shown to inhibit osteocalcin secretion and mineralisation and be cytotoxic over prolonged periods.[224,297,298] Information on *in vitro* toxicity has shown that aluminium, vanadium and titanium can bind to the thiol groups of bacterial cellular membrane proteins that result in an affected cell function, since such proteins are involved in the facilitation of metal transport across the cell membrane.[224,299,300]

Osteoblasts treated with sub-lethal levels of $Al_3{}^+$ and $Ti_4{}^+$ ions show little difference in alkaline phosphatase activity and cell proliferation. However, it has been shown that osteocalcin secretion is inhibited, as is mineralised matrix formation.[301] Work by Scotchford *et al.* carried out in 2002 using primary human osteoblasts on aluminium surfaces demonstrated that the metal affected cell morphology and cytoskeletal organisation, and caused inferior cell growth on the surfaces.[224] It was suggested that the cellular response to aluminium surfaces may be due to the oxide being a positively charged surface at pH 7.[224]

5.5.2.2 *Chromium*

Pure chromium is a human carcinogen in its chromium(VI) oxidation state and thus it is not widely used as a biomaterial. Chromium-based

Table 5.2 Biotoxicity information for metals.

Metal	Dose
Aluminium	The Environmental Protection Agency (EPA) has recommended a secondary maximum contaminant level (SMCL) of 0.05–0.2 mg L^{-1} for aluminium in drinking water. The Occupational Safety and Health Administration (OSHA) has set a legal limit of 15 mg m^{-3} (total dust) and 5 mg m^{-3} (respirable fraction) aluminium in dusts averaged over an 8 h work day.[385]
Bismuth	Animal studies have shown that IV injection of elemental bismuth (1.8 mg kg^{-1}) results in heart blockage and death. The acute toxicity of soluble bismuth pharmaceuticals results in lethal dose values ranging from 13–82 mg kg^{-1} body weight.[386]
Chromium	The lethal dose for chromium(VI) is in the range of 50–150 mg kg^{-1}. The World Health Organization (WHO) has recommended a maximum chromium(VI) limit of 0.05 mg L^{-1}. The OSHA has set a permissible exposure limit of 0.5 mg m^{-3} for chromium in air.[387,388]
Copper	The EPA has determined that drinking water should not contain more than 1.3 mg L^{-1} of copper. The OSHA has set a limit of 0.1 mg m^{-3} for copper fumes (vapour generated from heating copper) and 1.0 mg m^{-3} for copper dusts (fine metallic copper particles) and mists (aerosols of soluble copper) in workroom air.[388,389]
Gallium	Based on three published animal studies, the National Institute of Occupational Safety and Health (NIOSH) recommends the maximum exposure limit for gallium to be 2 μg m^{-3} in air.[390,391]
Iron	Humans experience iron toxicity above 20 mg kg^{-1} of body mass, and 60 mg kg^{-1} is considered a lethal dose. The NIOSH and OSHA have set a permissible exposure limit of 5 and 10 mg m^{-3} respectively for iron dust and fume in air.[392,393]
Molybdenum	Studies on rats show a median lethal dose (LD_{50}) of as low as 180 mg kg^{-1} for some molybdenum compounds. OSHA has set a limit of 5 mg m^{-3} for soluble molybdenum compounds and 15 mg m^{-3} for insoluble molybdenum compounds and total dust in workroom air. The NIOSH has established a level of 5000 mg m^{-3} for insoluble molybdenum compounds and 1000 mg m^{-3} for soluble molybdenum compounds that are considered immediately dangerous and likely to cause death.[303,394,395]
Nickel	In the US, the minimal risk level of nickel and its compounds is set to 0.2 mg m^{-3} for inhalation over 15–364 days. OSHA has set a limit of 1 mg m^{-3} for metallic nickel and in nickel compounds in workroom air. The EPA recommends that the drinking water levels for nickel should not be more than 0.1 mg L^{-1}.[396,397]
Niobium	Concentrations of niobium, niobium nitride and/or pentoxide in the air should not exceed 40 mg m^{-3}.
Palladium	Concentrations of palladium in the workplace for poorly soluble compounds (metals and their oxides) is recommended to be less than 0.1 mg m^{-3}.[398]
Platinum	The WHO recommends an exposure limit of 2 μg m^{-3} for platinum in workroom air over an 8 h work day.[399]
Silver	The EPA recommends that the concentration of silver in drinking water should not exceed 0.10 mg L^{-1} because of the skin discoloration that may occur. The OSHA and NIOSH limits silver in workplace air to 0.01 mg m^{-3} for an 8 h workday, for a 40 h working week. The American Conference of Government Industrial

Table 5.2 (*Continued*)

Metal	Dose
	Hygienists (ACGIH) recommends workplace air to contain no more than 0.1 mg m^{-3} of silver metal and 0.01 mg m^{-3} of soluble silver compounds.[400,401]
Stainless Steel	Studies on rats show that inhalation toxicity LD_{50} for stainless steel is higher than 5 mg L^{-1}.[402]
Tantalum	The OSHA and NIOSH have reported that the permissible exposure limit for an 8 h time weighted average is 5 mg m^{-3} for tantalum metal and oxide dust.
Thallium	The OSHA and NIOSH have reported that the permissible exposure limit for thallium is 0.1 mg m^{-3}.[398]
Titanium	When quantities of up to 0.8 mg of titanium were ingested by humans on a daily basis, most titanium was found to be excreted without being digested or absorbed.[403]
Tungsten	Studies in humans showed that exposure to tungsten at 0.1–0.2 mg m^{-3} for a period of 2.5 to 7 years caused pneumoconiosis.[404]
Vanadium	OSHA set a legal limit of 0.5 mg m^{-3} for vanadium pentoxide respirable dust as a maximum not to be exceeded during the work day. A limit of 0.1 mg m^{-3} for vanadium pentoxide fumes has also been established.[405,406]
Yttrium	The OSHA and NIOSH have reported that the permissible exposure limit is 1 mg m^{-3}.
Zinc	The EPA has stated that drinking water should contain no more than 5 mg mL^{-1} of zinc. The OSHA has set an average legal limit of 1 mg m^{-3} for zinc chloride fumes and 5 mg m^{-3} for zinc oxide (dusts and fumes) in workplace air.[407]
Zirconium	The OSHA has evaluated toxicity and defined long term and short term exposure limits for zirconium compounds to be at 5 mgZr m^{-3}. For 'inhalable particles', with a higher mean aerodynamic diameter, the value was set to 10 mg m^{-3}.[408]

compounds have been suggested to cause a number of physiological issues, including irritation to the mucous membranes, skin, eyes, irritant dermatitis and allergic contact dermatitis. Chronic exposure to such substances can cause permanent eye injury.[203] The toxicity and carcinogenic properties of hexavalent chromium(VI) are due to its strong oxidative properties.[103,302,303] However, chromium is widely used in biomaterials as a component in a range of alloys, including surgical steel and cobalt–chromium–vanadium alloys.[203] It is not surprising therefore that the release of CoCr alloys in a physiological environment has generated concerns. The Cr^{3+} and Cr^{6+} states are the most common oxidation states of chromium and the ions of CoCr (Co^{2+}, Cr^{3+} and Cr^{6+}) have been reported to result in cellular disfunction, including genotoxicity, necrosis and cell apoptosis.[118,304,305]

Orthopaedic biomaterials may naturally form a passive oxide layer and materials such as CoCrMo and Ti–6Al–4V form such a protective oxide layer. This layer will protect a surface from the dissolution effects of being in a physiological solution. However, when mechanical events at the surface occur, the oxide layer becomes disrupted, and continual disruption and reformation of the oxide layer have been shown to increase corrosion rates.[292,306–308] This may lead to the production of metal ions and wear debris. If patients have implants that have metal-on-metal systems, which may occur with some types of CoCrMo metals, the continued exposure of their bodies to wear products may result in chromium and cobalt release into the synovial fluid. The metal ions can then enter the blood and increased serum metal ion levels may result in adverse tissue reactions and teratogenic effects.[309–311]

5.5.2.3 Cobalt

Cobalt alloys are used in hip replacements, and the mechanical action of such replacements may result in wear of the implant and the production of metallic particles. A build-up of these metallic components in the localised tissue may damage the tissues surrounding the implant due to hypersensitivity reactions.[312] The toxicity of cobalt has been observed after the insertion of hip prostheses, with wearing of the prostheses releasing cobalt ions and nanoparticles that can affect liver cells, although the mechanisms that result in such reactions are not well understood.[313] The cytotoxic effect of cobalt ions was further supported by a study that demonstrated that cobalt ions altered osteocyte morphology and had a significant effect on osteocyte concentration.[314] In some patients, cobalt toxicity has also been reported to result in neurological and systemic symptoms after 4–5 years of implantation when the metal was used in alloys for metal-on-metal hip prostheses.[309]

5.5.2.4 Copper

Copper is an essential element for humans as it plays a role in many biological processes, including the catalytic activities of metabolic enzymes and proteins.[315,316] These processes are involved in the oxidative stress response, respiration, blood clotting and normal tissue function in tissues such as the nervous system or as part of the immune system. Copper is readily metabolised and utilised by the body when absorbed either orally or through tissues. Of the copper that is present within the body, the liver and the brain have the highest concentration, with most copper present in cupric form (Cu^{2+}), although this is not exclusive.[317] The majority of the copper in the body of normal healthy humans is linked to enzyme prosthetic groups or tightly bound to copper transport/chaperone proteins, which reduces the chance of unbound copper participating in redox reactions, thus ensuring that copper reaches specific target proteins.[318–321]

The use of copper in healthcare applications needs to ensure that the release of copper ions is low enough to be at levels safe for applications on or in the human body.[322] Generally, the risk of adverse reactions due to skin exposure is extremely low as skin is not generally sensitive to copper materials. Copper has been suggested for use in the antimicrobial treatment of wounds. However, the mode of antimicrobial action, may also have degenerative effects on human cells. For example, the presence of copper enables the production of ROS, which can damage DNA. Such modes of action can lead to carcinogenesis. In wounds where there is a reduced blood supply, such as in diabetic ulcers or pressure sores, the use of copper in wound dressings can lead to an accumulation of the metal, and this may be particularly problematic for patients with other comorbidities such as renal or liver malfunction.[323] Thus, copper can promote ROS formation by undergoing redox cycling reactions, which is thought to be copper's main cytotoxic property, since ROS production affects the structures of essential biomolecules.[324,325] Cu^{2+} can be reduced to Cu^{+}, which is capable of catalysing hydroxyl radical formation *via* Haber–Weiss reactions from hydrogen peroxide.[326–328] Copper can also affect several enzymes, such as glutathione reductase and glucose-6-phosphatase, as the copper binds to their sulfhydryl groups, which interferes with the cell's protection from free radical damage. However, there are other mechanisms that contribute to the cellular overload of copper and thus copper cytotoxicity, typically due to alterations in gene expression or molecules. Other mechanisms include altered lipid metabolism, with one study demonstrating links between copper status and lipid metabolism, and another demonstrating a link between cell cycle and cholesterol metabolism and copper imbalances.[329–331] Copper may also result in altered hepatic gene expression due to prolonged exposure, where alterations cause a shift towards anaerobic metabolism.[329] Further, copper may promote the apoptosis of cells *via* p53 dependent or independent pathways,[332] with one study demonstrating Cu^{2+} triggered hepatocyte apoptosis through acidic sphingomyelinase activation and ceramide release.[333]

Copper cytotoxicity can cause disruption in copper homeostasis, which has been associated with hepatic disorders, neurodegenerative disorders and other diseases.[334] It has been shown to induce a predominance of neutrophils, persistent secretion of pro-inflammatory cytokines, high degrees of cellular damage and a high and extended leukotactic response.[335–337] The US Environmental Protection Agency (EPA) gives copper a carcinogenic classification of D, meaning that there is not enough evidence to determine if copper is in fact carcinogenic, although some studies have suggested that it is.[338,339]

5.5.2.5 Gold

The biocompatibility of five metal coatings, and their effect on the soft tissue reaction in rats demonstrated that the addition of higher concentrations of gold and palladium to silver resulted in a thicker soft tissue capsule, and coatings with high concentrations of palladium induced the thickest fibrous capsule after 21 days of implantation.[340]

5.5.2.6 Molybdenum

Molybdenum toxicity is rare in humans, but it has been associated with issues due to dietary overdose, through nutritional supplementation. Over ingestion of molybdenum results in effects such as neurotoxicity and gout-like symptoms.[341]

5.5.2.7 Niobium

As a metal, niobium is generally considered nontoxic and is not involved in any known biological processes, and is therefore considered hypoallergenic.[342] The safety of niobium was indicated in one study, which demonstrated that the niobium ion safe concentration limit was greater than titanium when tested against osteoblast-like cells.[343] However, when a range of metal ions were tested for their ability to induce DNA damage, apoptosis, and necrosis in human (Jurkat) T-cells, cobalt and niobium were found to be the most toxic metals, although overall, the results indicated that soluble metal ions released from implants may not be preferentially genotoxic to lymphocytes.[344]

5.5.2.8 Nickel

Nickel is used in stainless steel and nickel–titanium orthodontic alloys, from which it is released.[345–347] Nickel is a known allergen and is highly toxic to the human body when implanted within bone or muscle nickel has resulted in necrosis or severe local tissue irritation. Nickel has been shown to be less harmful to osteoblasts than vanadium or cobalt, which are used in other alloys.[201,348,349] Nickel has also been found to be released and transported systemically to other organs, and it has also been found adjacent to tissues surrounding stainless steel implants.[350,351]

Since nickel has been demonstrated to be toxic to the body, there has been the development of austenitic stainless steels that are nickel free and high in nitrogen that can be used in medical applications. The nitrogen improves the structure and mechanical properties of the stainless steel.[103,352,353] Other nickel-containing alloys are now used only in short term implants, such as wires and bone plates, for example, Co–Ni–Cr–Mo–W–Fe and Co–20Cr–15W–alloys.[103]

5.5.2.9 Silver

Silver is widely used as a biomaterial, specifically due to its antimicrobial properties, although local and systemic toxicity effects of silver have been detailed.[230] The most common side effect associated with prolonged silver exposure is the development of an irreversible pigmentation of either the skin (argyria) or the eyes (argyrosis).[289] The affected areas have a bluish-grey or slate-grey pigmentation that is more prominent in areas of the body exposed to sunlight.[354] Argyria occurs due to silver deposits forming in the subdermal connective tissue, including the glands and vascular tissues.[355] This condition has previously resulted from the use of colloidal silver and silver-based medicinal products.[356–358] Argyria has also been a reported side effect of the use of silver-coated megaprostheses. In one study, 7 of the 34 patients with silver-coated megaprostheses implants developed localised argyria (23%) after a median time period of 25.7 months.[359] Interestingly, the duration of the implant did not influence local argyria development, whilst patients with and without local argyria had comparable silver concentrations in the blood and aspiration fluids.[359] Patients diagnosed with argyria had no neurological symptoms or evidence of renal or hepatic failure, therefore short term surveillance of blood silver levels was not required.[359] Silver has been reported to have an injurious effect on mammalian cells, including the central nervous tissue[360,361] and arteriolar vascular smooth muscle.[230,362] It has been reported that micromolar concentrations of silver are capable of causing significant cell toxicity in white blood cells.[363,364] Other symptoms related to silver accumulation include kidney and liver damage, irritation of the skin, respiratory and intestinal tract, and an alteration in blood cell morphology.[289]

In some cases, it has been suggested that silver has a similar toxicity to mercury.[365] A range of metals were analysed *in vitro* against human peripheral immune cells for their cytotoxicity. The work demonstrated that monocytes, B, and T lymphocytes were equally sensitive to all the metals tested and were ranked in the following order, $Hg \approx Ag \geq Cd \approx Cu \geq Pb \approx Zn$ of decreasing cytotoxic potency.[365] In another study, a silver surface was tested in conjunction with a hamster dorsal skinfold chamber model. This demonstrated that leukocytes are persistently activated along with disruption in the microvascular endothelial integrity, increase venular dilation and leukocyte extravasation.[230]

5.5.2.10 Stainless Steel

There are multiple issues with the use of stainless steel as a biomaterial. High material density and low biocompatibility can aggravate bone healing processes.[353] Also, medical stainless steels can corrode when in prolonged contact with serous fluid,[366] leading to device failure and the premature removal of medical implants.[367,368] Due to the toxicity of nickel ions from stainless steels, high-nitrogen nickel-free austenitic stainless steels are becoming more widely used as an alternative medical stainless steel.[369] Research has also found that titanium orthopaedic biomaterials are more stable and functional than those made from stainless steel.[370]

However, despite this, stainless steel is an alloy that is commonly utilised. Zinc, copper and chromium are released from stainless steel due to the electronic structure of such elements at the atomic level and the phase structure of the alloy.[99,371] Many stainless steels that are used for prolonged periods in the body demonstrate corrosion resistance, and can produce toxic reactions in the body and induce chronic allergies.[103,372,373] Stainless-steel implants incorporated in animal models have shown that although the animals demonstrated toxic effects due to the stainless steel, the animals tend to recover, whereas titanium implants were shown to have much higher cytotoxic effects on the animals.[230]

Leaching of nickel ions from stainless steel due to implant corrosion can cause inflammation, cytotoxicity and carcinogenicity.[374] In the oral cavity, the signs of nickel allergy may be extensive.[375] 316L stainless steel is not regularly used in implants since there have been reported toxicity and carcinogenicity issues due to released nickel and chromium.[103] However, stainless steel is low cost and thus it is still used in many temporary devices.[376,377]

Iron is a trace element that is found in almost all living organisms. However, if the corrosion of implants results in excessive iron release, issues such as long term organ damage, liver failure, coma, metabolic acidosis, shock, coagulopathy, adult respiratory distress syndrome, and ultimately death can occur if left untreated.[103,378]

5.5.2.11 Vanadium

Vanadium in the form of vanadate has close similarity to phosphate, which may explain some of its toxicological effects. Vanadium is commonly used in pharmaceuticals, and can accumulate in bone.[291,379] Vanadium compounds have been shown to produce significant cytotoxicity in various osteoblast cells.[224,380–382] Vanadium toxicity mainly originates from its compounds such as oxides.[383] The titanium–aluminium (6 wt%)–vanadium (4 wt%) (Ti–6Al–4V) alloy is widely used as a metallic implant material, which is considered as biocompatible and demonstrates high corrosion resistance.[291] However, it is thought that the release of titanium, vanadium and aluminium ions from the implants can potentially induce adverse health effects.[291] Both vanadium and aluminium released from Ti–6Al–4V alloys have been found to be linked to long-term health problems, such as osteomalacia, Alzheimer's disease and other neuropathies.[384]

5.6 Conclusions

This chapter highlights the complex and combined effects of surface properties and how they are linked to factors affecting microbial attachment, adhesion, retention and biofilm formation. Thus, the use of individual surface parameters to enhance the understanding of microbial binding to a surface is no longer adequate. Moreover, the effects of surface parameters on microbial retention should be considered individually for each bacterial species and environment. Of the metal ions that demonstrate consistent antimicrobial activity, titanium, zirconium, molybdenum, niobium, platinum, palladium and gold are the least toxic to human cells but demonstrate limited antimicrobial activity. This suggests that there must be a compromise between the biomaterial design and an acceptable, non-toxic release of antimicrobial ions if novel, non-cytotoxic, functional metal antimicrobial biomaterials are to be produced. The functionality and antimicrobial activity of such biomaterials must be tested under appropriate physiological solutions in order to determine their corrosion resistance and the effect of conditioning films on implant design.

References

1. A. M. Elsome, J. M. T. Hamilton-Miller, W. Brumfitt and W. C. Noble, *J. Antimicrob. Chemother.*, 1996, **37**, 911.
2. J. A. Lemire, J. J. Harrison and R. J. Turner, *Nat. Rev. Microbiol.*, 2013, **11**, 371.
3. G. Manivasagam, D. Dhinasekaran and A. Rajamanickam, *Recent Pat. Corros. Sci.*, 2010, **2**, 40.
4. A. Simchi, E. Tamjid, F. Pishbin and A. R. Boccaccini, *Nanomed-Nanotechnol.*, 2011, **7**, 22.
5. J. B. Park, *Biomaterials Science and Engineering*, Springer, Boston, 1984.
6. A. J. Slate, D. J. Wickens, M. El Mohtadi, N. Dempsey-Hibbert, G. West, C. E. Banks and K. A. Whitehead, *Sci. Rep.*, 2018, **8**, 1497.
7. C. A. Collinge, G. Goll, D. Seligson and K. J. Easley, *Orthopedics*, 1994, **17**, 445.
8. D. Gil, S. Shuvaev, A. Frank-Kamenetskii, V. Reukov, C. Gross and A. Vertegel, *Antimicrob. Agents Chemother.*, 2017, **61**, e00442.
9. C.-W. Chan, L. Carson, G. C. Smith, A. Morelli and S. Lee, *Appl. Surf. Sci.*, 2017, **404**, 67.
10. J. T. Kempthorne, R. Ailabouni, S. Raniga, D. Hammer and G. Hooper, *BioMed Res. Int.*, 2015, **2015**, 946215.
11. R. Zeng, M. Li, Q. Chen, L. Wang, P. Zhan, C. Wang, G. Lv, Y. Shen and W. Liu, *Antimicrob. Agents Chemother.*, 2014, **58**, 1443.
12. F. Mallard, V. Le Bourlot, C. Le Coeur, M. Avnaim, R. Péronnet, D. Claessen and T. Tully, *Funct. Ecol.*, 2020, **34**, 669.
13. L. G. Harris and R. G. Richards, *Injury*, 2006, **37**, 3.
14. K. C. Dee, D. A. Puleo and R. Bizios, *Cell. Mol. Biol.*, 2004, **8**, 419.
15. D. Campoccia, L. Montanaro and C. R. Arciola, *Biomaterials*, 2013, **34**, 8533.
16. I. Ktistakis, E. Guerado and P. V. Giannoudis, *Injury*, 2015, **46**, 35.
17. K. Hori and S. Matsumoto, *Biochem. Eng. J.*, 2010, **48**, 424.
18. P. S. Stewart and J. W. Costerton, *Lancet*, 2001, **358**, 135.
19. N. Høiby, T. Bjarnsholt, M. Givskov, S. Molin and O. Ciofu, *Int. J. Antimicrob. Agents*, 2010, **35**, 322.
20. F. Song, H. Koo and D. Ren, *J. Dent. Res.*, 2015, **94**, 1027.
21. U. Römling, S. Kjelleberg, S. Normark, L. Nyman, B. E. Uhlin and B. Åkerlund, *J. Intern. Med.*, 2014, **276**, 98.
22. J. W. Costerton, P. S. Stewart and E. P. Greenberg, *Science*, 1999, **284**, 1318.
23. N. Dror, M. Mandel, Z. Hazan and G. Lavie, *Sensors*, 2009, **9**, 2538.

24. S. J. McConoughey, R. Howlin, J. F. Granger, M. M. Manring, J. H. Calhoun, M. Shirtliff, S. Kathju and P. Stoodley, *Future Microbiol.*, 2014, **9**, 987.
25. C. Von Eiff, W. Kohnen, K. Becker and B. Jansen, *Artif. Organs*, 2005, **28**, 1146.
26. A. G. Gristina, M. Oga, L. X. Webb and C. D. Hobgood, *Science*, 1985, **228**, 990.
27. A. G. Gristina, *Science*, 1987, **237**, 1588.
28. A. Trampuz and W. Zimmerli, *Injury*, 2006, **37**, 59.
29. W. Zimmerli, *Best Pract. Res. Cl. Rh.*, 2006, **20**, 1045.
30. W. Zimmerli, A. Trampuz and P. E. Ochsner, *N. Engl. J. Med.*, 2004, **351**, 1645.
31. K. Gulati, M. S. Aw and D. Losic, *Nanoscale Res. Lett.*, 2011, **6**, 571.
32. C. A. Iobst, W. Mohammed and R. Colley, *Orthopedics*, 2017, **40**, 876.
33. J. C. R. Wormald, A. Jain, H. Lloyd-Hughes, S. Gardiner and M. D. Gardiner, *J. Plast. Reconstr. Aesthet. Surg.*, 2017, **70**, 1298.
34. A. Al-Ahmad, M. Wiedmann-Al-Ahmad, J. Faust, M. Bächle, M. Follo, M. Wolkewitz, C. Hannig, E. Hellwig, C. Carvalho and R. Kohal, *J. Biomed. Mater. Res. B. Appl. Biomater.*, 2010, **95**, 101.
35. H. Oliveira, S. Catros, O. Castano, S. Rey, R. Siadous, D. Clift, J. Marti-Munoz, M. Batista, R. Bareille, J. Planell and E. Engel, *Acta Biomater.*, 2017, **54**, 377.
36. J. E. Femino and K. Mueller, *Clin. Orthop. Relat. Res.*, 2001, **391**, 72.
37. M. Clauss, S. Graf, S. Gersbach, B. Hintermann, T. Ilchmann and M. Knupp, *Clin. Orthop. Relat. Res.*, 2013, **471**, 2312.
38. E. L. Cyphert and H. A. von Recum, *Exp. Biol. Med.*, 2017, **242**, 788.
39. S. M. Tambe, L. Sampath and S. M. Modak, *J. Antimicrob. Chemother.*, 2001, **47**, 589.
40. D. H. Nies, *Appl. Microbiol. Biotechnol.*, 1999, **51**, 730.
41. J. L. Hobman and L. C. Crossman, *J. Med. Microbiol.*, 2015, **64**, 471.
42. C. J. Van Oss, *Colloids Surf. B*, 1995, **5**, 91.
43. C. J. Van Oss and R. F. Giese, *Clays Clay Miner.*, 1995, **43**, 474.
44. M. Schauperl, M. Podewitz, B. J. Waldner and K. R. Liedl, *J. Chem. Theory Comput.*, 2016, **12**, 4600.
45. D. Ahmad, I. van den Boogaert, J. Miller, R. Presswell and H. Jouhara, *Energy Sources, Part A*, 2018, **40**, 2686.
46. J. M. Berg, L. G. T. Eriksson, P. M. Claesson and K. G. N. Borve, *Langmuir*, 1994, **10**, 1225.
47. S. J. Pogorzelski, A. Z. Mazurek and A. Szczepanska, *J. Marine Syst.*, 2013, **119**, 50.
48. C. I. Pereni, Q. Zhao, Y. Liu and E. Abel, *Colloids Surf. B*, 2006, **48**, 143.
49. M. N. Bellon-Fontaine, J. Rault and C. J. Van Oss, *Colloids Surf. B*, 1996, **7**, 47.
50. C. J. Van Oss and R. J. Good, *J. Protein Chem.*, 1988, **7**, 179.
51. R. M. Donlan and J. W. Costerton, *Clin. Microbiol. Rev.*, 2002, **15**, 167.
52. A. E. Zeraik and M. Nitschke, *Braz. Arch. Biol. Technol.*, 2012, **55**, 569.

53. A. Zita and M. Hermansson, *Appl. Environ. Microbiol.*, 1997, **63**, 1168.
54. E. Sinde and J. Carballo, *Food Microbiol.*, 2000, **17**, 439.
55. C. Faille, C. Jullien, F. Fontaine, M. N. Bellon-Fontaine, C. Slomianny and T. Benezech, *Can. J. Microbiol.*, 2002, **48**, 728.
56. M. L. Marin, Y. Benito, C. Pin, M. F. Fernandez, M. L. Garcia, M. D. Selgas and C. Casas, *Lett. Appl. Microbiol.*, 1997, **24**, 14.
57. M. S. Chae, H. Schraft, L. T. Hansen and R. Mackereth, *Food Microbiol.*, 2006, **23**, 250.
58. A. M. Gallardo-Moreno, M. A. Pacha-Olivenza, L. Saldaña, C. Pérez-Giraldo, J. M. Bruque, N. Vilaboa and M. L. González-Martín, *Acta Biomater.*, 2009, **5**, 181.
59. Y. Gao, Y. Liu, L. Zhou, Z. Guo, M. Rong, X. Liu, C. Lai and X. Ding, *PLoS One*, 2013, **8**, e68086.
60. T. Kawano, W. Prananingrum, Y. Ishida, T. Goto, Y. Naito, M. Watanabe, Y. Tomotake and T. Ichikawa, *PLoS One*, 2013, **8**, e84327.
61. A. Braem, L. Van Mellaert, D. Hofmans, E. De Waelheyns, J. Anné, J. Schrooten and J. Vleugels, *Powder Metall.*, 2013, **56**, 267.
62. D. Campoccia, L. Montanaro and C. R. Arciola, *Biomaterials*, 2013, **34**, 8018.
63. J. Verran, A. Packer, P. Kelly and K. A. Whitehead, *Lett. Appl. Microbiol.*, 2010, **50**, 258.
64. BS 1134:2010, *British Standards*, British Standards Institution, London, 2010, https://shop.bsigroup.com/ProductDetail/?pid=000000000030213633.
65. K. A. Whitehead and J. Verran, *Food Bioprod. Process.*, 2006, **84**, 253.
66. BS 1134-1:1988, *British Standards*, British Standards Institution, London, 1988, https://shop.bsigroup.com/ProductDetail/?pid=000000000000179371.
67. M. Sedlaček, B. Podgornik and J. Vižintin, *Wear*, 2009, **266**, 482.
68. P. D. Antonio, M. Lasalvia, G. Perna and V. Capozzi, *Biochim. Biophys. Acta Biomemb.*, 2012, **1818**, 3141.
69. E. S. Gadelmawla, M. M. Koura, T. M. A. Maksoud, I. M. Elewa and H. H. Soliman, *J. Mater. Process Technol.*, 2002, **123**, 133.
70. G. Aydin, I. Karakurt and K. Aydiner, *Int. J. Rock Mech. Min.*, 2013, **61**, 171.
71. Q. Zhao, Y. Liu, C. Wang, S. Wang, N. Peng and C. Jeynes, *Med. Eng. Phys.*, 2008, **30**, 341.
72. BS EN ISO 4287:1988 + A1:2009, *British Standards*, British Standards Institution, London, 2009, https://shop.bsigroup.com/ProductDetail/?pid=000000000030167230.
73. BS EN ISO 25187-2:2012, *British Standards*, British Standards Institution, London, 2012, https://shop.bsigroup.com/ProductDetail?pid=000000000030154352.
74. D. J. Wickens, S. Lynch, G. West, P. Kelly, J. Verran and K. A. Whitehead, *J. Microbiol. Methods*, 2014, **104**, 101.

75. C. Lüdecke, M. Roth, W. Yu, U. Horn, J. Bossert and K. D. Jandt, *Colloids Surf. B*, 2016, **145**, 617.
76. C. Faille, J. M. Membre, J. P. Ussier, M. N. Bellon-Fontaine, B. Carpentier, M. A. Laroche and T. Benezech, *Biofouling*, 2000, **15**, 261.
77. ISO TC 106 5, C. M. L. Bollen, B. N. A. Vandekerckhove and M. Quirynen, https://www.iso.org/organization/4381659.html.
78. E. Medilanski, K. Kaufmann, L. Y. Wick, O. Wanner and H. Harms, *Biofouling*, 2002, **18**, 193.
79. K. A. Whitehead, H. Li, P. J. Kelly and J. Verran, *J. Adhes. Sci. Technol.*, 2011, **25**, 2299.
80. A. Cunha, A. M. Elie, L. Plawinski, A. P. Serro, A. M. B. do Rego, A. Almeida, M. C. Urdaci, M. C. Durrieu and R. Vilar, *Appl. Surf. Sci.*, 2008, **360**, 485.
81. M. Y. Yüce and A. L. Demirel, *Eur. Phys. J. B*, 2008, **64**, 493.
82. A. J. Slate, K. A. Whitehead, D. A. C. Brownson and C. E. Banks, *Renew. Sustain. Energy Rev.*, 2019, **101**, 60.
83. K. A. Whitehead, J. Colligon and J. Verran, *Colloids Surf. B*, 2005, **41**(1), 29.
84. K. J. Edwards and A. D. Rutenberg, *Chem. Geol.*, 2001, **180**, 19.
85. M. Amin, S. Rowley-Neale, L. Shalamanova, S. Lynch, J. S. T. Wilson-Nieuwenhuis, M. El Mohtadi, C. E. Banks and K. A. Whitehead, *ACS Appl. Mater. Interfaces*, 2020, **12**, 21057.
86. D. Campoccia, L. Montanaro and C. R. Arciola, *Biomaterials*, 2006, **27**, 2331.
87. G. Colon, B. C. Ward and T. J. Webster, *J. Biomed. Mater. Res. A*, 2006, **78**, 595.
88. S. D. Puckett, E. Taylor, T. Raimondo and T. J. Webster, *Biomaterials*, 2010, **31**, 706.
89. D. C. Miller, K. M. Haberstroh and T. J. Webster, *J Biomed. Mater. Res. A*, 2005, **73**, 476.
90. A. Cohen, P. Liu-Synder, D. Storey and T. J. Webster, *Nanoscale Res. Lett.*, 2007, **2**, 385.
91. C. Yao, E. B. Slamovich and T. J. Webster, *J. Biomed. Mater. Res. A*, 2008, **85**, 157.
92. M. K. Akens, C. Chien, R. N. Katchky, H. J. Kreder, J. Finkelstein and C. M. Whyne, *BMC Musculoskelet. Disord.*, 2018, **19**, 260.
93. G. Bhardwaj and T. J. Webster, *Int. J. Nanomed.*, 2017, **12**, 363.
94. I. Yoda, H. Koseki, M. Tomita, T. Shida, H. Horiuchi, H. Sakoda and M. Osaki, *BMC Microbiol.*, 2014, **14**, 234.
95. K. A. Whitehead, S. Olivier, P. S. Benson, N. Arneborg, J. Verran and P. Kelly, *Int. J. Food Microbiol.*, 2015, **197**, 92.
96. C. L. Donaghy, R. McFadden, S. Kelaini, L. Carson, A. Margariti and C. W. Chan, *Opt. Laser Technol.*, 2020, **121**, 105793.
97. S. Svensson, M. Forsberg, M. Hulander, F. Vazirisani, A. Palmquist, J. Lausmaa, P. Thomsen and M. Trobos, *Int. J. Nanomed.*, 2014, **9**, 775.

98. T. F. Moriarty, L. Debefve, L. Boure, D. Campoccia, U. Schlegel and R. G. Richards, *Int. J. Artif. Organs*, 2009, **32**, 663.
99. S. Arango, A. Peláez-Vargas and C. García, *Coatings*, 2013, **3**, 1.
100. R. O. Darouiche, *Clin. Infect. Dis.*, 1999, **29**, 1371.
101. L. M. Coester, J. V. Nepola, J. Allen and J. L. Marsh, *Iowa Orthop. J.*, 2006, **26**, 48.
102. C. C. P. M. Verheyen, W. J. A. Dhert, J. M. A. de Blieck-Hogervorst, T. J. K. Van der Reijden, P. L. C. Petit and K. De Groot, *Biomaterials*, 1993, **14**, 383.
103. Q. Chen and G. A. Thouas, *Mater. Sci.*, 2015, **87**, 1.
104. R. O. Adams, *J. Vac. Sci. Technol.*, 1983, **1**, 12.
105. H. H. P. Fang, L.-C. Xu and K.-Y. Chan, *Water Res.*, 2002, **36**, 4709.
106. K. A. Whitehead and J. Verran, *J. Int. Biodeterior. Biodegrad.*, 2007, **60**, 74.
107. J. Lausmaa, *J. Electron Spectrosc.*, 1996, **81**, 343.
108. M. Y. H. Chin, A. Sandham, J. De Vries, H. C. van der Mei and H. J. Busscher, *Biomaterials*, 2007, **28**, 2032.
109. H. Ma, C. J. Winslow and B. E. Logan, *Colloids Surf. B*, 2008, **62**, 232.
110. H. E. Placko, S. Mishra, J. J. Weimer and L. C. Lucas, *Int. J. Oral Maxillofac. Implants*, 2000, **15**, 355.
111. D. V. Kilpadi, G. N. Raikar, J. Liu, J. E. Lemons, Y. Vohra and J. C. Gregory, *J Biomed. Mater. Res. A*, 1998, **40**, 646.
112. D. V. Kilpadi, J. J. Weimer and J. E. Lemons, *Colloids Surf., A*, 1998, **135**, 89.
113. J. Lausmaa, B. Kasemo and H. Mattsson, *Appl. Surf. Sci.*, 1990, **44**, 133.
114. J. Lausmaa, B. Kasemo, H. Mattsson and H. Odelius, *Appl. Surf. Sci.*, 1990, **45**, 189.
115. D. J. Blackwood and L. M. Peter, *Electrochim. Acta*, 1989, **34**, 1505.
116. C. Oldani and A. Dominguez, Titanium as a Biomaterial for Implants, *Rec. Adv. Arthroplasty*, 2012, **218**, 149.
117. M. H. Fathi, M. Salehi, A. Saatchi, V. Mortazavi and S. B. Moosavi, *Dent. Mater. J.*, 2003, **19**, 188.
118. B. Singh, G. Singh and B. S. Sidhu, *Surf. Coat. Technol.*, 2019, **377**, 124932.
119. A. Batal, R. Sammons and S. Dimov, *Mater. Sci. Eng., C*, 2019, **98**, 1005.
120. J. O. Noyce, H. Michels and C. W. Keevil, *Appl. Environ. Microbiol.*, 2006, **72**, 4239.
121. J. O. Noyce, H. Michels and C. W. Keevil, *J. Hosp. Infect.*, 2006, **63**, 289.
122. S. Mehtar, I. Wiid and S. D. Todorov, *J. Hosp. Infect.*, 2008, **68**, 45.
123. L. Weaver, H. T. Michels and C. W. Keevil, *J. Hosp. Infect.*, 2008, **68**, 145.
124. S. J. Dancer, *J. Hosp. Infect.*, 2004, **56**, 10.
125. A. Kramer, I. Schwebke and G. Kampf, *BMC Infect. Dis.*, 2006, **6**, 130.
126. A. L. Casey, D. Adams, T. J. Karpanen, P. A. Lambert, B. D. Cookson, P. Nightingale, L. Miruszenko, R. Shillam, P. Christian and T. S. J. Elliott, *J. Hosp. Infect.*, 2010, **74**, 72.

127. A. Mikolay, S. Huggett, L. Tikana, G. Grass, J. Braun and D. H. Nies, *Appl. Microbiol.*, 2010, **87**, 1875.
128. J. Elguindi, X. Hao, Y. Lin, H. A. Alwathnani, G. Wei and C. Rensing, *Appl. Microbiol.*, 2011, **91**, 237.
129. J. Elguindi, S. Moffitt, H. Hasman, C. Andrade, S. Raghavan and C. Rensing, *Appl. Microbiol.*, 2011, **89**, 1963.
130. J. C. Forero, E. Roa, J. G. Reyes, C. Acevedo and N. Osses, *Materials*, 2017, **10**, 1177.
131. A. Ito, H. Kawamura, M. Otsuka, M. Ikeuchi, H. Ohgushi, K. Ishikawa, K. Onuma, N. Kanzaki, Y. Sogo and N. Ichinose, *Mater. Sci. Eng., C*, 2002, **22**, 21.
132. T. J. Webster, C. Ergun, R. H. Doremus and R. Bizios, *J. Biomed. Mater. Res. A*, 2002, **59**, 312.
133. H.-J. Seo, Y.-E. Cho, T. Kim, H.-I. Shin and I.-S. Kwun, *Nutr. Res. Pract.*, 2010, **4**, 356.
134. A. Ewald, C. Käppel, E. Vorndran, C. Moseke, M. Gelinsky and U. Gbureck, *J. Biomed. Mater. Res. A*, 2012, **100**, 2392.
135. J. Zhou, F. Zhao, Y. Wang, Y. Zhang and L. Yang, *J. Lumin.*, 2007, **122**, 195.
136. Y. Zhu, Y. Qiu, R. Chen and L. Liao, *J. Nanosci. Nanotechnol.*, 2015, **15**, 5605.
137. C. R. Chitambar, *Int. J. Environ. Res. Public Health*, 2010, **7**, 2337.
138. S. Yamaguchi, *et al.*, *Nanomaterials*, 2019, **7**, 229.
139. M. D. Wadge, B. W. Stuart, K. G. Thomas and G. M. Grant, *Mater. Des.*, 2018, **155**, 264.
140. G. Kansu and A. K. Aydin, *Eur. J. Prosthodont. Restor. Dent.*, 1996, **4**, 129.
141. F. Novelli, M. Recine, F. Sparatore and C. Juliano, *Farmaco*, 1999, **54**, 232.
142. A. Berry, *Interdiscipl. Sci. Rev.*, 1992, **17**, 308.
143. D. Wiechmann, J. Gerß, T. Stamm and A. Hohoff, *Am. J. Orthod. Dentofac. Orthop.*, 2008, **133**, 359.
144. A. Plane and P. O. Teeth, *J. Esthet. Restor. Dent.*, 2014, **26**, 79–87.
145. D. Wiechmann, U. Meyer and A. Büchter, *Clin. Oral. Implants Res.*, 2007, **18**, 263.
146. I. Savvaidis, V. I. Karamushka, H. Lee and J. T. Trevors, *Biometals*, 1998, **11**, 69.
147. O. Saygun, *et al.*, *J. Surg. Res.*, 2006, **131**, 73.
148. R. Eisler, *Inflamm. Res.*, 2003, **52**, 487.
149. C. A. R. Chapman, *ACS Appl. Mater. Inter.*, 2015, **7**, 7093.
150. A. M. Ribeiro, T. H. S. Flores-Sahagun and R. C. A. Paredes, *J. Mater. Sci.*, 2016, **51**, 2806.
151. *Medical Applications of Titanium and Its Alloys: The Material and Biological Issues*, ed. S. A. Brown and J. E. Lemons, ASTM, West Conshohocken, PA, 1996.
152. W. F. Ho, S. C. Wu, S. K. Hsu, Y. C. Li and H. C. Hsu, *Mater. Sci. Eng.*, 2012, **C 32**, 517.
153. L. Trentani, *et al.*, *Biomaterials*, 2002, **23**, 2863.

154. D. M. Gordin, *et al.*, *J. Mater. Sci.: Mater. Med.*, 2004, **15**, 885.
155. P. Kovacs and J. Davidson, *European Pat.*, EP 0520721, 1992.
156. Y. L. Hao, S. J. Li, S. Y. Sun, C. Y. Zheng and R. Yang, *Acta Biomater.*, 2007, **3**, 277.
157. M. Niinomi, *Mater. Sci. Eng., A*, 1998, **243**, 231.
158. B. Woodward, *Platin. Met. Rev.*, 2012, **56**, 213.
159. M. J. Capeness, M. C. Edmundson and L. E. Horsfall, *New Biotechnol.*, 2015, **32**, 727.
160. A. Cowley, *Platin. Met. Rev.*, 2011, **55**, 98.
161. J. L. Clement and P. S. Jarrett, *Met. Based Drugs*, 1994, **1**, 467.
162. S. Marin, *et al.*, *Curr. Top. Med. Chem.*, 2015, **15**, 1596.
163. Y. Liu, *et al.*, *Biomaterials*, 2012, **33**, 8745.
164. G. Tokmaji, H. Vermeulen, M. C. A. Müller, P. H. S. Kwakman, M. J. Schultz and S. A. J. Zaat, *Cochrane Database Syst. Rev.*, 2015, **2015**, CD009201.
165. S. A. Hardman, *et al.*, *Ann. Vasc. Surg.*, 2004, **18**, 308.
166. A. Srinivasan, T. Karchmer, A. Richards, X. Song and T. Perl, *Infect. Control Hosp. Epidemiol.*, 2006, **27**, 38.
167. F. Donati, G. Di Giacomo, S. D'Adamio, A. Ziranu, S. Careri, M. A. Rosa and G. Maccauro, *BioMed Res. Int.*, 2016, **2016**, 9079041.
168. J. Mattos Corrêa, M. Mori, H. Lajas Sanches, A. Dibo da Cruz, E. Poiate and I. A. Venturini Pola Poiate, *Int. J. Biomater.*, 2015, **2015**, 485275.
169. L. Boulané-Petermann, *Biofouling*, 1996, **10**, 275.
170. A. Bekmurzayeva, W. J. Duncanson, H. S. Azevedo and D. Kanayeva, *Mater. Sci. Eng., C*, 2018, **93**, 1073.
171. X. Y. Fang, W. G. Wang, G. U. O. Hong, X. Zhang and B. X. Zhou, *J. Iron Steel Res. Int.*, 2007, **14**, 339.
172. H. Chai, *et al.*, *J. Mater. Sci.: Mater. Med.*, 2011, **22**, 2525.
173. S. Dittrick, V. K. Balla, S. Bose and A. Bandyopadhyay, *Mater. Sci. Eng., C*, 2011, **31**, 1832.
174. E. E. Nikishina, D. V. Drobot and E. N. Lebedeva, *Russ. J. Non-Ferr. Met.*, 2014, **55**, 130.
175. V. K. Balla, S. Bose, N. M. Davies and A. Bandyopadhyay, *JOM*, 2010, **62**, 61.
176. V. K. Balla, S. Banerjee, S. Bose and A. Bandyopadhyay, *Acta Biomater.*, 2010, **6**, 2329.
177. V. K. Balla, S. Bodhak, S. Bose and A. Bandyopadhyay, *Acta Biomater.*, 2010, **6**, 3349.
178. K. Prasad, *et al.*, *Sci. Rep.*, 2017, **7**, 1.
179. A. Rodríguez-Contreras, J. Guillem-Marti, O. Lopez, J. M. Manero and E. Ruperez, *Colloids Surf. B*, 2019, **182**, 110317.
180. A. H. Andrews Jr and M. J. Tamari, *Illinois Med. J.*, 1953, **103**, 175.
181. R. Molicki, B. Kłaptocz, M. Sypniewski and E. Wierzyński, *Czas. Stomatol.*, 1979, **32**, 595.
182. T. V. Kalinina, *Khirurgiia*, 1968, **44**, 97.
183. M. F. Vyrzhikovskaia, *Eksp. Khir.*, 1959, **4**, 38.

184. W. T. Huo, *et al.*, *Sci. Rep.*, 2017, **7**, 40868.
185. World Health Organisation, *Prevention of Coronary Heart Disease: Report of a WHO Expert Committee*, 1982.
186. B. D. Boyan, *et al.*, *J. Orthop. Res.*, 2003, **21**, 638.
187. X. He, X. Zhang, X. Wang and L. Qin, *Coatings*, 2017, **7**, 45.
188. S. Arens, *et al.*, *J. Bone Joint Surg.*, 1996, **78**, 647.
189. Y. Tian, H. Cao, Y. Qiao, F. Meng and X. Liu, *Acta Biomater.*, 2014, **10**, 4505.
190. R. Y. Pelgrift and A. J. Friedman, *Adv. Drug Deliv. Rev.*, 2013, **65**, 1803.
191. K. Whitehead, P. Kelly, H. Li and J. Verran, *J. Vac. Sci. Technol., B*, 2010, **28**, 180.
192. I. D. Akhidime, *et al.*, *Food Bioprod. Process.*, 2019, **113**, 68.
193. M. B. Guglielmotti, S. Renou and R. L. Cabrini, *Int. J. Oral Maxillofac. Implants*, 1999, **14**, 565.
194. C. Wetzel, F. Hollstein, R. H. W. Funk, N. Ozkucur and T. K. Monsees, *Surf. Coat. Technol.*, 2008, **202**, 5728.
195. C. K. Mueller, *et al.*, *Oral Surg, Oral Med. Oral Pathol. Oral Radiol.*, 2013, **116**, e1.
196. C. W. Chan, L. Carson and G. C. Smith, *Surf. Coat. Technol.*, 2018, **349**, 488.
197. M. Navarro, A. Michiardi, O. Castano and J. A. Planell, *J. R. Soc., Interface*, 2008, **5**, 1137.
198. B. Kasemo and J. Lausmaa, *CRC Crit. Rev. Clin. Neurobiol.*, 1986, **4**, 365.
199. H. Olin, B. O. Aronsson, B. Kasemo, J. Lausmaa and M. Rodahl, *Ultramicroscopy*, 1992, **42**, 567.
200. G. Rådegran, J. Lausmaa, L. Mattsson, U. Rolander and B. Kasemo, *J. Electron Microsc. Tech.*, 1991, **19**, 99.
201. H. J. Raithel, G. Ebner, K. H. Schaller, B. Schellmann and H. Valentin, *Am. J. Ind. Med.*, 1987, **12**, 55.
202. C. V. Vidal and A. Muñoz, *Electrochim. Acta*, 2009, **54**, 1798.
203. J. Levon, *et al.*, *J. Biomed. Mater. Res. A*, 2010, **92**, 1606.
204. P. Francois, P. Vaudaux and P. D. Lew, *Ann. Vasc. Surg.*, 1998, **12**, 34.
205. R. E. Baier, A. E. Meyer, J. R. Natiella, R. R. Natiella and J. M. Carter, *J. Biomed. Mater. Res.*, 1984, **18**, 337.
206. J. Dankert, *CRC Crit. Rev. Biocompat.*, 1986, **2**, 219.
207. M. Habash and G. Reid, *J. Clin. Pharmacol.*, 1999, **39**, 887.
208. A. J. Slate, *et al.*, *Colloids Surf. B*, 2019, **173**, 303.
209. L. Boulange-Petermann, B. Baroux and M. N. Bellon-Fontaine, *J. Adhes. Sci. Technol.*, 1993, **7**, 221.
210. K. A. Whitehead, P. S. Benson, L. A. Smith and J. Verran, *Biofouling*, 2009, **25**, 749.
211. K. A. Whitehead, P. S. Benson and J. Verran, *J. Biofouling*, 2011, **27**, 907.
212. D. S. Jones, J. G. McGovern, C. G. Adair, A. D. Woolfson and S. P. Gorman, *J. Mater. Sci.: Mater. Med.*, 2001, **12**, 399.
213. H. J. Busscher, *et al.*, *Sci. Transl. Med.*, 2012, **4**, 153rv110.

214. W. M. Dunne, *Clin. Microbiol. Rev.*, 2002, **15**, 155.
215. D. P. Gnanadhas, M. B. Thomas, R. Thomas, A. M. Raichur and D. Chakravortty, *Antimicrob. Agents Chemother.*, 2013, **57**, 4945.
216. E. M. C. X. Lima, H. Koo, A. M. Vacca Smith, P. L. Rosalen and A. A. Del Bel Cury, *Clin. Oral. Implants Res.*, 2008, **19**, 780.
217. C. R. Arciola, D. Campoccia, P. Speziale, L. Montanaro and J. W. Costerton, *Biomaterials*, 2012, **33**, 5967.
218. T. J. Kinnari, *et al.*, *Otol. Neurotol.*, 2005, **26**, 380.
219. J. W. Wilson, *et al.*, *Postgrad. Med. J.*, 2002, **78**, 216.
220. V. Solanki, M. Tiwari and V. Tiwari, *Int. J. Biol. Macromol.*, 2018, **112**, 54.
221. P. Klemm and M. A. Schembri, *Int. J. Med. Microbiol.*, 2000, **290**, 27.
222. S. Patel, *Gene Rep.*, 2017, **7**, 127.
223. A. Tristan, *et al.*, *J. Clin. Microbiol.*, 2003, **41**, 4465.
224. C. A. Scotchford, C. P. Gilmore, E. Cooper, G. J. Leggett and S. Downes, *J. Biomed. Mater. Res.*, 2002, **59**, 84.
225. Y. H. An, B. K. Blair, K. L. Martin and R. J. Friedman, Handbook of Bacterial Adhesion, in *Macromolecule Surface Coating for Preventing Bacterial Adhesion*, ed. R. J. Friedman and Y. H. An, Humana Press, NJ, US, 1st edn, 2000, pp. 609–625.
226. M. Fletcher, *Microbiology*, 1976, **94**, 400.
227. D. Eroshenko, I. Morozov and V. Korobov, *Curr. Microbiol.*, 2015, **70**, 846.
228. Y. H. An, J. Bradley, D. L. Powers and R. J. Friedman, *J. Bone Joint Surg. BR*, 1997, **79**, 816.
229. M. Herrmann, *et al.*, *J. Infect. Dis.*, 1988, **158**, 693.
230. C. N. Kraft, M. Hansis, S. Arens, M. D. Menger and B. Vollmar, *J. Biomed. Mat. Res.*, 2000, **49**, 192.
231. A. M. Pitz, G. W. Park, D. Lee, Y. L. Boissy and J. Vinjé, *Gut. Microbes.*, 2015, **6**, 93.
232. C. Selvi and D. Nartop, *Spectrochim. Acta, Part A*, 2012, **95**, 165.
233. S. L. Warnes and C. W. Keevil, *Appl. Environ. Microbiol.*, 2011, **77**, 6049.
234. M. Hans, *et al.*, *Langmuir*, 2013, **29**, 16160.
235. L. Weaver, J. O. Noyce, H. T. Michels and C. W. Keevil, *J. Appl. Microbiol.*, 2010, **109**, 2200.
236. C. E. Santo, *et al.*, *Appl. Environ. Microbiol.*, 2011, **77**, 794.
237. C. E. Santo, N. Taudte, D. H. Nies and G. Grass, *Appl. Environ. Microbiol.*, 2008, **74**, 977.
238. G. Grass, C. Rensing and M. Solioz, *Appl. Environ. Microbiol.*, 2011, **77**, 1541.
239. H. Irving and R. J. P. Williams, *J. Chem. Soc.*, 1953, **1953**, 3192.
240. M. Yasuyuki, *et al.*, *Biofouling*, 2010, **26**, 851.
241. F. Minandri, C. Bonchi, E. Frangipani, F. Imperi and P. Visca, *Future Microbiol.*, 2014, **9**, 379.
242. A. Cochis, *et al.*, *Biomaterials*, 2016, **80**, 80.
243. T. M. Auschill, *et al.*, *Eur. J. Oral Sci.*, 2002, **110**, 48.
244. M. P. Dittmer, *et al.*, *Head Face Med.*, 2015, **11**, 10.

245. H. Kawakami, K. Yoshida, Y. Nishida, Y. Kikuchi and Y. Sato, *ISIJ Int.*, 2008, **48**, 1299.
246. W. C. Witzleb, J. Ziegler, F. Krummenauer, V. Neumeister and K. P. Guenther, *Acta Orthop.*, 2006, **77**, 697.
247. J. Antoniou, *et al.*, *JBJS*, 2008, **90**, 142.
248. N. Wu, Y. Yu, T. Li, X. Ji, L. Jiang, J. Zong and H. Huang, *PLoS One*, 2016, **11**, 0167245.
249. W. Zhang, *et al.*, *Nanoscale*, 2016, **8**, 11642.
250. N. Qureshi, *et al.*, *Appl. Nanosci.*, 2015, **5**, 331.
251. C. Zollfrank, K. Gutbrod, P. Wechsler and J. P. Guggenbichler, *Mater. Sci. Eng., C*, 2012, **32**, 47.
252. K. Krishnamoorthy, M. Veerapandian, K. Yun and S. J. Kim, *Colloids Surf. B*, 2013, **112**, 521.
253. C. C. Mardare and A. W. Hassel, *ACS Comb. Sci.*, 2014, **16**, 631.
254. N. Gugala, J. A. Lemire and R. J. Turner, *J. Antibiot.*, 2017, **70**, 775.
255. M. Y. Vaidya, A. J. McBain, J. A. Butler, C. E. Banks and K. A. Whitehead, *Sci. Rep.*, 2017, **7**, 1.
256. M. A. Ali, *et al.*, *J. Inorg. Biochem.*, 2002, **92**, 141.
257. M. Trop, *et al.*, *J. Trauma Acute Care Surg.*, 2006, **60**, 648.
258. B. S. Atiyeh, M. Costagliola, S. N. Hayek and S. A. Dibo, *Burns*, 2007, **33**, 139.
259. Y. K. Jo, B. H. Kim and G. Jung, *Plant Dis.*, 2009, **93**, 1037.
260. V. Edwards-Jones, *Lett. Appl. Microbiol.*, 2009, **49**, 147.
261. M. Rai, A. Yadav and A. Gade, *Biotechnol. Adv.*, 2009, **27**, 76.
262. N. Durán, *et al.*, *Nanomed. Nanotechnol. Biol. Med.*, 2016, **12**, 789.
263. P. Velusamy, G. V. Kumar, V. Jeyanthi, J. Das and R. Pachaiappan, *Toxicol. Res.*, 2016, **32**, 95.
264. A. B. G. Lansdown, *J. Wound Care*, 2002, **11**, 125.
265. R. M. Goulter, I. R. Gentle and G. A. Dykes, *Lett. Appl. Microbiol.*, 2009, **49**, 1.
266. J. C. Ontong, S. Paosen, S. Shankar and S. P. Voravuthikunchai, *J. Microbiol. Methods*, 2019, **165**, 105692.
267. N. Stobie, *et al.*, *J. Colloid Int. Sci.*, 2010, **345**, 286.
268. O. Gordon, *et al.*, *Antimicrob. Agents Chemother.*, 2010, **54**, 4208.
269. Q. L. Feng, *et al.*, *J. Biomed. Mat. Res.*, 2000, **52**, 662.
270. Y. Y. Chang, H. L. Huang, H. J. Chen, C. H. Lai and C. Y. Wen, *Surf. Coat. Technol.*, 2014, **259**, 193.
271. C. Balagna, S. Spriano and M. G. Faga, *Mater. Sci. Eng. C*, 2012, **32**, 1868.
272. K. J. Welldon, G. J. Atkins, D. W. Howie and D. M. Findlay, *J. Biomed. Mater. Res. A*, 2008, **84**, 691.
273. A. Raghunath and E. Perumal, *Int. J. Antimicrob. Agents*, 2017, **49**, 137.
274. M. Vaidya, A. McBain, C. E. Banks and K. A. Whitehead, *Int. Biodeter. Biodegr.*, 2019, **141**, 39.
275. N. Gugala, D. Vu, M. D. Parkins and R. J. Turner, *Antibiotics*, 2019, **8**, 51.

276. A. J. Slate, L. Shalamanova, I. D. Akhidime and K. A. Whitehead, *Lett. Appl. Microbiol.*, 2019, **69**, 168.
277. M. A. Wassall, M. Santin, C. Isalberti, M. Cannas and S. P. Denyer, *J. Biomed. Mat. Res.*, 1997, **36**, 325.
278. B. Jansen, M. Rinck, P. Wolbring, A. Strohmeier and T. Jahns, *J. Biomat. Appl.*, 1994, **9**, 55.
279. L. Vroman, A. L. Adams, G. C. Fischer and P. C. Munoz, *Blood*, 1980, **55**, 156.
280. L. Vroman and A. L. Adams, *Surf. Sci.*, 1969, **16**, 438–446.
281. J. Jones-Carson, J. Laughlin, M. A. Hamad, A. L. Stewart, M. I. Voskuil and A. Vázquez-Torres, *PLoS One*, 2008, **3**, e1976.
282. M. Long and H. J. Rack, *Biomaterials*, 1998, **19**, 1621.
283. D. Kent, G. Wang, Z. Yu and M. S. Dargusch, *Mat. Sci. Eng. A*, 2010, **527**, 2246.
284. A. Misra, H. Kung, D. Hammon, R. G. Hoagland and M. Nastasi, *Int. J. Damage Mech.*, 2003, **12**, 365.
285. K. S. Egorova and V. P. Ananikov, *Organometallics*, 2017, **36**, 4071.
286. J. C. Hornez, A. Lefevre, D. Joly and H. F. Hildebrand, *Biomol. Eng.*, 2002, **19**, 103.
287. A. Shivaram, S. Bose and A. Bandyopadhyay, *Acta Biomater.*, 2017, **58**, 550.
288. N. R. Panyala, E. M. Peña-Méndez and J. Havel, *J. Appl. Biomed.*, 2008, **6**, 558.
289. P. L. Drake and K. J. Hazelwood, *Ann. Occ. Hyg.*, 2005, **49**, 575.
290. M. Bosetti, A. Massè, E. Tobin and M. Cannas, *Biomaterials*, 2002, **23**, 887.
291. Y. S. Hedberg and I. Odnevall Wallinder, *Biointerphases*, 2016, **11**, 018901.
292. J. B. Mistry, M. Chughtai, R. K. Elmallah, A. Diedrich, S. Le, M. Thomas and M. A. Mont, *J. Orthop. Trauma*, 2016, **17**, 1.
293. K. M. Ismail, A. Jayaraman, T. K. Wood and J. C. T. Earthman, *Electrochim. Acta*, 1999, **44**, 4685.
294. Y. Zhang, Y. Zheng, Y. Li, L. Wang, Y. Bai, Q. Zhao, X. Xiong, Y. Cheng, Z. Tang, Y. Deng and S. Wei, *PLoS One*, 2015, **10**, e0130774.
295. P. J. Chuang, V. Swaminathan, L. Pavlovsky, L. Marquez-Catral, D. L. Jones and L. Song, *J. Biomed. Mat. Res. Part A*, 2019, **107**, 2556.
296. C. C. Willhite, N. A. Karyakina, R. A. Yokel, N. Yenugadhati, T. M. Wisniewski, I. M. F. Arnold, F. Momoli and D. Krewski, *Crit. Rev. Toxicol.*, 2014, **44**, 1.
297. C. G. Bellows, J. N. M. Heersche and J. E. Aubin, *Calcified Tissue Inter.*, 1999, **65**, 59.
298. K. H. W. Lau, A. Yoo and S. P. Wang, *Mol. Cell. Biochem.*, 1991, **105**, 93.
299. P. Fanti, M. S. Kindy, S. Mohapatra, J. Klein, G. Colombo and H. H. Malluche, *Am. J. Phys. Endocrin. Metab.*, 2006, **263**, E1113.

300. H. Köpf and P. Köpf-Maier, in *Tumor Inhibition by Metallocene Dihalides of Early Transition Metals*, American Chemical Society, ACS Symposium, ACS Publications, 1983, vol. 209, 16, p. 315.
301. G. J. Thompson and D. A. Puleo, *Biomaterials*, 1996, **17**, 1949.
302. A. D. Dayan and A. J. Paine, *Hum. Exp. Toxicol.*, 2001, **20**, 439.
303. D. G. Barceloux and D. Barceloux, *J. Toxicol., Clin. Toxicol.*, 1999, **37**, 231.
304. D. Pradhan, A. W. Wren, S. T. Misture and N. P. Mellott, *Mater. Sci. Eng. C*, 2016, **58**, 918.
305. L. Luo, A. Petit, J. Antoniou, D. J. Zukor, O. L. Huk, R. C. W. Liu, F. M. Winnik and F. Mwale, *Biomaterials*, 2005, **26**, 5587.
306. E. Rudolf, *Acta Medica (Hradec Kralove). Suppl.*, 1998, **41**, 55.
307. J. L. Gilbert, C. A. Buckley and J. J. Jacobs, *J. Biomed. Mat. Res.*, 1993, **27**, 1533.
308. J. J. Jacobs, A. K. Skipor, L. M. Patterson, N. J. Hallab, W. G. Paprosky, J. Black and J. O. Galante, *JBJS*, 1998, **80**, 1447.
309. X. Mao, A. A. Wong and R. W. Crawford, *Med. J. Aust.*, 2011, **194**, 649.
310. U. E. Pazzaglia, P. Apostoli, T. Congiu, S. Catalani, M. Marchese and G. Zarattini, *Arch. Ortho Trauma Surg.*, 2011, **131**, 1299.
311. M. G. Zywiel, S. A. Sayeed, A. J. Johnson, T. P. Schmalzried and M. A. Mont, *Clin. Ortho. Related Res.*, 2011, **469**, 1536.
312. D. A. Jones, H. K. Lucas, M. O. Driscoll, C. H. G. Price and B. Wibberley, *J. Bone Joint Surg.*, 1975, **57**, 289.
313. Y. K. Liu, X. X. Deng and H. L. Yang, *Bone Joint Res.*, 2016, **5**, 461.
314. A. Kanaji, V. Orhue, M. S. Caicedo, A. S. Virdi, D. R. Sumner, N. J. Hallab, T. Yoshiaki and K. Sena, *J. Ortho. Surg. Res.*, 2014, **9**, 91.
315. M. Bonham, J. M. O'Connor, B. M. Hannigan and J. J. Strain, *Br. J. Nutr.*, 2002, **87**, 393.
316. J. Y. Uriu-Adams and C. L. Keen, *Mol. Aspects Med.*, 2005, **26**, 268.
317. J. R. Turnlund, W. R. Keyes, G. L. Peiffer and K. C. Scott, *Am. J. Clin. Nutrit.*, 1998, **67**, 1219.
318. P. J. Evans and B. Halliwell, in *Methods in Enzymology*, vol. 233, p. 82, Elsevier, 1994.
319. M. J. Burkitt, *Arch. Biochem. Biophys.*, 2001, **394**, 117.
320. A. K. Boal and A. C. Rosenzweig, *Chem. Rev.*, 2009, **109**, 4760.
321. J. R. Prohaska, *Am. J. Clin. Nutrit.*, 2008, **88**, 826S.
322. L. Y. Chen, Z. Xue, Z. J. Xu, J. Q. Chen, R. X. He, X. P. Nie, Q. P. Cao, X. D. Wang, S. Q. Ding and J. Z. Jiang, *Adv. Eng. Mat.*, 2012, **14**, 195.
323. G. Borkow, N. Okon-Levy and J. Gabbay, *Wounds*, 2010, **22**, 301.
324. L. M. Gaetke, H. S. Chow-Johnson and C. K. Chow, *Arch. Toxicol.*, 2014, **88**, 1929.
325. S. J. Lippard, *Science*, 1999, **284**, 748.
326. M. B. Kadiiska, P. M. Hanna, S. J. Jordan and R. P. Mason, *Mol. Pharmacol.*, 1993, **44**, 222.
327. I. Bremner, *Am. J. Clin. Nutrit.*, 1998, **67**, 1069S.

328. M. E. Letelier, S. Sánchez-Jofré, L. Peredo-Silva, J. Cortés-Troncoso and P. Aracena-Parks, *Chem.-Biol. Interact.*, 2010, **188**, 220.
329. E. M. Santos, J. S. Ball, T. D. Williams, H. Wu, F. Ortega, R. van Aerle, I. Katsiadaki, F. Falciani, M. R. Viant, J. K. Chipman and C. R. Tyler, *Environ. Sci. Technol.*, 2010, **44**, 820.
330. J. L. Burkhead, L. W. Gray and S. Lutsenko, *Biometals*, 2011, **24**, 455.
331. D. Huster and S. Lutsenko, *Mol. BioSyst.*, 2007, **3**, 816.
332. S. V. S. Rana, *J. Trace Elements Med. Biol.*, 2008, **22**, 262.
333. F. Lang, S. Ullrich and E. Gulbins, *Expert. Opin. Ther. Targets*, 2011, **15**, 1061.
334. J. R. Turnlund, W. R. Keyes, S. K. Kim and J. M. Domek, *Am. J. Clin. Nutr.*, 2005, **81**, 822.
335. F. Suska, M. Esposito, C. Gretzer, M. Källtorp, P. Tengvall and P. Thomsen, *Biomaterials*, 2003, **24**, 461.
336. F. Suska, C. Getzer, M. Esposito, L. Emanuelsson, A. Wennerberg, P. Tengval and P. Thomsen, *Biomaterials*, 2005, **26**, 519.
337. F. Suska, C. Gretzer, M. Esposito, P. Tengvall and P. Thomsen, *Biomaterials*, 2005, **26**, 5942.
338. T. Theophanides and J. Anastassopoulou, *Crit. Rev. Oncol. Hematol.*, 2002, **42**, 57.
339. N. Leone, D. Courbon, P. Ducimetiere and M. Zureik, *Epidemiology*, 2006, 308.
340. F. Suska, S. Svensson, A. Johansson, L. Emanuelsson, H. Karlholm, M. Ohrlander and P. Thomsen, *J. Biomed. Mat. Res. Part B: Appl. Biomat.*, 2010, **92**, 86.
341. A. Vyskočil and C. Viau, *J. Appl. Toxicol.*, 1999, **19**, 185.
342. N. A. Dsouki, M. P. de Lima, R. Corazzini, T. M. Gáscon, L. A. Azzalis, V. B. Campos Junqueira, D. Feder and F. L. A. Fonseca, *J. Mat. Sci.: Mat. Med.*, 2014, **25**, 1301.
343. Y. Li, C. Wong, J. Xiong, P. Hodgson and C. Wen, *J. Dent. Res.*, 2010, **89**, 493.
344. M. Caicedo, J. J. Jacobs, A. Reddy and N. J. Hallab, *J. Biomed. Mat. Res. A*, 2008, **86**, 905.
345. M. Ehrnrooth and H. Kerosuo, *Angle Orthod.*, 2009, **79**, 1194.
346. H.-H. Huang, Y.-H. Chiu, T.-H. Lee, S.-C. Wu, H.-W. Yang, K.-H. Su and C.-C. Hsu, *Biomat.*, 2003, **24**, 3585.
347. N. Jacobsen and A. Hensten-Pettersen, *Eur. J. Orthod.*, 2003, **25**, 591.
348. A. Yamamoto, R. Honma and M. Sumita, *J. Biomed. Mat. Res.*, 1998, **39**, 331.
349. A. Muñoz and M. Costa, *Toxicol. Appl. Pharmacol.*, 2012, **260**, 1.
350. M. Bergman, B. Bergman and R. Söremark, *J. Oral Rehabil.*, 1980, **7**, 325.
351. R. Michel and J. Zilkens, *Zeitsch. Orthop. Grenzg.*, 1978, **116**, 666.
352. K. Yang, Y. Ren and P. Wan, *Sci. China Technol. Sci.*, 2012, **55**, 329.
353. K. Yang and Y. Ren, *Sci. Tech. Adv. Mat.*, 2010, **11**, 014105.

354. H. B. Kwon, J. H. Lee, S. H. Lee, A. Y. Lee, J. S. Choi and Y. S. Ahn, *Ann. Dermatol.*, 2009, **21**, 308.
355. A. O. Gettler, C. P. Rhoads and S. Weiss, *Am. J. Pathol.*, 1927, **3**, 631.
356. N. S. Tomi, B. Kränke and W. Aberer, *Lancet*, 2004, **363**, 532.
357. L. Jonas, C. Bloch, R. Zimmermann, V. Stadie, G. E. Gross and S. G. Schäd, *Ultrastructural Pathol.*, 2007, **31**, 379.
358. A. Wadhera and M. A. Fung, *Dermatol. Online J.*, 2005, **11**, 12.
359. M. Glehr, A. Leithner, J. Friesenbichler, W. Goessler, A. Avian, D. Andreou, W. Maurer-Ertl, R. Windhager and P.-U. Tunn, *Bone Joint J.*, 2013, **95**, 988.
360. R. G. Bickford, G. Fischer and G. P. Sayre, *Proc. Staff. Meet. Mayo. Clin.*, 1957, **9**, 14.
361. T. L. Babb and W. Kupfer, *Exp. Neurol.*, 1984, **86**, 171.
362. W. F. Jackson and B. R. Duling, *Cir. Res.*, 1983, **53**, 105.
363. R. Gamelli, T. P. Paxton and M. O'Reilly, *Surg. Gynecol. Obstetrics*, 1993, **177**, 115.
364. R. L. Zapata-Sirvent and J. F. Hansbrough, *J. Burn Care Rehab.*, 1993, **14**, 132–140.
365. I. L. Steffensen, O. J. Mesna, E. Andruchow, E. Namork, K. Hylland and R. A. Andersen, *Gen. Pharma.: Vas. Sys.*, 1994, **25**, 1621.
366. K.-K. Chew, S. H. S. Zein and A. L. Ahmad, *Nat. Sci.*, 2012, **4**, 184.
367. J. Schalamon, T. Petnehazy, H. Ainoedhofer, E. B. Zwick, G. Singer and M. E. Hoellwarth, *J. Ped. Surg.*, 2007, **42**, 1584.
368. C. J. Kirkpatrick, V. Krump-Konvalinkova, R. E. Unger, F. Bittinger, M. Otto and K. Peters, *Biomol. Eng.*, 2002, **19**, 211.
369. N. Maruyama, H. Kawasaki, A. Yamamoto, S. Hiromoto, H. Imai and T. Hanawa, *Mater. Trans.*, 2005, **46**, 1588.
370. A. T. Mahar, S. S. Lee, F. D. Lalonde, T. Impelluso and P. O. Newton, *J. Pediatr. Orthop.*, 2004, **24**, 638.
371. G. Schmalz and D. Arenholt-Bindslev, *Biocompatibility of Dental Materials*, Springer, 2009, vol. 1.
372. J. J. Jacobs, C. Silverton, N. J. Hallab, A. K. Skipor, L. Patterson, J. Black and J. O. Galante, *Clin. Orthop. Relat. Res.*, 1999, 173.
373. M. Brayda-Bruno, M. Fini, G. Pierini, G. Giavaresi, M. Rocca and R. Giardino, *Int. J. Artif. Organs*, 2001, **24**, 41.
374. T. Eliades, H. Pratsinis, D. Kletsas, G. Eliades and M. Makou, *Am. J. Orthod. Dentofacial Orthop.*, 2004, **125**, 24.
375. J. Noble, S. I. Ahing, N. E. Karaiskos and W. A. Wiltshire, *Brit. Dental J.*, 2008, **204**, 297.
376. A. D. Mazzocca, M. P. Cote, C. L. Arciero, A. A. Romeo and R. A. Arciero, *Am. J. Sports Med.*, 2008, **36**, 1922.
377. A. P. Davies, H. G. Willert, P. A. Campbell, I. D. Learmonth and C. P. Case, *JBJS*, 2005, **87**, 18.
378. K. Cheney, C. Gumbiner, B. Benson and M. Tenenbein, *J. Toxicol. Clin. Toxicol.*, 1995, **33**, 61.
379. J. L. Domingo, *Mol. Cell. Biochem.*, 2000, **203**, 185.

380. G. C. McKay, R. Macnair, C. MacDonald and M. H. Grant, *Biomaterials*, 1996, **17**, 1339.
381. A. M. Cortizo, L. Bruzzone, S. Molinuevo and S. B. A. Etcheverry, *Toxicology*, 2000, **147**, 89.
382. V. C. Sálice, A. M. Cortizo, C. L. G. Dumm and S. B. Etcheverry, *Mol. Cell. Biochem.*, 1999, **198**, 119.
383. M. Altamirano-Lozano, L. Alvarez-Barrera, F. Basurto-Alcántara, M. Valverde and E. Rojas, *Terato. Carcino. Muta.*, 1996, **16**, 7.
384. S. Nag, R. Banerjee, J. Stechschulte and H. L. Fraser, *J. Mat. Sci.: Mat. Med.*, 2005, **16**, 679.
385. D. Krewski, D. R. A. Yokel, E. Nieboer, D. Borchelt, J. Cohen, J. Harry, S. Kacew, J. Lindsay, A. M. Mahfouz and V. Rondeau, *J. Toxicol. Environ. Health B*, 2007, **10**, 1.
386. S. Ivankovic and R. Preussmann, *Food Cosmet. Toxicol.*, 1975, **13**, 347.
387. I. Al-Saleh and I. Al-Doush, *Sci. Total Environ.*, 1988, **216**, 181.
388. S. B. Goldhaber, *Regul. Toxicol. Pharmacol.*, 2003, **38**, 232.
389. A. Dorsey and L. Ingerman, in *Toxicological Profile for Copper*, 1st edn, 2004, Agency for Toxic Substances and Disease Registry, p. 121.
390. D. R. Webb, S. E. Wilson and D. E. Carter, *Toxicol. Appl. Pharmacol.*, 1986, **82**, 405.
391. H. Yamauchi, K. Takahashi, Y. Yamamura and B. A. Fowler, *Toxicol. Appl. Pharmacol.*, 1992, **116**, 66.
392. J. W. Goller and N. W. Paik, *Am. Ind. Hyg. Assoc. J.*, 1985, **46**, 89.
393. C. Esteve, E. Alcaide and R. Ureña, *Aquat. Toxicol.*, 2012, **109**, 176.
394. S. C. Kerwien, in *Toxicity of Tungsten, Molybdenum, and Tantalum and the Environmental and Occupation Laws Associated with Their Manufacture, Use, and Disposal*, Army Armament Research Development and Engineering Center, Picatinny Arsenal, NJ, USA, 1996.
395. G. D. Todd, in *Toxicological Profile for Molybdenum: Agency for Toxic Substances and Disease Registry*, Atlanta, USA, 2017.
396. K. Hughes, M. E. Meek, P. K. L. Chan, J. Shedden, S. Bartlett and L. J. Seed, *J. Environ. Sci. Health C*, 1994, **12**, 417.
397. T. Kawamoto, T.-T.-P. Pham, T. Matsuda, T. Oyama, M. Tanaka, H.-S. Yu and I. Uchiyama, *Int. J. Hyg. Environ. Health*, 2011, **214**, 296.
398. K. J. Schulz, N. M. Piatak, & J. F. Papp, in *Niobium and Tantalum. Report No. 1411339916*, US Geological Survey, USA, 2017.
399. J. T. Hartmann and H.-P. Lipp, *Expert Opin. Pharmacother.*, 2003, **4**, 889.
400. Y. S. Kim, J. S. Kim, H. S. Cho, D. S. Rha, J. M. Kim, J. D. Park, B. S. Choi, R. Lim, H. K. Chang, Y. H. Chung, I. H. Kwon, J. Jeong, B. S. Han and I. J. Yu, *Inhalation Toxicol.*, 2008, **20**, 575.
401. S. Martin and W. Griswold, *Environ. Sci. Technol. Briefs Citizens*, 2009, **15**, 1.
402. T. Santonen, H. Stockmann-Juvala and A. Zitting, *Finnish Institute of Occupational Health*, 2010.

403. S. Yaghoubi, C. W. Schwietert and J. P. McCue, *Biol. Trace Element Res.*, 2000, **78**, 205.
404. S. Keith, in *Toxicological Profile for Tungsten*, Agency for Toxic Substances and Disease Registry, 2005.
405. J. L. Domingo, M. Gomez, D. J. Sanchez, J. M. Llobet and C. L. Keen, *Mol. Cell. Biochem.*, 1995, **153**, 233.
406. B. Gummow, C. J. Botha, J. Noordhuizen and J. A. P. Heesterbeek, *Prev. Vet. Med.*, 2005, **72**, 281.
407. A. El Safty, K. El Mahgoub, S. Helal and N. A. Maksoud, *Ann. NY Acad. Sci.*, 2008, **1140**, 256.
408. C. J. Spiegl, M. C. Calkins, J. J. DeVoldre, J. K. Scott, L. T. Steadman and H. E. Stokinger, in *Inhalation Toxicity of Zirconium Compounds. I. Short-term Studies*, Atomic Energy Project, Univ. of Rochester, NY, USA, 1956.

Subject Index

Note: elements are indexed under their full form, *e.g.* ‘cobalt’ rather than ‘Co’.